LINEAR ALGEBRA AND ITS APPLICATIONS
Volume 1: A First Course

WEBSITE

WWW. REED.CO.UK

MATHEMATICS AND ITS APPLICATIONS

Series Editor: G. M. BELL, Professor of Mathematics,
King's College London (KQC), University of London

NUMERICAL ANALYSIS, STATISTICS AND OPERATIONAL RESEARCH

Editor: B. W. CONOLLY, Emeritus Professor of Mathematics (Operational Research),
Queen Mary College, University of London

Mathematics and its applications are now awe-inspiring in their scope, variety and depth. Not only is there rapid growth in pure mathematics and its applications to the traditional fields of the physical sciences, engineering and statistics, but new fields of application are emerging in biology, ecology and social organization. The user of mathematics must assimilate subtle new techniques and also learn to handle the great power of the computer efficiently and economically.

The need for clear, concise and authoritative texts is thus greater than ever and our series will endeavour to supply this need. It aims to be comprehensive and yet flexible. Works surveying recent research will introduce new areas and up-to-date mathematical methods. Undergraduate texts on established topics will stimulate student interest by including applications relevant at the present day. The series will also include selected volumes of lecture notes which will enable certain important topics to be presented earlier than would otherwise be possible.

In all these ways it is hoped to render a valuable service to those who learn, teach, develop and use mathematics.

Mathematics and its Applications

Series Editor: G. M. BELL, Professor of Mathematics, King's College London (KQC), University of London

Series continued at back of book

LINEAR ALGEBRA AND ITS APPLICATIONS
Volume 1: A First Course

D. H. GRIFFEL, B.Sc., Ph.D.
School of Mathematics
University of Bristol

ELLIS HORWOOD LIMITED
Publishers · Chichester

Halsted Press: a division of
JOHN WILEY & SONS
New York · Chichester · Brisbane · Toronto

First published in 1989 by
ELLIS HORWOOD LIMITED
Market Cross House, Cooper Street,
Chichester, West Sussex, PO19 1EB, England
The publisher's colophon is reproduced from James Gillison's drawing of the ancient Market Cross, Chichester.

Distributors:

Australia and New Zealand:
JACARANDA WILEY LIMITED
GPO Box 859, Brisbane, Queensland 4001, Australia
Canada:
JOHN WILEY & SONS CANADA LIMITED
22 Worcester Road, Rexdale, Ontario, Canada
Europe and Africa:
JOHN WILEY & SONS LIMITED
Baffins Lane, Chichester, West Sussex, England
North and South America and the rest of the world:
Halsted Press: a division of
JOHN WILEY & SONS
605 Third Avenue, New York, NY 10158, USA
South-East Asia
JOHN WILEY & SONS (SEA) PTE LIMITED
37 Jalan Pemimpin # 05–04
Block B, Union Industrial Building, Singapore 2057
Indian Subcontinent
WILEY EASTERN LIMITED
4835/24 Ansari Road
Daryaganj, New Delhi 110002, India

© **1989 D.H. Griffel/Ellis Horwood Limited**

British Library Cataloguing in Publication Data
Griffel, D.H.
Linear algebra and its applications.
Vol. 1: A First course
1. Linear algebra
I. Title
512′.5

Library of Congress Card No. 88–13650

ISBN 0–85312–946–0 (Ellis Horwood Limited — Library Edn.)
ISBN 0–7458–0571–X (Ellis Horwood Limited — Student Edn.)
ISBN 0–470–21242–X (Halsted Press)

Printed in Great Britain by The Camelot Press, Southampton

Table of Contents

PART II   VECTOR SPACES AND LINEAR EQUATIONS

Table of contents

Contents of Volume 2

Preface

This book is intended

- to be intelligible and helpful to the novice, while including more advanced material for the more demanding reader;
- to present the core of vector space theory in a straightforward and reasonably concise way;
- to include many other theoretical and computational aspects of the subject, and applications in the sciences;
- to give the beautiful and important subject of eigenvalues its rightful place, as a key to understanding how a linear map behaves, and as a central concept in mechanics; and
- to outline very briefly the fundamental role of linear algebra in modern physics.

The unstarred sections give a self-contained course in basic linear algebra with no frills. The starred sections branch off in various directions, and may be omitted if desired.

In general, the theory is developed in full detail. But some topics require long and detailed proofs, making it hard for beginners to see the wood for the trees. Such proofs have been transferred to the ends of the chapters, so that the main ideas can stand out as clearly as possible.

PREREQUISITES

Readers should have met elementary calculus, and preferably complex numbers; but very little mathematical sophistication is required. Appendices to Volume 1 offer guidance to readers unfamiliar with abstract mathematics, and give some basic information about sets, functions, complex numbers, and other such topics.

STRUCTURE OF VOLUME 1

This volume can be used for a basic course on the most useful parts of linear algebra. The unstarred sections of Chapters 1 to 7 give a self-contained introduction to matrices, real and complex vector spaces, linear maps, and the eigenvalue problem. The starred sections develop the main theme in three directions: algebraic theory, scientific applications, and numerical computation. But they can all be skipped; they are not needed for understanding the unstarred sections.

Chapter 8 gives a stripped-down version of diagonalisation for matrices and quadratic forms, without using the idea of matrix representations. Readers intending to proceed to Volume 2 may skip Chapter 8; in Chapter 10 they will meet diagonalisation in its natural context: matrix representations of linear transformations.

STRUCTURE OF VOLUME 2

The unstarred sections continue the development of the unstarred sections of Volume 1, covering matrix representations of linear maps, diagonalisation from this point of view, inner product spaces, and the spectral theorem for self-adjoint operators. Again, many optional sections are grouped around the main stem. And the last two chapters consist entirely of more advanced material: least squares approximation, the singular-value decomposition, and duality.

PROBLEMS AND EXERCISES FOR THE READER

At the end of each chapter there are many problems, of varying difficulty; some of them give useful extensions or applications of the theory.

There are also exercises included in the body of the text. They are quite easy, and are intended to be worked as soon as they are encountered, as an aid to getting to grips with the text. Full solutions to the exercises are given at the back of the book.

ACKNOWLEDGEMENTS

I have inflicted earlier versions of this material on several hundred students; I am grateful to them for finding many mistakes, and for encouraging me to think that writing yet another account of linear algebra may not be a complete waste of time.

A number of colleagues have kindly read parts of the manuscript and made helpful contributions. It is a pleasure to acknowledge my gratitude to Arthur Chatters, Michael Drazin, Philip Drazin, John Pryce, Andy Vince, Andy Wathen, and particularly to Geoffrey Goodhill whose perceptive comments have improved several chapters of this volume.

As well as helpful colleagues, I am blessed with an ideal publisher. Quite apart from his patience when typescript fails to arrive on time, it has been a consistent pleasure to work with Ellis Horwood and his family and friends during the production of this book.

Despite the efforts of all these people, some of my mistakes are bound to have survived. I will be most grateful to any readers kind enough to point them out.

School of Mathematics   David Griffel
University Walk
Bristol, BS8 1TW

Part I

Vectors and matrices

The central ideas of our subject are contained in Parts II to V. Part VI deals with some more advanced topics, and this first part contains an elementary treatment of vector and matrix algebra. A firm grasp of this material is essential for the more abstract theory of Parts II and III.

1

Elementary vector algebra

Linear algebra is an abstract structure built on a geometric foundation. This first chapter lays the foundations, and is essential reading, though some of it may already be familiar to you. The first five sections set out the ideas of vector algebra in two and three dimensions, and the last section defines vectors in n-space, the first step towards the general abstract theory of Part II.

1A INTRODUCTION TO VECTORS

Mathematics is useful because it is related to the world in which we live. Elementary mathematics deals with numbers; anything which has a magnitude, or size, can be expressed mathematically in terms of numbers. But not everything can be described adequately in this way. For example, the strength of the wind is a magnitude and is described by a number; but the wind also has a direction which is just as important. Vectors are the mathematical objects used to describe things like the wind, having a direction as well as a magnitude. This chapter sets out the algebra of vectors; most of linear algebra is based on generalisations of these ideas.

A vector, then, is a combination of a magnitude and a direction. Vectors have no other properties: the magnitude and direction specify a vector completely. We print them in bold type; thus **v** denotes a vector (in handwriting, underlining is more convenient). We can represent **v** on a diagram by drawing an arrow in the direction of **v**, with length equal to the magnitude of **v** on some agreed scale of measurement. For example, if **v** is the vector of magnitude 2 in the northwest direction, it is represented by any of the arrows of Fig. 1.1 (with the conventions that north is upwards and that magnitude 1 is to be represented by a length of 0.5 cm). Note that the position of the arrow on the paper is immaterial; all arrows with the same direction and length are equally good representations of the vector **v**.

Fig. 1.1 – Arrows which all represent the same vector v.

Vectors are mathematical objects of a new kind. We can take nothing for granted; we must be careful to define the meaning of any mathematical operation to be applied to them. In principle we can define the meaning of addition and multiplication for vectors in any reasonable way that we like. In practice, the definitions in this chapter are those that have been found most useful.

Definition 1. The **sum** of two vectors **u** and **v** is defined as follows. Draw arrows representing **u** and **v**, with their tails at the same point (see Fig. 1.2(a)). Construct the parallelogram with these arrows as two of its sides. Draw the diagonal arrow as in Fig. 1.2(b). Then **u** + **v** is the vector represented by this diagonal arrow.   □

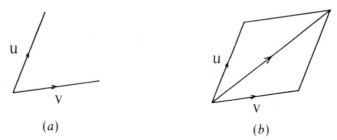

Fig. 1.2 – Adding two vectors.

This may at first sight seem an odd way of defining addition. But consider the following example.

Example 1. A **displacement** is one of the basic examples of a vector. It means the operation of moving things a certain distance in a certain direction: thus 'two metres northwards' specifies a certain displacement. Let **n** be a vector of magnitude 2 north, and **s** a vector of magnitude 4 southwest. If an object at a point P is displaced by vector **n**, it moves to Q (see Fig. 1.3). If it is now displaced by vector **s**, its final position is R. The net effect of the two displacements is to move the object from P to R, and this is just what would be achieved by a single displacement through the vector **n** + **s**. Our definition of vector addition thus fits in with the idea of combining displacements.

Example 2. Velocities are closely related to displacements: velocity is displacement per unit time. We have seen that displacements are combined by the parallelogram rule of vector addition. Velocities add in the same way.

Suppose, for example, an aircraft is flying in a wind of strength and direction given

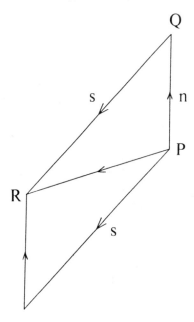

Fig. 1.3 – Combining two displacements.

by the vector **w**. The aircraft's instruments give its speed through the air, and the direction in which it is heading. We combine these two things into a vector **a**, with magnitude equal to the speed through the air, and in the direction in which the aircraft is heading. It is shown in books on mechanics that the vector sum **w** + **a** gives the magnitude and direction of the total velocity of the aircraft, including the effect of the wind.

Example 3. A force on an object is a push or a pull of a certain strength in a certain direction. It can therefore be represented by a vector. If two forces F and G are applied to an object simultaneously, experiment shows that the net effect is the same as that of a single force equal to the vector sum of F and G. In other words, forces add by the parallelogram rule of vector addition.   □

The brief accounts above cannot do justice to the subtlety of the concepts of relative velocity and force. The main point is that vectors, and the addition rule of Definition 1, are extremely useful in mechanics. The details are interesting, but part of another story (see Feynman *et al.* (1964), for example).

The triangle rule

There is another version of the parallelogram rule, as follows. To find **u** + **v**, draw an arrow representing **u**, then draw an arrow representing **v** with its tail at the head of the **u** arrow (see Fig. 1.4). Then **u** + **v** is given by joining the tail of the **u** arrow to the head of the **v** arrow. In short, it is the third side of the triangle.

The triangle rule is really the same as the parallelogram rule. Fig. 1.5 shows the

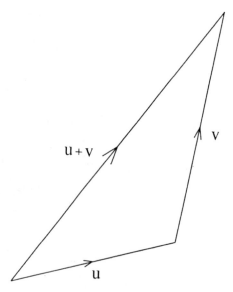

Fig. 1.4 – The addition triangle.

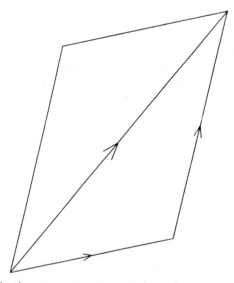

Fig. 1.5 – The addition triangle as part of the addition parallelogram.

triangle as part of the parallelogram picture. But the triangle diagram is less cluttered and, as we shall see, more convenient for some purposes.

1B   ELEMENTARY VECTOR ALGEBRA

This section studies the algebraic properties of vectors.

Addition

It is obvious from the definition that the sum of two vectors does not depend on their order: $\mathbf{u} + \mathbf{v} = \mathbf{v} + \mathbf{u}$ for all vectors $\mathbf{u}$ and $\mathbf{v}$. It is not quite so obvious that when three vectors are added together, the order in which the additions are performed makes no difference. But it is easily proved.

Theorem 1. For all vectors $\mathbf{u}$, $\mathbf{v}$ and $\mathbf{w}$,

$$\mathbf{u} + (\mathbf{v} + \mathbf{w}) = (\mathbf{u} + \mathbf{v}) + \mathbf{w}$$

Proof. In Fig. 1.6 the sum $\mathbf{v} + \mathbf{w}$ is given by the diagonal of the parallelogram OBQC, namely OQ. The sum of $\mathbf{u}$ and $\mathbf{v} + \mathbf{w}$ is the diagonal of the parallelogram OAZQ, namely OZ. The sum of $\mathbf{u} + \mathbf{v}$ and $\mathbf{w}$ is the diagonal of the parallelogram OPZC, which is OZ again.   ☐

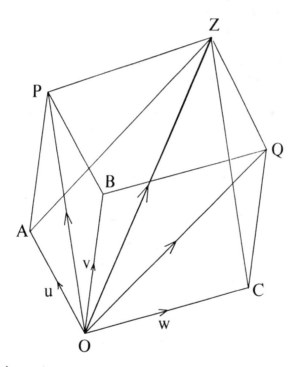

Fig. 1.6 – Adding three vectors.

Remark. You may or may not like this proof. It consists of a diagram, not a chain of logical reasoning. This is inevitable; any proof must start from the definition of addition, which is itself geometrical. If you find this approach woolly and illogical, I must ask you to be patient. We shall construct a rigorously algebraic theory of vectors later. But that theory is much easier to understand if one has an intuitive grasp of vector geometry.   ☐

We have now shown that adding vectors is rather like adding numbers; three vectors can be added in any order to give the same answer. We now proceed to extend other aspects of ordinary arithmetic to vectors.

The zero vector

In ordinary arithmetic there is a special number, called zero, which can be added to any number without changing it. Similarly there is a special vector, called the zero vector, which can be added to any other vector without changing it. It has magnitude zero, and is denoted by $\mathbf{0}$.

An awkward question now arises. What is the direction of the zero vector? Consider the examples in the last section. A displacement of magnitude zero does not change anything: it is the leave-it-alone operation, and has no direction. Similarly, zero velocity means standing still, and zero force means that there is no push or pull applied, so there is no direction involved. We are forced to conclude that the zero vector has no direction. Its arithmetic is specified by the rule that $\mathbf{0} + \mathbf{a} = \mathbf{a}$ for every vector $\mathbf{a}$; this, together with the fact that the magnitude of $\mathbf{0}$ is 0, is all that one needs to know about the zero vector.

Objection! 'You said in the last section that a vector has magnitude and direction. Now you say that the zero vector has no direction. You are contradicting yourself!'

Reply. 'Use your common sense. We began with vectors as things with magnitude and direction. Then we introduced a new object, of a different kind, with no direction. There is no real contradiction here. Nonzero vectors have magnitude and direction, and add by the parallelogram law; the set of all vectors is the set of nonzero vectors together with a new object, called the zero vector, which has zero magnitude and no direction.'

Subtraction

Subtracting $\mathbf{v}$ from $\mathbf{u}$ means finding a vector which when added to $\mathbf{v}$ gives $\mathbf{u}$. Draw arrows representing $\mathbf{u}$ and $\mathbf{v}$, with their tails together (see Fig. 1.7); then $\mathbf{u} - \mathbf{v}$ is given by the arrow from the head of $\mathbf{v}$ to the head of $\mathbf{u}$.

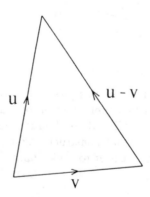

Fig. 1.7 – Subtraction.

We must now ask whether subtraction has the same properties as in ordinary algebra.

Exercise 1. Verify, by drawing diagrams, that for any vectors **u**, **v**, **w** we have $(\mathbf{u} - \mathbf{v}) - \mathbf{w} = \mathbf{u} - (\mathbf{v} + \mathbf{w})$.

Note. When exercises appear in the middle of the text like this, they are intended to be done before you read on. There are solutions at the back of the book. ☐

Other similar relations can be verified in the same way. The general conclusion is that addition and subtraction works for vectors in the same way as for numbers.

Multiplication

For any real number a we have $a + a = 2a$. This equation is so simple that it must surely apply to vectors too. But if **a** is a vector, the expression 2**a** is meaningless at this stage; we have not yet defined what it means to multiply a vector by a number.

It is fairly obvious how to do this. The vector **a** + **a** is in the same direction as **a** and has twice its magnitude; this follows from the triangle rule for addition, as shown in Fig. 1.8. In general, for any positive number k we define $k\mathbf{a}$ to be the vector in the same direction as **a** and with k times its magnitude. In the special case of the zero vector, we define $k\mathbf{0}$ to be **0**, in analogy with the rule $k0 = 0$ of ordinary algebra. Similarly, for any vector **a** we define 0**a** (where 0 denotes the number zero) to be the zero vector.

$$a \qquad a$$

Fig. 1.8 – Doubling a vector.

We have now defined the product of a vector with a positive number or zero. What about negative multiples? Common sense suggests that $-k\mathbf{a} + k\mathbf{a}$ should equal the zero vector. The triangle rule for addition shows that the sum of two vectors is zero only when they have the same magnitude and opposite directions, as in Fig. 1.9. So we define the product of a vector **a** with a negative number $-k$ to be the vector with magnitude k times the magnitude of **a**, and in the opposite direction.

Fig. 1.9 – Two vectors adding to zero.

This completes the definition of what it means to multiply a vector by a number. It is easy to show that the usual rules of algebraic manipulation are valid.

Theorem 2. For any vectors **u** and **v**, and any real numbers k and m, we have
(a) $k(m\mathbf{u}) = (km)\mathbf{u}$;
(b) $(k + m)\mathbf{u} = k\mathbf{u} + m\mathbf{u}$;
(c) $k(\mathbf{u} + \mathbf{v}) = k\mathbf{u} + k\mathbf{v}$;
(d) $\mathbf{u} + (-1)\mathbf{v} = \mathbf{u} - \mathbf{v}$.

Proof. (a) Multiplying the magnitude of **u** by m and then multiplying the result by k gives a vector in the same direction as **u** but of magnitude km times the magnitude of **u**. This is the same as (km)**u**.

(b) See Fig. 1.10 for the case where k and m are positive; similar diagrams apply when one or both are negative. If one is zero the result is trivial.

Fig. 1.10 – Adding two multiples of the same vector.

(c) The parallelogram for adding k**u** to k**v** is similar to that for adding **u** to **v**, but scaled by k. See Fig. 1.11.

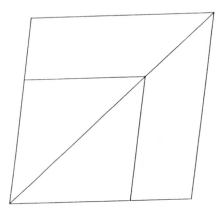

Fig. 1.11 – Adding multiples of two vectors.

(d) We must show that $[\mathbf{u}+(-1)\mathbf{v}]+\mathbf{v}=\mathbf{u}$; this holds because (b) implies that $(-1)\mathbf{v}+\mathbf{v}=[-1+1]\mathbf{v}=\mathbf{0}$. □

Properties of the magnitude

We denote the magnitude of a vector **v** by $\|\mathbf{v}\|$. It is a positive number unless $\mathbf{v}=\mathbf{0}$, in which case $\|\mathbf{0}\|=0$.

Theorem 3. For any vectors **u** and **v** and any real number k,
(a) $\|\mathbf{u}+\mathbf{v}\|\leq\|\mathbf{u}\|+\|\mathbf{v}\|$;
(b) $\|k\mathbf{u}\|=|k|\,\|\mathbf{u}\|$.

Proof. (a) follows at once from the triangle rule for addition; one side of a triangle cannot be longer than the sum of the lengths of the other two sides.

(b) follows at once from the definitions of k**u** in the three cases $k>0$, $k=0$, $k<0$. □

This completes our account of vector addition and multiplication by real numbers.

Position vectors

Vectors are very useful in geometry. It is easy to relate the abstract concept of a vector (magnitude and direction) to the geometrical concept of a point. Begin by choosing a certain point in space as the 'origin' or 'reference point'. Then for any other point P, define the **position vector** of P with respect to the origin O to be the vector with magnitude and direction equal to the length and direction of the line OP (see Fig. 1.12).

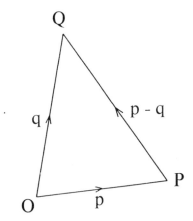

Fig. 1.12 – Position vector of a point.

Although the position vector is defined in terms of an arbitrarily chosen origin, its usefulness hinges on the fact that the choice of origin does not really matter. For example, consider two points P and Q, with position vectors **p** and **q** with respect to some origin O. Then the vector **p** − **q** has the magnitude and direction of the line from P to Q. It is independent of the choice of O.

Applications of position vectors are given in the problems.

1C   TWO-DIMENSIONAL VECTORS: BASIS AND COMPONENTS

We have been discussing vectors in terms of direction, implying that they exist in space. In this section we assume that the space is two-dimensional; that is, we assume that all our vectors lie in a fixed plane.

Given two different directions in the plane, every vector **x** in the plane can be expressed as a sum of two vectors in the two given directions. The procedure is as follows. Draw an arrow representing the given vector **x**; draw lines in the two given directions through the head and through the tail (see Fig. 1.13(a)). The points of intersection of the lines are the heads of the arrows representing vectors **a** and **b** in the two given directions, such that **a** + **b** = **x** (see Fig. 1.13(b)). This procedure is called **resolving** the vector **x** along the two given directions.

We often think of forces in this way. For example, when an aircraft is flying horizontally, the air exerts a force on it. We think of this force as made up of a horizontal air resistance (bad!) and a vertical lifting force which stops the aircraft

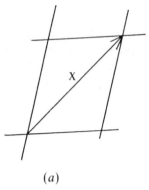
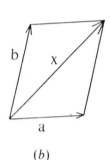

(a)   (b)

Fig. 1.13 – Resolving a two-dimensional vector.

falling (good!). In other words, we resolve the force in the horizontal and vertical directions. Procedures like this are often useful.

Basis vectors

Consider two nonzero nonparallel vectors **u** and **v** (lying in a fixed plane, like all the vectors in this section). Every vector **x** in the plane can be expressed as a sum of vectors parallel to **u** and **v** respectively. A vector parallel to **u** is of the form $a\mathbf{u}$ for some number a: similarly for **v**. Thus every **x** can be expressed as $a\mathbf{u} + b\mathbf{v}$ for some numbers a and b (which will depend on **x**, of course). The pair of vectors **u**, **v** is called a basis set, because every vector in the plane can be expressed in terms of the basic vectors **u** and **v**.

Definition 1. A **basis** for the vectors in a plane is a pair of nonzero vectors in different directions.

Note. The phrase 'in different directions' is to be interpreted as 'pointing along different lines'. Two vectors pointing in precisely opposite directions, such as **v** and $-\mathbf{v}$, are along the same line, and are not regarded as being in different directions for the purposes of Definition 1.   □

Basis sets of two vectors express in mathematical terms the idea that planes are two-dimensional. In the next section we shall see that three-dimensional space requires basis sets of three vectors, and in later chapters we shall define higher-dimensional (even infinite-dimensional) spaces using similar ideas.

Example 1. In the horizontal plane, let **u**, **v**, **w** be vectors of magnitude 1, 3 and 3 respectively, pointing north, northeast and south respectively. Then $\{\mathbf{u}, \mathbf{v}\}$ is a basis for the set of horizontal vectors. Since **u** and **w** point along the same line (see the note above), $\{\mathbf{u}, \mathbf{w}\}$ is not a basis.

Definition 2. An expression of the form $a\mathbf{u} + b\mathbf{v}$ is called a **linear combination** of **u** and **v**.
                                                                        □

Thus every vector in the plane is a linear combination of the basis vectors. The coefficients a and b are called 'components'.

Definition 3. The **components** of a vector **x** with respect to a basis $\{u, v\}$ are the numbers a and b such that $x = au + bv$.   □

The components of a vector with respect to a given basis contain all the information about the vector; if the components of **x** are known, then **x** can easily be written down: $x = au + bv$.

Now, there are infinitely many different pairs of basis vectors. Any of these pairs can be used to express a given vector in terms of components. In practice one chooses a basis which makes the work as simple as possible. For example, when an aircraft is flying horizontally it is natural to take a basis consisting of horizontal and vertical vectors, but when the aircraft is climbing it may be better to take basis vectors parallel and perpendicular to the aircraft. In general, the choice of basis depends on the details of the problem at hand.

The algebra of components

We have seen that the components contain all the information about the vector. In fact, calculations with vectors can be completely replaced by calculations with components. The advantage of this is that components are numbers, and adding numbers is much simpler than adding geometrical vectors by the parallelogram rule.

In order to use this method, we must be sure that each vector corresponds to a unique set of components.

Theorem 1 (uniqueness of components). With respect to a given basis, each vector has one and only one pair of components.

Proof. We must show that there cannot be two different pairs of components for the same vector. In other words, if

$$x = au + bv \quad \text{and} \quad x = Au + Bv \tag{1}$$

we must show that $a = A$ and $b = B$. This is easy. Subtracting the two equations in (1) gives $0 = (a - A)u + (b - B)v$ or

$$(A - u)u = (b - B)v \tag{2}$$

We wish to prove that $A - a = B - b = 0$. It follows from (2) that if one vanishes, then the other does. If they are both bonzero, then (2) says that a vector parallel to **u** equals a vector parallel to **v**, which is impossible because **u** and **v** are in different directions (by Definition 1). We have now ruled out all possibilities except $A - a = B - b = 0$, so $A = a$ and $B = b$, as required.   □

The phrase 'uniqueness of components' must be interpreted with care. Components with respect to a given basis are unique, but there are many different bases, and a given vector has different components with respect to different bases.

We shall now consider the effect on the components of performing algebraic operations on the vectors. We need a more systematic notation to deal with more complicated algebraic expressions.

Notation. $\{\mathbf{b}_1, \mathbf{b}_2\}$ will denote a pair of basis vectors. The components of a vector $\mathbf{v}$ with respect to this basis will be written V_1 and V_2. Thus for any vector $\mathbf{v}$ we have

$$\mathbf{v} = V_1 \mathbf{b}_1 + V_2 \mathbf{b}_2 \tag{3}$$

Theorem 2 (algebra in components). The components of a linear combination of vectors are the corresponding combinations of their components. In symbols: for any numbers c and d and any vectors $\mathbf{u}$ and $\mathbf{v}$, the components of $c\mathbf{u} + d\mathbf{v}$ are $cU_1 + dV_1$ and $cU_2 + dV_2$.

Proof. Applying (3) to $\mathbf{u}$ and $\mathbf{v}$ we have

$$c\mathbf{u} + d\mathbf{v} = cU_1 \mathbf{b}_1 + cU_2 \mathbf{b}_2 + dV_1 \mathbf{b}_1 + dV_2 \mathbf{b}_2 = (cU_1 + dV_1)\mathbf{b}_1 + (cU_2 + dV_2)\mathbf{b}_2$$

Thus the components of $c\mathbf{u} + d\mathbf{v}$ are $cU_1 + dV_1$ and $cU_2 + dV_2$.   □

Theorem 2 shows that algebraic operations on vectors are exactly mirrored by operations on their components. Hence geometrical vector algebra can be replaced by calculations with numbers, the components of the vectors.

Finally we express the magnitude of a vector in terms of its components. In our standard notation we have

$$\mathbf{v} = V_1 \mathbf{b}_1 + V_2 \mathbf{b}_2$$

If θ is the angle between $\mathbf{b}_1$ and $\mathbf{b}_2$ (see Fig. 1.14), trigonometry gives

$$\|\mathbf{v}\|^2 = \|V_1 \mathbf{b}_1\|^2 + \|V_2 \mathbf{b}_2\|^2 - 2\|V_1 \mathbf{b}_1\|\|V_2 \mathbf{b}_2\| \cos\theta$$
$$= (V_1\|\mathbf{b}_1\|)^2 + (V_2\|\mathbf{b}_2\|)^2 - 2|V_1 V_2|\|\mathbf{b}_1\|\|\mathbf{b}_2\| \cos\theta$$

This is an ugly formula. It becomes simpler if the basis vectors have magnitude 1.

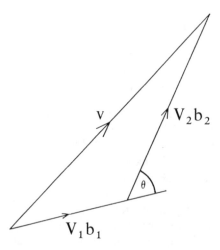

Fig. 1.14 – A two-dimensional vector resolved into components.

Definition 4. A **unit vector** is a vector with magnitude 1.   □

The magnitude of a vector is given in terms of its components with respect to a basis consisting of unit vectors by

$$\|\mathbf{v}\| = [(V_1)^2 + (V_2)^2 - 2|V_1 V_2|\cos\theta]^{1/2}$$

The expression simplifies further if the basis vectors are perpendicular. Then $\cos\theta = 0$, and we have

$$\|\mathbf{v}\| = [(V_1)^2 + (V_2)^2]^{1/2} \tag{4}$$

if V_1 and V_2 are the components of $\mathbf{v}$ with respect to perpendicular unit vectors. This formula can be interpreted as a form of Pythagoras' theorem – see Fig. 1.15.

   This completes our discussion of two-dimensional vectors. The next section extends the ideas to three dimensions.

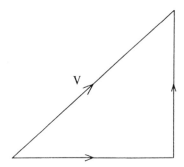

Fig. 1.15 – A two-dimensional vector resolved into components along perpendicular unit vectors.

1D   THREE-DIMENSIONAL VECTORS

The two-dimensionality of a plane means that plane vectors can be expressed as linear combinations of a basis of two vectors. In three-dimensional space, one expects a basis to consist of three vectors. However, some care is needed.

   A basis for a set S of vectors must, by definition, have the property that each vector in S is a linear combination of the basis vectors. In the last section we saw that if S is the set of vectors in a plane, then every pair of nonzero vectors in different directions is a basis. Is every set of three nonzero vectors in different directions in three-dimensional space a basis?

Exercise 1. Find a set of three nonzero vectors, in different directions, which does not satisfy the condition of the paragraph above. (Reminder: exercises like this, in the middle of a section, are intended to be done before you read on. Answers are at the back of the book.)   □

We shall define a basis for three-dimensional vectors in a way that excludes the awkward case of Exercise 1.

Definition 1. A basis for three-dimensional vectors is a set of three nonzero vectors which do not all lie in a single plane.   □

If a basis $\{\mathbf{u}, \mathbf{v}, \mathbf{w}\}$ is given, any vector $\mathbf{x}$ can be expressed as a linear combination of the basis vectors. The procedure for doing this is as follows.

Draw arrows representing the vectors, as in Fig. 1.16(a). The two basis vectors $\{\mathbf{u}, \mathbf{v}\}$ define a plane through the origin containing $\mathbf{u}$ and $\mathbf{v}$. There are two other planes through the origin containing the other two pairs of basis vectors. Draw planes parallel to these planes, passing through the tip of the $\mathbf{x}$ arrow. The six planes form a parallelepiped (that is, a box whose angles need not be right angles). The three edges of the parallelepiped give vectors $\mathbf{p}$, $\mathbf{q}$, $\mathbf{r}$ parallel to $\mathbf{u}$, $\mathbf{v}$ and $\mathbf{w}$ respectively, such that $\mathbf{x} = \mathbf{p} + \mathbf{q} + \mathbf{r}$ (see Fig. 1.16(b)). The vector $\mathbf{x}$ is thus resolved in the directions of the three given basis vectors. Since $\mathbf{p}$ is parallel to $\mathbf{u}$, there is a number a such that $\mathbf{p} = a\mathbf{u}$. Similarly there are numbers b and c such that $\mathbf{q} = b\mathbf{v}$ and $\mathbf{r} = c\mathbf{w}$, and we have $\mathbf{x} = a\mathbf{u} + b\mathbf{v} + c\mathbf{w}$. In this way, every vector can be expressed in terms of a basis set, and the coefficients are called the **components** of $\mathbf{x}$ with respect to the basis $\{\mathbf{u}, \mathbf{v}, \mathbf{w}\}$.

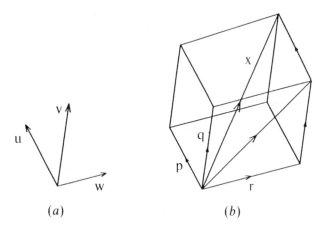

(a)   (b)

Fig. 1.16 – Resolving a three-dimensional vector.

We shall now show that components in three dimensions behave much as in the plane case.

Theorem 1 (uniqueness of components). With respect to a given basis, a given vector has a unique set of components.

Proof. Let $\{\mathbf{b}_1, \mathbf{b}_2, \mathbf{b}_3\}$ be a basis, and suppose that

$$\mathbf{x} = a_1\mathbf{b}_1 + a_2\mathbf{b}_2 + a_3\mathbf{b}_3$$

and

$$\mathbf{x} = A_1\mathbf{b}_1 + A_2\mathbf{b}_2 + A_3\mathbf{b}_3$$

so that a_1, a_2, a_3 and A_1, A_2, A_3 are two sets of components of $\mathbf{x}$. We must show that the a's equal the A's.

Subtracting the equations above gives

$$(a_1 - A_1)\mathbf{b}_1 + (a_2 - A_2)\mathbf{b}_2 + (a_3 - A_3)\mathbf{b}_3 = 0$$

Suppose that one of the a's is different from the corresponding A, say $(a_2 - A_2) \neq 0$. Then we have

$$\mathbf{b}_2 = c_1\mathbf{b}_1 + c_3\mathbf{b}_3 \qquad (1)$$

where $c_i = (A_i - a_i)/(a_2 - A_2)$ for $i = 1$ and 3. But (1) implies that $\mathbf{b}_2$ is a linear combination of $\mathbf{b}_1$ and $\mathbf{b}_3$ and therefore lies in the same plane. This is impossible because three basis vectors cannot lie in the same plane (Definition 1). Hence no a can be different from the corresponding A; there cannot be two different sets of components. □

In the same way as in section 1C, one can show that the components of a linear combination of vectors are combinations of their components. The magnitude of a vector is expressed in terms of components by a complicated formula, as in the two-dimensional case; but if the basis consists of unit vectors perpendicular to each other, then

$$\| \mathbf{v} \| = [(V_1)^2 + (V_2)^2 + (V_3)^2]^{1/2}$$

This can be interpreted as a three-dimensional version of Pythagoras' theorem: the square of the length of the diagonal of a rectangular box equals the sum of the squares of the three edges. It can be proved by two applications of the ordinary Pythagoras theorem (see Fig. 1.17).

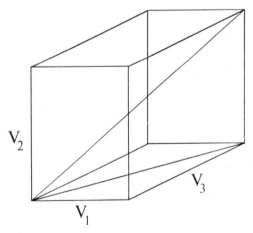

Fig. 1.17 – A three-dimensional vector resolved into components along perpendicular unit vectors.

1E   THE DOT PRODUCT

We have defined the product of a vector and a number. What about the product of two vectors? There are two useful ways of multiplying two vectors, one of which gives a vector as the product, the other a scalar. The vector product is rather specialised; it works in three-dimensional space but not in higher dimensions. We shall consider only the scalar product.

The product of two vectors which gives a scalar is usually denoted by $\mathbf{a} \cdot \mathbf{b}$, and is therefore called the dot product.

Definition 1. The **dot product** of two vectors $\mathbf{a}$ and $\mathbf{b}$ is the number $\mathbf{a} \cdot \mathbf{b} = \| \mathbf{a} \| \, \| \mathbf{b} \| \cos \theta$. Here θ is the angle between $\mathbf{a}$ and $\mathbf{b}$, defined as in Fig. 1.18 to be between 0 and π inclusive. □

Fig. 1.18 – Angle between two vectors.

The dot product of two parallel vectors is just the product of their magnitudes, because $\cos(0) = 1$. In particular, for any vector $\mathbf{a}$ we have

$$\mathbf{a} \cdot \mathbf{a} = \| \mathbf{a} \|^2 \tag{1}$$

If two vectors are at right angles, the cosine of the angle between them is zero, so their dot product is zero. Vectors at right angles are said to be **orthogonal** to each other. Orthogonality means vanishing of the dot product.

Warning 1. The example above illustrates a peculiar feature of vector algebra. In ordinary algebra, the product of two numbers is zero only if one of them is zero. That is why a nonzero factor can be cancelled from both sides of an equation: if $xy = xz$, then $x(y - z) = 0$, so either $x = 0$ or $y = z$. The rule 'if $xy = xz$ and $x \neq 0$ then $y = z$' is called the **cancellation law**.

In vector algebra it does not hold. The equation $\mathbf{x} \cdot \mathbf{y} = \mathbf{x} \cdot \mathbf{z}$ is equivalent to $\mathbf{x} \cdot (\mathbf{y} - \mathbf{z}) = 0$, or $\| \mathbf{x} \| \, \| \mathbf{y} - \mathbf{z} \| \cos \theta = 0$. This will be true, regardless of the magnitudes of $\mathbf{x}$, $\mathbf{y}$ and $\mathbf{z}$, if $\cos \theta = 0$, that is. the angle between $\mathbf{x}$ and $\mathbf{y} - \mathbf{z}$ is a right angle. The cancellation law does not hold for dot products. □

The $\cos \theta$ factor in Definition 1 is responsible for the failure of the cancellation law. But is this factor really necessary? Why not define the product of two vectors simply as the product of their magnitudes?

Exercise 1. Define a new kind of product by $\mathbf{a} * \mathbf{b} = \| \mathbf{a} \| \, \| \mathbf{b} \|$. Show that $\mathbf{a} * (\mathbf{b} + \mathbf{c})$ is not always equal to $\mathbf{a} * \mathbf{b} + \mathbf{a} * \mathbf{c}$. (Hint: consider the case $\mathbf{b} = -\mathbf{c}$.) □

Exercise 1 shows that omitting the $\cos \theta$ term leads to breakdown of the usual algebraic rules of manipulation. That is why the $\cos \theta$ term is there. We shall soon see another advantage of Definition 1: it has a simple geometrical meaning.

It is clear from Exercise 1 that we cannot take the properties of multiplication for granted; they need proof. Since the dot product has been defined geometrically, the proofs must be geometrical. They depend on the following idea.

Definition 2. Given two nonzero vectors **a** and **b**, the **projection** of **a** on **b** is the number $\|\mathbf{a}\| \cos \theta$, where θ is the angle between **a** and **b**. If $\mathbf{a} = \mathbf{0}$, its projection on to any nonzero vector is 0. Projection on to the zero vector is not defined.   □

To interpret this definition, draw arrows representing **a** and **b** with tails at a point O (see Fig. 1.19). Draw a line from the head of the **a** arrow perpendicular to **b**, meeting the line of the **b** arrow at P. Simple trigonometry shows that the projection of **a** on **b** equals the length of OP if OP is in the same sense as the **b** arrow, as in Fig. 1.19, and minus the length if not, as in Fig. 1.20. Note that if $\mathbf{b} = \mathbf{0}$ then it has no direction, and this construction cannot be carried out.

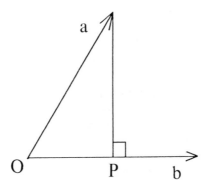

Fig. 1.19 – Projection of **a** on **b**: first case.

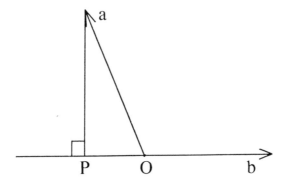

Fig. 1.20 – Projection of **a** on **b**: second case.

Why is it called the 'projection'? If a beam of light shines vertically downwards in the plane of Fig. 1.19, the **a** arrow casts a shadow on the **b** arrow. The projection of **a** on **b** equals the length of the shadow cast, or projected, by **a** on to a line along the direction of **b**.

Exercise 2. Does the projection of **a** on **b** equal the projection of **b** on **a**?

Proposition 1. If $\mathbf{b} \neq \mathbf{0}$, then $\mathbf{a} \cdot \mathbf{b}$ equals $\|\mathbf{b}\|$ times the projection of **a** on **b**.

Proof. It follows at once from the definitions.   □

If you are not familiar with the term 'proposition', consult Appendix A. You might like to read it in any case.

The interpretation of the dot product in terms of projections is the key to its algebraic properties.

Theorem 1 (properties of the dot product). For any vectors **a**, **b**, **c** and any number k, we have

(i) $\mathbf{a} \cdot \mathbf{b} = \mathbf{b} \cdot \mathbf{a}$;

(ii) $(k\mathbf{a}) \cdot \mathbf{b} = k(\mathbf{a} \cdot \mathbf{b})$;

(iii) $\mathbf{a} \cdot (\mathbf{b} + \mathbf{c}) = \mathbf{a} \cdot \mathbf{b} + \mathbf{a} \cdot \mathbf{c}$;

(iv) $|\mathbf{a} \cdot \mathbf{b}| \leq \|\mathbf{a}\| \, \|\mathbf{b}\|$.

Proof. (i) follows at once from the definition, which is completely symmetrical between **a** and **b**.

(ii) Suppose $k \geq 0$. Then $k\mathbf{a}$ is a vector in the direction of **a** and k times its magnitude, so its projection on **b** is k times the projection of **a**. If $k < 0$ then $k\mathbf{a}$ is in the opposite direction to **a**, so its projection has the opposite sign, agreeing with $k(\mathbf{a} \cdot \mathbf{b})$.

(iii) Using (i), we have $\mathbf{a} \cdot (\mathbf{b} + \mathbf{c}) = (\mathbf{b} + \mathbf{c}) \cdot \mathbf{a}$. Fig. 1.21 shows that the projection of **b** + **c** on **a** equals the sum of the projections of **b** and **c**. The figure is drawn for the case where all the projections are positive; the other cases work in a similar way. Of course, if **a** = **0** then the projections are not defined, but then (iii) holds because all the dot products are zero.

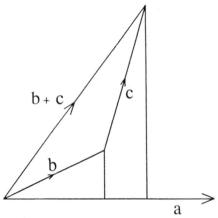

Fig. 1.21 – Projection of the sum of two vectors.

(iv) This follows at once from the definition and the fact that $|\cos \theta| \leq 1$ for all θ.   □

Theorem 1 implies that the usual rules of algebraic manipulation apply to the dot product. For example, brackets can be removed in the usual way:

$$(k\mathbf{a} + m\mathbf{b}) \cdot (n\mathbf{c} + p\mathbf{d}) = kn\mathbf{a} \cdot \mathbf{c} + mn\mathbf{b} \cdot \mathbf{c} + kp\mathbf{a} \cdot \mathbf{d} + mp\mathbf{b} \cdot \mathbf{d}$$

This is proved by using (iii) to express the left-hand side as $(k\mathbf{a} + m\mathbf{b})\cdot(n\mathbf{c}) +$
$(k\mathbf{a} + m\mathbf{b})\cdot(p\mathbf{d})$, using (i) to reverse the products, then using (iii) again and finally (ii).
The details are very boring. Manipulations of this sort can generally be taken for
granted now that Theorem 1 has been established.

Warning 2. Apart from the cancellation law, most of the rules of algebra apply to
vector addition and dot products. But care is needed. Vector equations contain both
vectors and numbers. You may be surprised to learn that beginners sometimes write
things like $\mathbf{a} + \mathbf{b}\cdot\mathbf{c}$, which is nonsense, of course; a vector cannot be added to a
number. One should always check that equations make sense. It should be as
instinctive as, say, a driver checking the road ahead and behind; failure to do so can
be fatal.
    Again, in ordinary algebra one can multiply three or more numbers together. But
$\mathbf{a}\cdot\mathbf{b}\cdot\mathbf{c}$ does not make sense in vector algebra. Since $\mathbf{a}\cdot\mathbf{b}$ is a scalar, one cannot take
its dot product with $\mathbf{c}$. One can multiply the scalar $\mathbf{a}\cdot\mathbf{b}$ by the vector $\mathbf{c}$; this must be
written $(\mathbf{a}\cdot\mathbf{b})\mathbf{c}$. Then $(\mathbf{a}\cdot\mathbf{b})\mathbf{c} \neq \mathbf{a}(\mathbf{b}\cdot\mathbf{c})$; the first vector is parallel to $\mathbf{c}$, the second to $\mathbf{a}$.

Warning 3. We have defined vector addition and multiplication, but not division.
You might care to think about what obstacles lie in the way of defining division. But
all that you really need to know is that our theory of vectors does not include division,
so:

DIVIDING BY A VECTOR IS NONSENSE

Orthogonality and components

We have seen that three-dimensional vectors can be described completely by their
components with respect to a basis. The dot product should therefore be expressible
in terms of components. If vectors $\mathbf{u}$ and $\mathbf{v}$ have components U_i and V_i $(i = 1, 2, 3)$
with respect to a basis $\{\mathbf{b}_i: i = 1, 2, 3\}$, then

$$\mathbf{u}\cdot\mathbf{v} = (U_1\mathbf{b}_1 + U_2\mathbf{b}_2 + U_3\mathbf{b}_3)\cdot(V_1\mathbf{b}_1 + V_2\mathbf{b}_2 + V_3\mathbf{b}_3) \tag{2}$$

The product can be expanded into nine terms involving all possible dot products of
pairs of the $\mathbf{b}_i$. In general, this is not very useful.
    But suppose that the basis vectors are orthogonal, that is, perpendicular to each
other like the usual Cartesian x, y, z axes. Then $\mathbf{b}_1\cdot\mathbf{b}_2 = \mathbf{b}_1\cdot\mathbf{b}_3 = \ldots = 0$, and only
three terms survive from the expansion of (2):

$$\mathbf{u}\cdot\mathbf{v} = U_1V_1\mathbf{b}_1\cdot\mathbf{b}_1 + U_2V_2\mathbf{b}_2\cdot\mathbf{b}_2 + U_3V_3\mathbf{b}_3\cdot\mathbf{b}_3$$

The formula is simpler still if $\mathbf{b}_1$, $\mathbf{b}_2$ and $\mathbf{b}_3$ are unit vectors, so that $\mathbf{b}_1\cdot\mathbf{b}_1 = \ldots = 1$
(cf. (1)). Then

$$\mathbf{u}\cdot\mathbf{v} = U_1V_1 + U_2V_2 + U_3V_3 \tag{3}$$

Definition 3. An **orthonormal basis** is a basis consisting of orthogonal unit vectors.
$\square$

Orthonormal bases allow one to calculate dot products by the simple formula (3). But orthonormality has another advantage.

In section 1D we showed how to find the components of a vector with respect to a given basis. The method involved drawing planes in three dimensions, not a very practical proposition. But if the basis is orthonormal, there is a much simpler way.

Let $\{e_1, e_2, e_3\}$ be an orthonormal basis. The components of the vector e_1 with respect to this basis are 1, 0, 0 because $e_1 = 1e_1 + 0e_2 + 0e_3$. For any vector u, equation (3) therefore gives

$$u \cdot e_1 = U_1$$

Thus the first component of u can be found by taking its dot product with the first basis vector. In the same way we have

$$u \cdot e_i = U_i \quad \text{for } i = 1, 2, 3 \tag{4}$$

Warning 4. This method of finding components works for orthonormal bases only.
□

It is often convenient to express (4) in the form

$$U_i = \|u\| \cos \theta_i \tag{5}$$

where θ_i is the angle between u and the ith orthonormal basis vector. The component of any vector in any direction equals the magnitude of the vector times the cosine of the angle between the direction and the vector.

Example 1. Suppose an aircraft is descending at an angle θ to the horizontal. There is an air resistance force of magnitude R, say, in the direction opposite to its motion. Question: how much does this help to balance the force of gravity? Answer: the resistance force is at an angle $\pi/2 - \theta$ to the vertical, hence by (5) its vertical component is $R \cos(\pi/2 - \theta) = R \sin \theta$, acting against gravity.

1F n-VECTORS

We began this chapter with geometrical reasoning. But as the subject develops, the geometry becomes overlaid with an algebraic formalism based on components. In this section we shall take the algebraic point of view and extend it to n dimensions. The theory does not depend logically on geometrical intuition. But geometry is not far from the surface; it underlies and motivates the algebraic theory.

The algebraic approach to vectors is based on components. If a basis is given, a vector in three-space is represented by a set of three numbers, its components. This set of three numbers will be called a 3-vector. In a similar way we can consider sets of n numbers for any n.

Definition 1. For any positive integer n, an n-vector is an ordered set of n numbers. (See Appendix C for an account of ordered sets.)

Notation. The *n*-vector consisting of the numbers $a_1, a_2, \ldots, a_n$ is denoted by $(a_1, a_2, \ldots, a_n)$, or by the single letter a.   □

The number a_r (where r is a number from 1 to n) is called the *r*th **entry** of the vector; the idea is that the numbers are entered into the format (, ,...,). We use ordinary type for *n*-vectors, reserving bold type for geometric vectors.

Now, suppose that we have a basis in 3-space. Each geometric vector corresponds to a certain 3-vector, containing its components with respect to that basis. Changing to a new basis will give different components, and thus different 3-vectors corresponding to the same geometric vector. But for a fixed basis, each geometric vector corresponds to a unique 3-vector.

Vectors in two-dimensional space have pairs of components, and correspond to 2-vectors. We shall shortly meet an example of an *n*-vector for $n > 3$. But first we lay down the rules for adding and multiplying *n*-vectors. We simply generalise the rules for 2-vectors and 3-vectors given earlier in this chapter.

Definition 2. The **sum** of two *n*-vectors a and b is the *n*-vector $a + b = (a_1 + b_1, a_2 + b_2, \ldots, a_n + b_n)$. In other words: to add *n*-vectors, just add the entries.

Definition 3. If a is an *n*-vector and k is a number, the **product** of k and a is the *n*-vector $ka = (ka_1, ka_2, \ldots, ka_n)$.

Definition 4. The **dot product** of *n*-vectors a and b is the number

$$a \cdot b = a_1 b_1 + a_2 b_2 + \ldots + a_n b_n.$$ □

Theorem 1C-2 shows that Definitions 2 and 3 correspond to the rules for adding geometric vectors and multiplying them by numbers; the 2-vector of components of $\mathbf{u} + \mathbf{v}$ is the sum of the 2-vectors of components of $\mathbf{u}$ and $\mathbf{v}$, and so on. Similarly for three dimensions.

In the same way, equation (3) of section 1E shows that the dot product of two three-dimensional vectors equals the dot product of their component 3-vectors – provided that the basis is orthonormal. As far as the dot product is concerned, there is a simple relation between geometric and algebraic vectors only when the basis is orthonormal, whereas for addition and multiplication by a scalar, any basis can be used. For this reason, the abstract structure of linear algebra, set out in Part II, is based on addition and multiplication by scalars only; dot products appear at a much later stage.

It is easy to prove that addition and multiplication of *n*-vectors obey the usual algebraic rules: $a + b = b + a$, $k(a + b) = ka + kb$, and so on. These relations hold for *n*-vectors because they hold for the individual numbers of which they are composed. The dot product too has the usual properties, such as $a \cdot (b + c) = a \cdot b + a \cdot c$. This is easily proved by writing out both sides as sums of products of components, and showing that they are equal.

We can define the magnitude of an *n*-vector by means of equation (1) of section 1E.

Definition 5. The **magnitude** of an *n*-vector x is the number $\|x\| = \sqrt{(x \cdot x)}$.   □

Note that the symbol $\sqrt{\ }$ denotes the positive square root of a positive number (except that $\sqrt{0}=0$, of course). It is impossible for $x\cdot x$ to be negative, because it is a sum of squares: $(x_1)^2+(x_2)^2+\dots$. It is easy to verify that the magnitude of an n-vector has the obvious properties: the magnitude of $2x$ is twice the magnitude of x, and so on. We shall give a systematic treatment in Chapter 11.

The dot product can also be used to define angles in n-space. This is done by taking over the formula in Definition 1E-1, but turning the logic upside down. In section 1E, the notion of angle in three-dimensional space was regarded as intuitively clear, and we defined the dot product in terms of it. Here, in n-space, there is no intuitive idea of angle; we define it in terms of the dot product given in Definition 4.

Definition 6. The **angle** between two nonzero n-vectors x and y is defined as

$$\cos^{-1}\left[\frac{x\cdot y}{\|x\|\,\|y\|}\right]$$

The case of orthogonal vectors (where $\theta=90°$) is by far the most important.

Exercise 1. Show that each of the n-vectors $(1,0,0,\dots,0)$, $(0,1,0,0,\dots,0),\dots,$ $(0,0,\dots,0,1)$ has magnitude 1, and each is orthogonal to the others.

Remark. Definition 6 raises a question: what if the number in square brackets turns out to be greater than 1?

Answer: it never does. This will be proved in Chapter 11. (Until then we shall never need to assume that there is a well-defined real angle between every two nonzero vectors, so no logical inconsistency arises.)

Example 1. Shopping lists are n-vectors. Suppose I buy a_1 kg of peanuts, a_2 kg of betel nuts, a_3 kg of walnuts, and so on, down to a_n kg of hazelnuts. Let a be the n-vector with entries $a_1,\dots,a_n$, and let p be the n-vector whose ith entry is the price per kg of the ith variety of nut. Then the total cost is $a\cdot p$. Furthermore, if there are several transactions, represented by n-vectors $a,b,\dots,e$, then the overall cost is $(a+b+\dots+e)\cdot p$. Examples of this kind are discussed in Chapter 14.

PROBLEMS FOR CHAPTER 1

Hints and answers for problems marked [a] will be found on page 277.

Sections 1A, 1B

[a]1. Let u and v be the position vectors of points U and V. Show that for any number t, the vector $tu+(1-t)v$ is the position vector of a point on the line joining U and V. What values of t correspond to U? to V? to their midpoint? to points lying between U and V?

2. A, B, C, D are four points in space, and P, Q, R, S are the midpoints of the lines AB, BC, CD, DA. Is it obvious that the points P, Q, R, S all lie in one plane? This problem proves that PQRS is a parallelogram (and therefore lies in a plane).

Write down expressions for the position vectors of P, Q, R, S in terms of those of A, B, C, D. Hence show that the vectors corresponding to the lines PQ and RS are equal and parallel. Similarly for the other two sides of PQRS.

[a]3. Under what conditions are the following true?

  (a) $\|\mathbf{x} + \mathbf{y}\| = \|\mathbf{x}\| + \|\mathbf{y}\|$;

  (b) $\|\mathbf{x} + \mathbf{y}\| = \|\mathbf{x}\| - \|\mathbf{y}\|$;

  (c) $\|\mathbf{x} - \mathbf{y}\| + \|\mathbf{y} - \mathbf{z}\| = \|\mathbf{x} - \mathbf{z}\|$.

[a]4. A point P has position vector $\mathbf{p}$; a line L passes through P in the direction of a vector $\mathbf{d}$. Write down the position vector of a point on L at a distance r from P.

[a]5. $\mathbf{a}$ and $\mathbf{b}$ are vectors in different directions, and $\mathbf{p}$ is the position vector of a point P. Describe geometrically the set of all points with position vectors $\mathbf{p} + s\mathbf{a} + t\mathbf{b}$, where s and t are arbitrary numbers.

  (Hint: look at Problem 4.)

[a]6. $\mathbf{a}$, $\mathbf{b}$ and $\mathbf{c}$ are vectors in different directions, and $\mathbf{p}$ is the position vector of a point P. Describe geometrically the set of all points with position vectors $\mathbf{p} + s\mathbf{a} + t\mathbf{b} + u\mathbf{c}$, where s, t and u are arbitrary numbers.

  (Hint: look at Problem 5.)

Sections 1C, 1D

[a]7. Vectors $\mathbf{p}$ and $\mathbf{q}$ are horizontal, of magnitudes 1 and 1.3 respectively. The vector $\mathbf{p}$ points due north, and $\mathbf{q}$ points 60° east of north.

  (a) Make an accurate drawing of the triangle for adding $\mathbf{p}$ and $\mathbf{q}$. What is the magnitude and the direction of $\mathbf{p} + \mathbf{q}$?

  (b) Find the components of $\mathbf{p}$ and $\mathbf{q}$ with respect to a basis consisting of unit vectors north and east. Hence find the components of $\mathbf{p} + \mathbf{q}$. Deduce its magnitude and direction, checking that your answers agree with part (a).

[a]8. Take three points A, B, C, not lying on a straight line, and a point O not in their plane. Let $\mathbf{a}$, $\mathbf{b}$, $\mathbf{c}$ be the position vectors of A, B, C with respect to O. They form a basis for three-dimensional vectors. Let X be a point with position vector $\mathbf{x}$ having components s, t, u with respect to this basis.

  If X lies in the plane of A, B, C, what can you say about s, t, u? What if X lies inside the triangle ABC?

  (Hint: see Problem 1.)

Section 1E

[a]9. Given a vector $\mathbf{a} \neq \mathbf{0}$ and a number $k \neq 0$, show that the set of points with position vectors $\mathbf{r}$ satisfying $\mathbf{r} \cdot \mathbf{a} = k$ is a plane. What is the distance of the plane from the origin? Describe its orientation.

[a]10. Find all vectors $\mathbf{x}$ satisfying $\mathbf{x} + \mathbf{a}(\mathbf{x} \cdot \mathbf{b}) = \mathbf{0}$, where $\mathbf{a}$ and $\mathbf{b}$ are given vectors.

  (Warning: be careful – the answer depends on the value of $\mathbf{a} \cdot \mathbf{b}$ and you should consider all possible cases.)

11. Find all vectors x satisfying $x + a(x \cdot b) = c$, where a, b and c are given vectors. (Hint: do Problem 10 first.)

Section 1F

[a]12. Find (i) the set of all 4-vectors orthogonal to $(1, 1, 1, 1)$; (ii) the set of all 4-vectors orthogonal to $(1, 0, 0, 0)$; (iii) the intersection of the above two sets.

[a]13. (a) Define the '2-dimensional projection' of a 3-vector (x_1, x_2, x_3) to be the 2-vector (x_1, x_2) obtained by ignoring the third component. The projection of a set of 3-vectors is thus a set of 2-vectors.

What is the 2-dimensional projection of the unit sphere $S = \{x \in \mathbb{R}^3 : x \cdot x = 1\}$? (Note: 'sphere' means the surface, not the solid interior.)

For any number t, define the 't-slice' of the sphere S as the set of all points x in S such that $x_3 = t$. For example, the 0-slice is a circle of radius 1 in the xy plane. If $|t| > 1$, the t-slice does not exist (more correctly, it is the empty set). Describe the 2-dimensional projection of the t-slice for values of t ranging from -1 to 1.

(b) The unit sphere in $\mathbb{R}^n$ is the set of all n-vectors x such that $x \cdot x = 1$. Define the '3-dimensional projection' of an n-vector $(x_1, \dots, x_n)$ to be the 3-vector (x_1, x_2, x_3) obtained by ignoring all but the first three components.

(i) Describe the 3-dimensional projection of the unit sphere in $\mathbb{R}^n$ where $n > 3$.

(ii) Define the t-slice of the unit sphere in $\mathbb{R}^4$ as the set of all vectors on the sphere with 4-coordinate t. Describe the 3-dimensional projection of the t-slice for all possible values of t.

(For a readable and wide-ranging discussion, see Rucker (1986).)

14. If each of a, b, c can be either 0 or 1, there are eight possible 3-vectors (a, b, c). They are the position vectors of the corners of a cube. Similarly, the 4-vectors (a, b, c, d), where each of a, b, c, d can be either 0 or 1, are the corners of the '4-dimensional cube' in $\mathbb{R}^4$, called a hypercube or a tesseract. A cube can be drawn on paper; the picture contains two squares and four distorted squares all joined together. Similarly, a tesseract can be represented by a three-dimensional structure consisting of cubes and distorted cubes joined together. This structure itself can be drawn on paper. Try and visualise it.

(See Rucker (1986).)

2

Matrices

This chapter continues the train of thought of section 1F. An n-vector is a set of numbers arranged in a row. Section 2A shows that it is sometimes more natural to arrange numbers in a rectangle; this arrangement is called a matrix (plural: matrices). The rest of the chapter develops their properties.

This material is essential for what follows. But it is not all needed immediately; if you like, you can read the first two sections and then proceed to Chapter 3, coming back to this chapter when necessary.

2A   BASIC IDEAS

One of the main problems of algebra is solving sets of linear equations, such as

$$\left. \begin{array}{r} 2x - 3y + 2z = 1 \\ x + 2y - 3z = -1 \\ -x + 5y - 6z = 2 \end{array} \right\} \tag{1}$$

In this chapter we develop an efficient notation for dealing with them – and, as often happens in mathematics, the notation develops a life of its own, and turns into a major branch of mathematics in its own right. This is how the subject of linear algebra began (see Appendix H).

There is clearly scope for streamlining the notation of the equations above. There is no need to write each variable three times. It is the numerical coefficients that really matter. We write them down in a table, thus:

$$\begin{pmatrix} 2 & -3 & 2 \\ 1 & 2 & -3 \\ -1 & 5 & -6 \end{pmatrix}$$

The numbers on the right-hand sides are equally important; it is natural to arrange them in a vertical column, in the way that they appear in (1) above. We write down the names of the unknowns in a similar column, giving the formula

$$
\begin{pmatrix} 2 & -3 & 2 \\ 1 & 2 & -3 \\ -1 & 5 & -6 \end{pmatrix} \begin{pmatrix} x \\ y \\ z \end{pmatrix} = \begin{pmatrix} 1 \\ -1 \\ 2 \end{pmatrix}
\tag{2}
$$

This is simply another notation for equations (1). For three equations, it is not much shorter than the original version; but for larger systems, the saving of effort is considerable.

The general system of p equations for q unknowns is written in the form

$$
\left.\begin{array}{l} a_{11}x_1 + a_{12}x_2 + \ldots + a_{1q}x_q = b_1 \\ a_{21}x_1 + a_{22}x_2 + \ldots + a_{2q}x_q = b_2 \\ \cdots\cdots\cdots\cdots\cdots\cdots\cdots\cdots\cdots\cdots \\ a_{p1}x_1 + a_{p2}x_2 + \ldots + a_{pq}x_q = b_p \end{array}\right\}
\tag{3}
$$

The unknowns $x_1, x_2, \ldots, x_q$ are to be determined in terms of $b_1, \ldots, b_p$ and the pq numbers a_{11}, a_{12}, etc. Later chapters will deal with solving these equations; here we consider matters of notation only.

We abbreviate the system (3) in the same way as above:

$$
\begin{pmatrix} a_{11} & a_{12} & \cdots & a_{1q} \\ a_{21} & a_{22} & \cdots & a_{2q} \\ \cdot & \cdot & \cdots & \cdot \\ \cdot & \cdot & \cdots & \cdot \\ a_{p1} & a_{p2} & \cdots & a_{pq} \end{pmatrix} \begin{pmatrix} x_1 \\ x_2 \\ \cdot \\ \cdot \\ x_q \end{pmatrix} = \begin{pmatrix} b_1 \\ b_2 \\ \cdot \\ \cdot \\ b_p \end{pmatrix}
\tag{4}
$$

Each group of numbers enclosed in brackets here is called a matrix.

Definition 1. A **matrix** is a set of numbers arranged in a rectangle. The individual numbers are called the **entries** or **elements** of the matrix. □

Thus equation (4) contains three matrices. The set of entries in one horizontal line is called a **row** of the matrix, and the set of entries in one vertical line is called a **column**. In equation (2), the first matrix has three rows and three columns, each consisting of three elements. The second matrix in (2) has one column, consisting of three entries, and three rows, each having one entry.

A matrix with the same number of rows and columns, like the first matrix in (2), is called **square**; otherwise it is **rectangular**.

Examples 1. (a)

$$
\begin{pmatrix} 1 & -2 & 3.1 \\ 0 & 0.3 & 9 \end{pmatrix}
$$

is a rectangular matrix with two rows and three columns.

(b) (-3) is a matrix: it has one row and one column.   □

Each entry of a matrix is labelled by the row and the column in which it appears. Thus a_{rs} denotes the entry in the rth row and the sth column of the first matrix in (4); it is referred to as the (r, s) entry. The $(3, 2)$ entry of the first matrix in equation (2) is 5.

The diagonal line of entries running downwards from the top left corner is called the **main diagonal** of the matrix. In (2), the first matrix has $2, 2, -6$ on its main diagonal. These are often referred to as the **diagonal entries**, the word 'main' being omitted. The diagonal entries of Example 1(a) are $1, 0.3$. In general, the main diagonal will end at the right-hand edge if the matrix has more rows than columns, otherwise it ends at the bottom as in Example 1(a).

A matrix consisting of a single column, like the second and third matrices in the equations above, is called a **column matrix**; its elements can be labelled by a single index, as in equation (4). Similarly, a matrix consisting of a single row is called a **row matrix**.

Example 2. The equation $ax + by + cz = 1$ can be written in matrix notation as

$$(a, b, c)\begin{pmatrix} x \\ y \\ z \end{pmatrix} = (1)$$

This is a special case of equation (4), with $p = 1$ and $q = 3$.   □

Row and column matrices are closely related to n-vectors as defined in section 1F. A row matrix with n entries is an n-vector written out horizontally; a column matrix with n entries is an n-vector written vertically. A given n-vector can be written as either a row or a column matrix; the two matrices are different, though they come from the same n-vector.

A matrix with r rows and k columns is called an **r by k matrix**; the three matrices in (4) are p by q, q by 1, and p by 1 respectively. An r by k matrix is short and fat if $r < k$, and tall and thin if $r > k$; row and column matrices are extreme examples of these types.

The matrix with entries a_{rs} can be denoted by (a_{rs}). Here r and s are 'dummy indices', that is, they can be replaced by any other letters: (a_{rs}) and (a_{ij}) mean exactly the same thing, namely, the matrix with $(1, 1)$ entry a_{11}, $(1, 2)$ entry a_{12}, etc. Using this notation, we can write (4) more concisely as

$$(a_{rs})(x_i) = (b_j)$$

We often use a single letter to stand for a matrix, usually a capital letter. The matrix (a_{rs}) is often denoted by A, and in this notation (4) becomes

$$AX = B \tag{5}$$

When a matrix is denoted by a single letter such as A, it is often convenient to denote its (r, s) entry by A_{rs}, so that $A = (A_{rs})$. In our example here, A_{rs} and a_{rs} stand for the same thing. You may find the number of different notations and technical terms daunting at first, but you will soon get used to them.

Note carefully that when we write A and X next to each other in equation (5), it does not mean 'multiply A by X'; the operation of multiplying matrices has not yet been defined. The string of symbols $AX = B$ is merely a shorthand notation for the equations (3).

However, we shall soon introduce algebraic operations on matrices, which will give a different point of view on equation (5). We begin in the next section by defining addition for matrices.

2B   MATRIX ALGEBRA

A matrix is a convenient way of writing down a set of numbers. We can use matrices to add and multiply sets of numbers *en masse*. Given two matrices of the same size and shape, their sum is defined in the natural way by adding corresponding entries of the two given matrices. For example,

$$\begin{pmatrix} 2 & 3 & 4 \\ 5 & 4 & 3 \end{pmatrix} + \begin{pmatrix} 0 & -1 & 2 \\ 2 & 1 & 0 \end{pmatrix} = \begin{pmatrix} 2 & 2 & 6 \\ 7 & 5 & 3 \end{pmatrix}$$

Similarly, if k is a number and A is a matrix, we define the product kA to be the matrix obtained by multiplying each entry of A by k; for example,

$$-2 \begin{pmatrix} 2 & 3 \\ 3 & -1 \\ 0 & 1 \end{pmatrix} = \begin{pmatrix} -4 & -6 \\ -6 & 2 \\ 0 & -2 \end{pmatrix}$$

We express these definitions formally as follows:

Definition 1. If A and B are two p by q matrices and k is a number, we define $A + B$ to be the matrix with (i,j) entry $A_{ij} + B_{ij}$, and we define kA to be the matrix with (i,j) entry kA_{ij} (where i ranges from 1 to p, and j from 1 to q).   □

Just as in ordinary algebra, the matrix $(-1)A$ is written $-A$, and subtraction of two matrices A and B is defined by $A - B = A + (-B)$.

Exercise 1. Show that $A - B$ is the matrix with entries $A_{rs} - B_{rs}$.

Example 1. A railway network joins n towns. We describe the network by an n by n matrix R with $R_{ij} = 1$ if there is a direct rail connection between town i and j, and $R_{ij} = 0$ if not.

There is also a bus service, described by a matrix B with $B_{ij} = 1$ if there is a direct bus service between towns i and j, and $B_{ij} = 0$ if not. To find out whether it is possible to travel directly between towns i and j using train or bus, look at $R + B$: its (i,j) entry is positive if and only if such a journey is possible.

Railways need maintenance. Define a matrix M with $M_{ij} = 1$ if the line between towns i and j is blocked by engineering work, and $M_{ij} = 0$ otherwise. The the matrix $R - M$ gives information on which direct journeys are possible, taking the maintenance work into account.

Indirect journeys, involving changing trains, can also be dealt with by matrices; see Example 2D-2.   □

If matrix algebra is like ordinary algebra, then $A - A$ should be zero. So we need a matrix version of the number 0. We define it in the natural way, as the matrix with all its entries zero. There are in fact many zero matrices, of different dimensions: for each pair of positive integers m and n, there is an m by n zero matrix. Strictly speaking we should use a notation which indicates the dimensions, such as $0(m, n)$, but this is not usually necessary. In the equation $A - A = 0$, for example, 0 obviously denotes the zero matrix with the same dimensions as A.

We can now add and subtract matrices and multiply them by numbers. The usual rules of algebra apply: $A + B = B + A$ because $A_{rs} + B_{rs} = B_{rs} + A_{rs}$ for all r, s; $(A + B) + C = A + (B + C)$ for a similar reason; $k(A + B) = kA + kB$; and so on. One could write down formal proofs of these results. But they are so easy that it is hardly worthwhile. We shall simply note that addition and multiplication work for matrices just like numbers (or vectors – see Theorem 1B-2), and proceed to consider another operation on matrices, this time one which is nothing like the operations of ordinary arithmetic.

A p by q matrix is a rectangular array of pq numbers, arranged in p rows of q numbers each. We sometimes wish to turn the array sideways and write it as q rows of p numbers, or, what is the same thing, p columns of q numbers. This operation is called transposing the matrix.

Definition 2. The **transpose** of a p by q matrix A is the q by p matrix A^{T} defined by $(A^{\mathrm{T}})_{ij} = A_{ji}$ for $i = 1, \ldots, q$ and $j = 1, \ldots, p$.

Example 2. If

$$A = \begin{pmatrix} 1 & 3 & 5 \\ 2 & 4 & 6 \end{pmatrix}$$

then

$$A^{\mathrm{T}} = \begin{pmatrix} 1 & 2 \\ 3 & 4 \\ 5 & 6 \end{pmatrix}$$

Exercise 2. Show that: (i) the transpose of a column matrix is a row matrix, and vice versa; (ii) transposing a matrix twice gives the original matrix; (iii) if $A + B = C$, then $A^{\mathrm{T}} + B^{\mathrm{T}} = C^{\mathrm{T}}$.   □

It is easy to generalise (iii): given any equation involving addition, subtraction, and multiplication by numbers, one can transpose all the matrices and the equation is still true.

Consider the matrix R of Example 1. If a journey is possible in one direction, it is also possible in the other. Hence $R_{ij} = R_{ji}$ for all i and j, so $R = R^{\mathrm{T}}$. Matrices of this type are particularly important.

Definition 3. A matrix A is **symmetric** if $A = A^{\mathsf{t}}$, that is, if $A_{ij} = A_{ji}$ for all i, j.   □

Note that this is only possible for square matrices. Otherwise A and A^{T} have different numbers of rows and cannot possibly be equal.

Example 3.

$$\begin{pmatrix} 1 & 2 & 4 \\ 2 & 0 & -1 \\ 4 & -1 & 0 \end{pmatrix}$$

is a symmetric matrix.   □

Symmetric matrices are so called because they have a diagonal symmetry property: each element equals its mirror image across the main diagonal. They have a special place in matrix theory; they occur naturally in many applications, and they have very useful theoretical properties, as we shall see in Chapter 8.

But meanwhile, the next section considers another operation on matrices: multiplying them together.

2C   MATRIX MULTIPLICATION

The obvious way to multiply two matrices A and B is to multiply each entry of A by the corresponding entry of B. But there is a much more useful way of defining matrix multiplication, arising from the following train of thought.

In the equation $AX = B$ of section 2A, we regarded the x_r as unknowns, to be determined in terms of the given numbers a_{rs} and b_r. But sometimes the x's are known and the b's are not; the equations then determine the b's in terms of the x's. We can think of $AX = B$ as transforming a set of variables X into a new set B.

It is natural to consider the effect of two such transformations in succession.

Example 1. A bakery makes biscuits of three types, G, M, D,* according to the following recipes:

	Flour	Sugar	Butter	Margarine	
Type G	0.5	0.1	0.35	0.05	
Type M	0.55	0.15	0.15	0.05	(1)
Type D	0.65	0.2	0.0	0.15	

The figures give the quantity of ingredients, in kilograms, for one kg of the biscuits.

The raw-material cost of the biscuits depends on the prices of the ingredients. If b_1, b_2, b_3 are the costs per kg of the three types of biscuit, and $p_1, \ldots, p_4$ are the

---

* Good, mediocre, and disgusting.

prices of a kg of flour, sugar, butter, and margarine respectively, then

$$\left.\begin{aligned}
b_1 &= 0.5p_1 + 0.1p_2 + 0.35p_3 + 0.05p_4 \\
b_2 &= 0.55p_1 + 0.15p_2 + 0.15p_3 + 0.05p_4 \\
b_3 &= 0.65p_1 + 0.2p_2 \qquad\quad\; + 0.15p_4
\end{aligned}\right\} \tag{2}$$

In the matrix notation of section 2A, we have

$$B = RP \tag{3}$$

where B and P are column matrices with three and four entries respectively, and R is the 3 by 4 recipe matrix in (1) above. The matrix R thus transforms P (raw-material prices) into B (biscuit costs).

The company sells three kinds of tins of assorted biscuits, made up as follows:

	Type G	Type M	Type D	
Assortment I	1.1	0.5	0.4	
Assortment II	0.5	1.0	0.5	(4)
Assortment III	0.0	0.3	0.7	

The table gives the amount of each type of biscuit, in kg, in each type of tin. The costs t_1, t_2, t_3 of the three tins are

$$t_1 = 1.1b_1 + 0.5b_2 + 0.4b_3$$
$$t_2 = 0.5b_1 + b_2 + 0.5b_3$$
$$t_3 = \qquad\; 0.3b_2 + 0.7b_3$$

These equations can be written

$$T = AB \tag{5}$$

where T is the column matrix with elements t_r, and A is the matrix in (4).

We have three columns of numbers, P (prices of ingredients), B (costs of the biscuits), and T (costs of the tins of biscuits). The matrix R transforms P into B according to (3), and the matrix A transforms B into T according to (5). It would be useful to be able to transform P directly into T, without calculating B. We shall now construct the mathematical apparatus for doing this. □

Consider then three sets of numbers X, Y, Z, related by sets of linear equations like (2). In our abbreviated notation we have

$$\left.\begin{aligned}
BX &= Y \\
AY &= Z
\end{aligned}\right\} \tag{6}$$

where A and B are matrices containing the coefficients of the equations. Writing out the general case in detail is very complicated. For simplicity we shall take the case where Z and X are 2-vectors and Y is a 3-vector; this case illustrates the essential idea.

The equation $BX = Y$ expresses the three components y_1, y_2 and y_3 in terms of the two components of X, giving

$$y_1 = B_{11}x_1 + B_{12}x_2$$
$$y_2 = B_{21}x_1 + B_{22}x_2$$
$$y_3 = B_{31}x_1 + B_{32}x_2$$

Similarly, $AY = Z$ stands for

$$z_1 = A_{11}y_1 + A_{12}y_2 + A_{13}y_3$$
$$z_2 = A_{21}y_1 + A_{22}y_2 + A_{23}y_3$$

To go directly from the x's to the z's, we must eliminate the y's. Thus

$$z_1 = A_{11}(B_{11}x_1 + B_{12}x_2) + A_{12}(B_{21}x_1 + B_{22}x_2) + A_{13}(B_{31}x_1 + B_{32}x_2)$$
$$= (A_{11}B_{11} + A_{12}B_{21} + A_{13}B_{31})x_1 + (A_{11}B_{12} + A_{12}B_{22} + A_{13}B_{32})x_2$$

A similar calculation gives

$$z_2 = (A_{21}B_{11} + A_{22}B_{21} + A_{23}B_{31})x_1 + (A_{21}B_{12} + A_{22}B_{22} + A_{23}B_{32})x_2$$

We thus have a set of equations relating the z's to the x's, which can be written

$$Z = CX$$

where

$$C_{11} = A_{11}B_{11} + A_{12}B_{21} + A_{13}B_{31}$$

and generally

$$C_{ij} = A_{i1}B_{1j} + A_{i2}B_{2j} + A_{i3}B_{3j} \tag{7}$$

for $i, j = 1, 2$.

Since $Z = AY$ and $Y = BX$, it is natural to write $Z = ABX$ and define the product AB to be the matrix C with entries defined by (7).

Example 2. Take

$$A = \begin{pmatrix} 1 & 2 & 3 \\ 4 & 5 & 6 \end{pmatrix} \qquad B = \begin{pmatrix} 7 & -8 \\ -5 & 6 \\ -3 & 4 \end{pmatrix}$$

Then

$$C_{11} = 1(7) + 2(-5) + 3(-3) \qquad C_{12} = 1(-8) + 2(6) + 3(4)$$
$$C_{21} = 4(7) + 5(-5) + 6(-3) \qquad C_{22} = 4(-8) + 5(6) + 6(4)$$

so

$$C = AB = \begin{pmatrix} 1 & 2 & 3 \\ 4 & 5 & 6 \end{pmatrix} \begin{pmatrix} 7 & -8 \\ -5 & 6 \\ -3 & 4 \end{pmatrix} = \begin{pmatrix} -12 & 16 \\ -15 & 22 \end{pmatrix} \qquad \square$$

We now generalise. Things work in the same way for matrices of any size, with one important proviso. The product matrix C is defined by (7), in which the second subscript of A takes the same values as the first subscript of B. Hence the second subscript of A must range over the same values as the first subscript of B. This means that the number of rows of A must equal the number of columns of B.

Definition 1. Two matrices C and D are said to be **conformable for multiplication**, in the order CD, if the number of columns of C equals the number of rows of D. If C and D are conformable, with C being m by n and D being n by p, then CD is the m by p matrix defined by

$$(CD)_{ij} = \sum_{k=1}^{n} C_{ik}D_{kj} \quad \text{for all } i,j \tag{8}$$

Of course, 'for all i,j' here means for all i from 1 to m and all j from 1 to p. □

Note that CD and DC are not the same thing in matrix multiplication. The order of the factors is vital.

Exercise 1. With A and B as in Example 2, calculate BA.

Example 3. In equation (3) above, RP is merely an abbreviation for equations (2). However, R is 3 by 4 and P is 4 by 1, so R and P are conformable according to our definition. Their product is the 3 by 1 matrix whose $(i, 1)$ entry is $\sum_k R_{ik}P_{k1}$. Here P_{k1} is the kth entry of the first (and only) column of P, so $P_{k1} = p_k$. Hence the ith entry of the column matrix RP is just b_i as given by equation (2). We can now interpret (3) as saying that B is the product of the matrices R and P. Our notation for sets of equations, as introduced in section 2A, can be consistently reinterpreted in terms of matrix products.

Example 4. A car hire firm has three depots. It is found that at the end of every month, a quarter of the cars which were in Cardiff at the beginning of the month are in London, and one fifth of them are in Manchester. Similarly, one third of the cars in London at the beginning of the month end up in Cardiff and one third in Manchester, and of the cars starting in Manchester, half end up in London and one sixth in Cardiff.

Let (C, L, M) be the numbers of cars in Cardiff, London and Manchester at the start of a month, and (C', L', M') the numbers at the start of the next month. Then C' equals the number left in Cardiff from the previous month plus the numbers arriving from London and Manchester, so $C' = (1 - 0.25 - 0.2)C + 0.33L + 0.17M$. There are similar expressions for L' and M'; they can be summarised as

$$\begin{pmatrix} C' \\ L' \\ M' \end{pmatrix} = \begin{pmatrix} 0.55 & 0.33 & 0.17 \\ 0.25 & 0.33 & 0.5 \\ 0.2 & 0.33 & 0.33 \end{pmatrix} \begin{pmatrix} C \\ L \\ M \end{pmatrix} \tag{9}$$

The entries of this matrix are between 0 and 1, and each column adds up to 1. A square matrix of this type is called a **stochastic matrix**. More precisely, an n by n matrix P is called stochastic if $0 \le P_{ij} \le 1$ for all i,j, and $\sum_i P_{ij} = 1$ for each j. Matrices of this kind describe stochastic (random) processes; here the destinations of the cars are random, but the overall proportions are predictable.

Let X', P and X denote the matrices in (9), so that $X' = PX$. Let X'' be the column matrix of numbers of cars at the end of two months. Then $X'' = PX' = PPX$, and similarly for longer times. Thus multiplying the matrix P by itself many times gives information on the long-term behaviour of the system.

This example is typical of a large class of situations called 'Markov processes'. The methods of linear algebra give a great deal of information about their behaviour, as we shall see in later chapters. □

The process of matrix multiplication may be clearer from the following point of view. An m by n matrix A can be viewed as a set of m rows, each of which is an n-vector. Likewise, an n by p matrix B can be viewed as a set of p columns, each of which is an n-vector. Then the (i,j) entry of AB is the dot product of the ith row of A with the jth column of B. In symbols,

$$(AB)_{ij} = \mathbf{a}_i \cdot \mathbf{b}_j$$

where $\mathbf{a}_i$ is the ith row of A and $\mathbf{b}_j$ is the jth column of B, both of which are n-vectors. If you write out this inner product you will see that it is simply another way of expressing equation (8).

This representation of AB leads to a technique for writing down matrix products. Write B above and to the right of A, as follows (cf. Example 2):

$$\begin{pmatrix} 1 & 2 & 3 \\ 4 & 5 & 6 \end{pmatrix} \begin{pmatrix} \boxed{\begin{matrix} 7 & -8 \\ -5 & 6 \\ -3 & 4 \end{matrix}} \\ \boxed{-15} \end{pmatrix}$$

This leaves rectangular spaces at the top left and the bottom right. The matrices are conformable if the top left space is square – here it is 3 by 3, so they are conformable. The size of the product matrix is given by the space at the bottom right – here it is 2 by 2. The product matrix is obtained by multiplying and adding corresponding entries of rows and columns as indicated: the $(2, 1)$ entry of this product is $4.7 + 5(-5) + 6(-3) = -15$.

This point of view on matrix multiplication is particularly fruitful in the special case of the product Ax where x is a column matrix. The first entry of Ax is then the sum of the first entries of the columns of A multiplied by the entries of x. In general, the ith entry of Ax is obtained by multiplying the ith entries of the columns of A by x_i and adding. In other words, Ax is a column matrix which equals the sum of the columns of A multiplied by the entries of x. This simple observation is very useful. We therefore set it out as a lemma.

Lemma 1 (the column lemma). Let A be a matrix, with columns $a_1, \ldots, a_q$. Then for any q-vector x, written as a column matrix, Ax is the column matrix $x_1 a_1 + \ldots + x_q a_q$.

Proof. $(Ax)_i = A_{i1}x_1 + \ldots + A_{iq}x_q$. Since A_{ij} is the ith entry of the jth column of A, this equals the ith entry of $(x_1a_1 + \ldots + x_qa_q)$. $\qquad\qquad\square$

There is a similar result for rows; see Problem 8. But vectors are written as columns more often than rows, and it is the column lemma which will be used in later chapters.

This completes our discussion of how to multiply matrices. The next section works out some consequences of the definition.

2D   THE ALGEBRA OF MATRIX MULTIPLICATION

Adding matrices is much like ordinary algebra, but multiplication is different in several ways. Given two matrices of the same dimensions, we can always add them, but we cannot always multiply them. If A is a p by q matrix, then AB makes sense only if B has q rows. So if A and B are two rectangular matrices of the same dimensions, they cannot be multiplied together. In particular, if A is rectangular, then the product AA does not exist. If A is square, however, then AA exists; we write it as A^2.

Even if AB exists, BA need not exist; consider, for example, the case where A is 2 by 2 and B is 2 by 3. And even if AB and BA both exist, they may not be equal. The order in which matrices are multiplied is crucial.

Example 1. Let

$$A = \begin{pmatrix} 1 & 1 \\ 1 & 1 \end{pmatrix} \qquad B = \begin{pmatrix} 1 & 1 \\ -1 & -1 \end{pmatrix}$$

Then

$$AB = \begin{pmatrix} 0 & 0 \\ 0 & 0 \end{pmatrix} \qquad BA = \begin{pmatrix} 2 & 2 \\ -2 & -2 \end{pmatrix} \qquad\square$$

We say that matrices A and B **commute** with each other if $AB = BA$. This is the exception rather than the rule; most pairs of matrices, like those in Example 1, do not commute; we say that matrix multiplication in general is **noncommutative**.

Exercise 1. Is the following statement true or false?: $(A + B)^2 = A^2 + 2AB + B^2$ for all n by n matrices A, B.

Example 2. In Example 2B-1 we set up an n by n matrix R with $R_{ij} = 1$ if there is a direct rail connection between towns i and j, and $R_{ij} = 0$ otherwise. Now, $(R^2)_{ij} = R_{i1}R_{1j} + \ldots + R_{in}R_{nj}$. Suppose this sum is nonzero, and that the kth term, say, is nonzero. Then R_{ik} and R_{kj} are both 1. This means that there are direct connections from i to k and from k to j. Thus $(R^2)_{ij}$ tells us whether it is possible to go from i to j changing trains once. In fact, $(R^2)_{ij}$ equals the number of different routes one can take by changing trains once between i and j.

A similar argument shows that $(RB)_{ij} = 0$ unless one can go from i to j starting on the train and then changing to a bus; similarly, the entries of BR tell us which journeys can be made starting on the bus and changing to a train. In this example, the noncommutativity of multiplication expresses the obvious fact that going by train followed by bus is different from going by bus followed by train. □

Although matrices do not commute in general, there is a special class of matrices which do.

Definition 1. A matrix is **diagonal** if the only nonzero entries are on the main diagonal. In other words, A is diagonal if $A_{ij} = 0$ whenever $i \neq j$. The square matrix with entries $a, b, \ldots, k$ on the main diagonal and zeros everywhere else is denoted by $\text{diag}(a, b, \ldots, k)$. Thus

$$\text{diag}(a, b, \ldots, k) = \begin{pmatrix} a & 0 & 0 & \ldots & 0 & 0 \\ 0 & b & 0 & \ldots & 0 & 0 \\ . & . & . & \ldots & . & . \\ . & . & . & \ldots & . & . \\ 0 & 0 & 0 & \ldots & 0 & k \end{pmatrix}$$ □

Note that the entries $a, b, \ldots$ on the diagonal may or may not be zero. For example, the zero matrix is a diagonal matrix; it satisfies the condition $A_{ij} = 0$ for $i \neq j$.

Exercise 2. Write down two 2 by 2 diagonal matrices and calculate their products in both orders. □

It is easy to verify that the product of two n by n diagonal matrices is diagonal, with entries obtained by multiplying corresponding entries of the two original matrices. It follows that any two n by n diagonal matrices commute.

We now return to general matrix algebra. Apart from noncommutativity, most of the other rules of elementary algebraic manipulation apply.

Theorem 1 (matrix manipulation). The following hold for all matrices A, B, C, D for which the products and sums exist:

(a) $(AB)C = A(BC)$;
(b) $A(B + C) = AB + AC$;
(c) $(A + B)(C + D) = AC + AD + BC + BD$;
(d) $A(kB) = k(AB)$ for all numbers k. □

The proof of these results is very uninteresting; it consists of working out both sides of the equations in (a) to (d) and showing that they are equal. If you work out some simple examples, with 2 by 2 matrices, you will see how a general proof could be written out.

The statement (a) is called the associative law of multiplication. It says that when three matrices are multiplied together, the order in which the two multiplications are performed does not matter. (Do not confuse the order in which two multiplications are performed, which makes no difference, with the order of the factors, which is crucial.) The associative law is not obvious: there are algebraic structures in which it does not hold (see Problem 21). However, matrix multiplication is associative, though not commutative.

Statements (b) and (c) are versions of what is called the distributive law. Statement (d) says that scalars commute with matrices, even though matrices do not commute with each other. The whole theorem can be summed up by saying that matrices can be manipulated like numbers, provided that the order of matrix factors is not changed.

Example 3. If A and B are square matrices of the same size, then

$$(A + B)^2 = (A + B)(A + B) = A^2 + AB + BA + B^2 \tag{1}$$

If BA were equal to AB, this could be written in the familiar form $A^2 + 2AB + B^2$. But BA and AB are different in matrix algebra, and (1) cannot be simplified further.

Equation (1) has a nice interpretation in our railway example. We saw in section 2B that $R + B$ gives the number of ways (if any) of travelling directly between two towns using either rail or bus. We saw above that R^2, B^2, RB and BR give the numbers of ways of making the journey in two stages using rail only, bus only, rail followed by bus, and bus followed by rail. The equation $(R + B)^2 = R^2 + RB + BR + B^2$ means that the total number of ways of making the journey in two stages is the sum of the four possible combinations of modes of transport. ☐

Theorem 1 explains how multiplication interacts with addition and scalar multiplication. In section 2B we introduced another matrix operation: transposition. We now ask what happens when you take the transpose of a product.

Theorem 2 (transpose of product). If A and B are conformable for multiplication, then

$$(AB)^\mathsf{T} = B^\mathsf{T} A^\mathsf{T}$$

Proof. We must first check that the product $B^\mathsf{T} A^\mathsf{T}$ exists. Suppose that A and B are conformable p by q and q by r matrices respectively. Then B^T is r by q and A^T is q by p; therefore $B^\mathsf{T} A^\mathsf{T}$ is defined, and for all i, j we have

$$(B^\mathsf{T} A^\mathsf{T})_{ij} = \sum_k (B^\mathsf{T})_{ik} (A^\mathsf{T})_{kj}$$

$$= \sum_k B_{ki} A_{jk}$$

$$= \sum_k A_{jk} B_{ki} = (AB)_{ji} = [(AB)^\mathsf{T}]_{ij}$$

We have shown that the (i, j) entries of the matrices $B^\mathsf{T} A^\mathsf{T}$ and $(AB)^\mathsf{T}$ are the same, for all i and j. This proves the theorem. ☐

Exercise 3. If X and Y are n by 1 matrices, that is, column vectors with n entries, then the dot product of X and Y, as defined in section 1E, equals $X^\mathsf{T} Y = Y^\mathsf{T} X$. On the other hand, XY^T is an n by n matrix.

Note. An exercise like this, in the form of a bald statement, is a challenge: prove it! Solutions are at the back of the book.

Example 4. The linear equations of section 2A can be written in the form $AX = Y$, where X and Y are column matrices. Theorem 2 gives the equivalent form

$$X^\mathsf{T} A^\mathsf{T} = Y^\mathsf{T}$$

The transpose of a column matrix is a row matrix. We see that a set of linear equations can be written in terms of row matrices, provided that the coefficient matrix multiplies the row of unknowns on the right. The row version and the column version are entirely equivalent.   □

We now proceed with the task of constructing a matrix analogue of ordinary algebra. In section 2B we defined the zero matrix. It is easy to see that $0A = 0$ for any matrix A, where 0 denotes an appropriate zero matrix. More precisely, if A is an s by t matrix, and n is any positive integer, then $0A = 0'$ where 0 and $0'$ denote the n by s and n by t zero matrices respectively.

The zero matrix behaves much like the number 0 in ordinary arithmetic. What about the number 1? We have $1a = a$ for every number a; it would be useful to have a matrix I, say, such that $IA = A$ for every matrix A. Such a matrix, unlike the zero matrix, must be square, otherwise IA and A would not have the same number of rows and could not be equal.

Definition 2. The n by n **unit matrix** I_n is defined by

$$I_n = \begin{pmatrix} 1 & 0 & 0 & . & . & 0 & 0 \\ 0 & 1 & 0 & . & . & 0 & 0 \\ 0 & 0 & 1 & . & . & 0 & 0 \\ . & & & . & & & . \\ . & & & . & & & . \\ . & & & . & . & 1 & 0 \\ 0 & 0 & 0 & . & . & 0 & 1 \end{pmatrix}$$

or $I_n = \operatorname{diag}(1, 1, \ldots, 1)$, in the notation of Definition 1.   □

For each value of n there is an n by n unit matrix I_n. Every matrix A can be multiplied by an appropriate unit matrix I_n with n chosen to make the matrices conformable.

Exercise 4. If A is a p by q matrix, then $I_p A = A I_q = A$.   □

We often omit the subscript when there is no danger of confusion, and write I for the unit matrix.

The case $q = 1$ of Exercise 4 shows that for every column matrix x we have $Ix = x$. We now show that I is the only matrix with this property.

Theorem 3 (characterisation of the unit matrix). If an n by n matrix A satisfies $Ax = x$ for all column n-vectors x, then A is the unit matrix.

Proof. We are given that $Ax = x$ for all x, but in fact to prove that $A = I$ we only need the special case

$$Ae_j = e_j \quad \text{for } j = 1, \ldots, n \tag{2}$$

where e_j is the jth column of I. Since all entries of e_j are 0 except for the jth, which is 1, it follows from the column lemma of section 2C that Ae_j is the jth column of A. Now (2) shows that the jth column of A equals e_j, which is the jth column of I. This holds for all j, so $A = I$. $\quad\square$

The entries of the unit matrix are sometimes written as δ_{ij}, and given a special name.

Definition 3. The **Kronecker delta symbol** δ_{ij} is defined by $\delta_{ij} = 0$ if $i \neq j$, and $\delta_{ij} = 1$ if $i = j$. $\quad\square$

Since $Ix = x$ for any n-vector x, it follows that

$$\sum_{j=1}^{n} \delta_{ij} x_j = x_i \quad \text{for } i = 1, \ldots, n \tag{3}$$

Indeed, this is obvious without using matrices: all terms in the sum vanish except the term with $j = i$.

The matrices I and 0 have many of the same properties as the numbers 1 and 0. But there is an important difference. Example 1 shows that the product of two nonzero matrices can be zero. In the same way as in section 1E, this leads to breakdown of the cancellation law. In other words, $AX = AY$ does not imply that either $X = Y$ or $A = 0$.

Exercise 5. Construct nonzero matrices A, X, Y such that $AX = AY$ but $X \neq Y$. $\quad\square$

We have now found two novel features of matrix algebra: noncommutativity and the failure of the cancellation law. There is a third. For every nonzero number x there is a reciprocal x^{-1} such that $xx^{-1} = 1$. This is not true for matrices, as we shall see in the next section.

2E   INVERSES

In ordinary algebra, for every nonzero number a there is one and only one number b satisfying $ab = 1$; it is called the reciprocal or inverse of a. The matrix analogue of 1 is I, so it is natural to ask if, given a matrix $A \neq 0$, there is a unique B satisfying $AB = I$.

Exercise 1. Let

$$A = \begin{pmatrix} 1 & 2 & 3 \\ 3 & 2 & 1 \end{pmatrix} \qquad B = \begin{pmatrix} -1/2 & 1/2 \\ 3/4 & -1/4 \\ 0 & 0 \end{pmatrix} \qquad B' = \begin{pmatrix} 1/2 & 3/2 \\ -5/4 & 9/4 \\ 1 & 1 \end{pmatrix}$$

Work out the products AB and AB', showing that they both equal I. $\quad\square$

Thus there can be two different matrices B satisfying $AB = I$ for the same A. Here there are infinitely many: given any values for the entries of the last row of B, one can find values for the other rows such that $AB = I$.

It can also happen that for a given nonzero A, there is no matrix B satisfying $AB = I$. The transpose of the matrix A in Exercise 1 is such a matrix. To add to the confusion, observe that we could consider the equation $BA = I$ instead of $AB = I$. The subject of matrix inverses is not simple, and we must proceed carefully, by way of the following definitions.

Definition 1. If $AB = I$, we say that B is a **right inverse** of A, and A is a **left inverse** of B.

Example 1. In Exercise 1, B and B' are right inverses of A. Transposing the equation $AB = AB' = I$ gives (using Theorem 2D-2) $B^T A^T = B'^T A^T = I^T = I$. Hence A^T has at least two left inverses.

Example 2. Let

$$A = \begin{pmatrix} 1 & 1 \\ 1 & 1 \end{pmatrix}$$

If A has a right inverse B, say, then B must be 2 by 2, otherwise AB would not be square. Let a, b, c, d be the unknown elements of B. Then

$$\begin{pmatrix} 1 & 1 \\ 1 & 1 \end{pmatrix}\begin{pmatrix} a & b \\ c & d \end{pmatrix} = \begin{pmatrix} 1 & 0 \\ 0 & 1 \end{pmatrix}$$

giving $a + c = 1, b + d = 0, a + c = 0, b + d = 1$. These equations contradict each other. Therefore there cannot be any numbers a, b, c, d satisfying them. Hence there is no matrix B satisfying $AB = I$: A has no right inverse. A similar argument shows that A has no left inverse either.

Example 3. Applying the method of Example 2 to the matrix

$$A = \begin{pmatrix} 1 & 1 \\ 2 & 1 \end{pmatrix}$$

gives a set of equations for a, b, c, d which can easily be solved by elimination, giving just one solution: the right inverse of A is

$$\begin{pmatrix} -1 & 1 \\ 2 & -1 \end{pmatrix}$$

It is easy to show in the same way that there is just one left inverse, which is the same as the right inverse. This matrix has a single inverse, which works on both sides. $\square$

If the left and right inverses of A are the same, there is no need to specify left or right, and we can simply refer to the inverse.

Definition 2. If $AB = BA = I$, then we say B is the **inverse** of A. If a matrix has an inverse, it is said to be **invertible**. The inverse as defined here is sometimes called a **two-sided inverse** when it is desired to contrast it with the left and right inverses. $\square$

This definition leads easily to the following results.

Theorem 1 (properties of the inverse)
    (a) Every invertible matrix is square.
    (b) If B is an inverse of A, then A is an inverse of B.
    (c) If B is an inverse of A, then B is a left inverse and a right inverse of A.
    (d) If A is invertible, then so is its transpose A^T. If B is an inverse of A, then B^T
is an inverse of A^T.

Proof. (a) If $AB = BA = I$, and A is p by q and B is r by s, then $q = r$, for otherwise
the product AB would not exist. Similarly $s = p$ because the product BA exists. Thus
AB is p by p and BA is q by q; but these matrices are equal, hence $p = q$ and A is square.
    (b) and (c) follow immediately from the definitions.
    (d) If $AB = BA = I$, then by Theorem 2D-2, $B^T A^T = A^T B^T = I^T = I$. This is the
condition for B^T to be the inverse of A^T.   □

Inverses are very useful. For example, a set of linear equations for unknowns x
can be written

$$Ax = c \qquad\qquad\qquad (1)$$

If A has a left inverse B, then $BA = I$, and multiplying (1) on the left by B gives

$$x = Bc$$

Thus every left inverse B of A immediately gives a solution Bc of (1). Now, there can
be many left inverses of a given matrix A; they give rise to many different solutions
of (1). But for two-sided inverses things are different.

Theorem 2. An invertible matrix has only one inverse.

Proof. Suppose that B and C are two inverses for A, so that $AB = BA = AC = CA = I$.
Then

$$\begin{aligned}
C = IC &= (BA)C \quad \text{(since } BA = I) \\
&= B(AC) \quad \text{(since multiplication is associative)} \\
&= BI \qquad\;\; \text{(since } AC = I) \\
&= B
\end{aligned}$$

Thus any two inverses for A must be the same; the inverse is unique.   □

    We denote the inverse of A by A^{-1}. We could not use this notation before proving
Theorem 2, because there might have been several possible matrices, all with an equal
claim to be called A^{-1}. The notation would have been ambiguous, and therefore
unacceptable.
    The uniqueness of inverses is related to the theory of linear equations. If A is
invertible it must be square, so $Ax = b$ is a set of n equations with n unknowns, and
has a unique solution. Rectangular matrices correspond to sets with more unknowns
than equations, or vice versa. If there are more unknowns than equations, then the

equations do not determine the unknowns completely, and there are likely to be many solutions. This is related to the nonuniqueness of left inverses; the full story is told in Chapter 4.

For real numbers, the two conditions (i) having an inverse, and (ii) being nonzero are equivalent. We have seen that for matrices they are different; (ii) does not imply (i). Which of these conditions is the 'natural generalisation' to matrices of the nonzero numbers? The answer is very often (i). For example, the product of two nonzero numbers is nonzero; we shall now show that the product of two invertible matrices is invertible. We saw in section 2D that this last statement becomes false if 'invertible' is replaced by 'nonzero'.

Theorem 3 (product of invertible matrices). If A and B are invertible n by n matrices, then AB is invertible and

$$(AB)^{-1} = B^{-1}A^{-1}$$

Proof. We have

$$(B^{-1}A^{-1})AB = B^{-1}(A^{-1}A)B$$

$$= B^{-1}IB = B^{-1}B = I$$

Similarly $AB(B^{-1}A^{-1}) = AA^{-1} = I$. Thus $B^{-1}A^{-1}$ satisfies the conditions defining the inverse of AB.   □

Exercise 2. If A, B, C are invertible matrices of the same size, then their product is invertible, and its inverse is the product of their inverses in the reverse order. (Reminder: an exercise in the form of a statement is a challenge: prove it!)   □

This can obviously be generalised to the product of any number of invertible matrices.

The results of this section have mainly been of the form 'if A is invertible, then...'. Since invertible matrices have nice properties, it would be useful to be able to see whether a given matrix A has an inverse or not. There is no easy way of doing this at the present stage. We will develop a method in Chapter 4, based on the general theory of Chapter 3.

2F   PARTITIONED MATRICES

This section deals with a technical matter which will not be needed until the middle of Chapter 4. It can be omitted on a first reading.

Consider the matrix

$$\begin{pmatrix} 1 & 0 & 0 & 0 & 0 \\ 0 & 1 & 0 & 0 & 0 \\ 0 & 0 & 1 & 0 & 0 \\ 0 & 0 & 0 & 2 & 3 \\ 0 & 0 & 0 & 3 & 2 \end{pmatrix} \tag{1}$$

It is natural to describe it as consisting of a 3 by 3 unit matrix with a block of zeros to the right, a block of zeros underneath, and a 2 by 2 matrix in the bottom right corner. A matrix described in these terms is said to be **partitioned**, and can be written compactly as

$$\left(\begin{array}{c|c} A & B \\ \hline C & D \end{array}\right)$$

where A, B, C and D are matrices. We think of it as a 2 by 2 matrix whose entries are themselves matrices. In the example above, A is the 3 by 3 unit matrix, B and C are zero matrices, and D is 2 by 2.

This idea can be extended to matrices with several blocks, provided that the partitioning lines extend across the full width and height of the matrix. The first matrix below is properly partitioned, but the second is not:

$$\left(\begin{array}{cc|cc|c} 1 & 3 & 4 & 2 & 1 \\ 2 & 3 & 0 & 1 & 3 \\ 1 & 2 & 0 & 4 & 1 \\ \hline 0 & 2 & 1 & 3 & 0 \end{array}\right) \qquad \left(\begin{array}{c|cc} 1 & 0 & 0 \\ 3 & 0 & 0 \\ \hline 1 & 1 & 0 \\ 0 & 0 & 1 \end{array}\right)$$

The second matrix here has three submatrices arranged in a triangle; it cannot be regarded as a rectangular array of submatrices in the way that the first matrix can; it is not a properly partitioned matrix.

Block notation. The submatrices (or blocks) into which a matrix A is partitioned are often denoted by A_{ij}; thus A_{ij} denotes the ith block down and jth block along. In (1), for example, A_{12} is the 3 by 2 zero matrix. Note that the number of rows in the block A_{ij} depends only on j, being the same for all i; similarly the number of columns in the block depends only on i.   □

We now consider the algebra of partitioned matrices. If two matrices are partitioned in the same way, then they can be added block by block. Thus, for example,

$$\left(\begin{array}{c|c|c} A & B & C \\ \hline D & E & F \end{array}\right) + \left(\begin{array}{c|c|c} A' & B' & C' \\ \hline D' & E' & F' \end{array}\right) = \left(\begin{array}{c|c|c} A+A' & B+B' & C+C' \\ \hline D+D' & E+E' & F+F' \end{array}\right)$$

provided that A has the same dimensions as A', etc. (otherwise $A + A'$ would not make sense). This follows immediately from the definition of matrix addition. Multiplication by a number works in the same kind of way.

Multiplication of partitioned matrices is more useful, and, as you might expect, a little more intricate.

Definition 1. Consider matrices A and B which are conformable for multiplication in the order AB. They are **partitioned conformably** if
  (i) the number of blocks in a row of A equals the number of blocks in a column of B, and
  (ii) the matrix A_{ij} is conformable with the matrix B_{jk} for all i, j, k (using the notation introduced above for submatrices).

Theorem 1 (block multiplication). Let A and B be conformably partitioned matrices, with blocks A_{ij} and B_{ij}. Then the (i,j) block of the matrix AB is given by

$$(AB)_{ij} = \sum_k A_{ik}B_{kj} \qquad \square \quad (2)$$

The beauty of this is that block multiplication works exactly as if the blocks were just elements of a matrix. The only difference is that each term on the right of (2) is itself a matrix product.

Example 1

$$\left(\begin{array}{c|c} A & B \\ \hline C & D \end{array}\right)\left(\begin{array}{c} E \\ \hline F \end{array}\right) = \left(\begin{array}{c} AE + BF \\ \hline CE + DF \end{array}\right) \qquad (3)$$

where $A, \ldots, F$ are matrices satisfying the above conditions. Take

$$A = \begin{pmatrix} 1 \\ 2 \end{pmatrix} \quad B = \begin{pmatrix} 1 & 3 \\ 0 & 2 \end{pmatrix} \quad C = (-1) \quad D = (3 \quad -1) \quad E = (2) \quad F = \begin{pmatrix} 1 \\ -1 \end{pmatrix}$$

It is easy to muliply out the matrices on each side of (3), and verify that they are equal.
$\square$

If you work out the details of this example, you will understand the block multiplication theorem. To write out a general proof in detail would need notation so complicated as to obscure the basic idea, which is more clearly shown by the example.

The idea of partitioning leads us to perceive a matrix in different ways. A p by q matrix is a rectangular array of pq numbers. It can be seen as a set of q columns, each of which is a p-vector, or as a set of p rows, each of which is a q-vector. These are particularly useful ways of partitioning a matrix.

Consider, for example, a matrix product AB, where A is p by q and B is q by r. Partitioning A into its row and B into its columns gives

$$AB = \begin{pmatrix} a_1 \\ \hline a_2 \\ \hline \vdots \\ \hline a_p \end{pmatrix} (b_1 \mid b_2 \mid \ldots \mid b_r)$$

where a_i is a row matrix, the ith row of A, and b_i is a column matrix, the ith column of B. The (i,j) entry of AB is the product a_ib_j of a 1 by q matrix with a q by 1 matrix, giving a 1 by 1 matrix, that is, a single number. We have

$$(AB)_{ij} = a_ib_j \qquad (4)$$

The number a_ib_j is the sum of the products of corresponding components of the two q-vectors a_i and b_j; in the language of vector algebra, it is their dot product. Thus the (i,j) entry of AB is the dot product of the ith row of A with the jth column of B. We have come back, by a different route, to the result reached near the end of section 2C.

Example 2. The neighbouring countries of Franconia and Gaul contain respectively f and g towns with railway stations. The Franconian rail network can be represented in the manner of section 2B by an f by f matrix F, and the Gallic network by a g by g matrix G.

There are also lines which cross the border. We define a cross-border matrix C, of dimensions f by g, by $C_{ij} = 1$ if there is a line joining the ith town in Franconia with the jth town in Gaul, and 0 otherwise. Now the complete Franco–Gallic railway system can be described by a single matrix R, as follows. Renumber the towns in Gaul by adding f to their original number; then all the railway towns in the two countries are numbered in a single sequence from 1 to $f + g$. Define an $(f + g)$ by $(f + g)$ matrix R by $R_{ij} = 1$ if there is a line between towns i and j, and 0 otherwise. Then a few moments' thought will show that

$$R = \left(\begin{array}{c|c} F & C \\ \hline C^{\mathrm{T}} & G \end{array}\right)$$

This partitioning of the combined rail network matrix reflects the natural division into purely Franconian lines, purely Gallic lines, and cross-border lines.

In Example 2D-2 we saw that squaring the network matrix gives information about journeys with one change of train. Using the block multiplication rule gives

$$R^2 = \left(\begin{array}{c|c} F^2 + CC^{\mathrm{T}} & FC + CG \\ \hline C^{\mathrm{T}}F + GC^{\mathrm{T}} & C^{\mathrm{T}}C + G^2 \end{array}\right)$$

The top left block here says that to travel between two Franconian towns changing trains once, you can either take two Franconian trains or travel across the border to Gaul and then back across the border again; the top right block says that to go from Franconia to Gaul changing trains once, you can either first take a Franconian train and then cross the border, or cross first and then take a Gallic train; and so on. The partitioned multiplication rule reflects the natural structure of the problem under discussion.

PROBLEMS FOR CHAPTER 2

Hints and answers for problems marked [a] will be found on page 278.

Sections 2A, 2B

[a]1. Let

$$A = (2, 1, 0) \qquad B = (-1, 3, 2) \qquad C = \begin{pmatrix} 2 & 1 & 0 \\ 0 & 2 & 1 \end{pmatrix}$$

$$D = \begin{pmatrix} 1 \\ 0 \\ 1 \end{pmatrix} \qquad E = \begin{pmatrix} 0 & 1 \\ 1 & 2 \\ 2 & 1 \end{pmatrix} \qquad F = \begin{pmatrix} 1 & 2 \\ 2 & -1 \end{pmatrix}$$

Which of the following make sense? Evaluate those that do.
(a) $2A - 3B$; (b) $B + D$; (c) $C + E$; (d) $C + E^{\mathrm{T}}$; (e) $B^{\mathrm{T}} + D$; (f) $C + F$.

a2. Are the following statements true or false?
(a) All the rows of a given matrix have the same number of entries.
(b) Two matrices can be added if they have the same number of columns.
(c) For a 50 by 60 matrix, the rows have more entries than the columns.
(d) The main diagonal of a matrix has the same number of entries as each row.
Note: questions of the form 'true or false?' should be answered by a logical argument: either a proof that it is true, or a counterexample showing that it is false (see Appendix A). Sometimes (as in (a) here) the question will be so simple that a plain yes or no will suffice; but usually a proof is needed.

3. Example 2B-1 shows how a network corresponds to a matrix with entries 0 or 1. Draw the network corresponding to

$$\begin{pmatrix} 0 & 1 & 1 & 0 & 0 \\ 1 & 0 & 0 & 1 & 1 \\ 1 & 0 & 0 & 1 & 0 \\ 0 & 1 & 1 & 0 & 0 \\ 0 & 1 & 0 & 0 & 0 \end{pmatrix}$$

4. A matrix A is called **skew-symmetric** or **antisymmetric** if $A^T = -A$.
(a) Show that every diagonal entry of a skew-symmetric matrix is zero.
(b) Write down a 3 by 3 skew-symmetric matrix.
(c) Show that if a matrix is both symmetric and skew-symmetric, then it is zero.
(d) Prove that every square matrix can be expressed as a symmetric matrix plus a skew-symmetric matrix.
(See also Chapter 4, Problem 26.)

5. The matrix

$$\begin{pmatrix} 1 & 2 & 3 & 4 \\ 4 & 1 & 2 & 3 \\ 3 & 4 & 1 & 2 \\ 2 & 3 & 4 & 1 \end{pmatrix}$$

is called a **circulant**. In general, a circulant is a square matrix for which each row is obtained from the row above by moving the last entry to the beginning and moving all the others along. Thus a circulant is completely specified when the first row is given.
(a) Show that the transpose of the matrix above is also a circulant. Is this a coincidence?
(b) Let A be an n by n circulant with first row $a_1, \ldots, a_n$. Show that $A_{ij} = a_{j-i+1}$ for all i and j from 1 to n, where a_i for $i \le 0$ is defined by $a_i = a_{i+n}$. Deduce that each column of A is obtained from the previous column by moving the last entry to the top and moving all the others down. In other words, A^T is a circulant if A is.
(See also Chapter 3, Problems 29, 30 and Chapter 7, Problem 8.)

Sections 2C, 2D

ᵃ6. This is a continuation of Problem 1. (g) AB; (h) BD; (i) DA; (j) $3AD - 4$; (k) $D^T E$;
(l) FC; (m) FCE; (n) FE; (o) C^2; (p) F^2.

ᵃ7. True or false?
  (a) A^2 makes sense if and only if A is square.
  (b) $AB = 0$ only if $A = 0$ or $B = 0$.
  (c) $AB = 0$ if $A = 0$ or $B = 0$.
  (d) $A^2 - B^2 = (A - B)(A + B)$ for all n by n matrices A, B.

ᵃ8. Formulate and prove an analogue of the column lemma that applies to rows.

9. Consider the matrices

$$\begin{pmatrix} 0 & 1 \\ 1 & 0 \end{pmatrix} \quad \begin{pmatrix} 0 & -i \\ i & 0 \end{pmatrix} \quad \begin{pmatrix} 1 & 0 \\ 0 & -1 \end{pmatrix}$$

Show that the square of each one equals I, that the product of any two is $\pm i$ times the third, and that any two of them satisfy the relation $AB + BA = 0$.
  (They are called the 'Pauli spin matrices', and are used in physics to describe electrons. $AB + BA = 0$ is called the 'anticommutation relation', because of its similarity to the condition $AB - BA = 0$ for matrices to commute.)

ᵃ10. Consider the statement 'If A is a 2 by 2 matrix with $A^3 = 0$, then $A^2 = 0$'. Try to decide whether it is true.
  (For an answer using subtle theoretical methods, see Chapter 10 (Volume 2), Problem 15.)

ᵃ11. Find all matrices which commute with

$$\begin{pmatrix} 1 & 2 \\ 3 & 4 \end{pmatrix}$$

ᵃ12. Set

$$R(t) = \begin{pmatrix} \cos t & -\sin t & 0 \\ \sin t & \cos t & 0 \\ 0 & 0 & 1 \end{pmatrix}$$

for all real t. Show that $R(x)R(-x) = R(0) = I$ and $R(x)R(y) = R(x + y)$ for all x, y.
  (The meaning of these equations will be discussed in section 5I.)

ᵃ13. True or false? 'If A and B are square matrices of the same size, and n is a positive integer, then $(A + B)^n = A^n + nA^{n-1}B + [n(n - 1)/2!]A^{n-2}B^2 + \ldots + B^n$.'

ᵃ14. True or false? 'For any n by n matrices A and B, $(A + B)^3 = A^3 + A^2B + ABA + BA^2 + AB^2 + BAB + B^2A + B^3$.' Generalise.

15. Find as many real 2 by 2 matrices A and B as you can which satisfy $A^2 = I$ and $B^2 = 0$.

16. Let

$$E(1,1) = \begin{pmatrix} 1 & 0 \\ 0 & 0 \end{pmatrix} \qquad E(1,2) = \begin{pmatrix} 0 & 1 \\ 0 & 0 \end{pmatrix} \qquad E(2,1) = \begin{pmatrix} 0 & 0 \\ 1 & 0 \end{pmatrix} \qquad E(2,2) = \begin{pmatrix} 0 & 0 \\ 0 & 1 \end{pmatrix}$$

Thus $E(i,j)$ is the 2 by 2 matrix with (i,j) entry 1 and all other entries 0, for i and $j = 1$ or 2.

Verify that $E(i,j)E(k,l) = 0$ if $j \neq k$, and $E(i,j)E(j,k) = E(i,k)$ for any i, j, k.

Try to describe in words the effect on a matrix A of multiplying it by $E(i,j)$. There are two cases to consider: $E(i,j)A$ and $AE(i,j)$.

17. Problem 16 generalises as follows. For any n, define the **n by n matrix units** (not to be confused with the unit matrix) as follows: for each i and j from 1 to n, $E(i,j)$ is the n by n matrix with (i,j) entry 1 and all other entries 0. The 2 by 2 matrix units are given in Problem 16.

Verify that for any r and s from 1 to n, the (r,s) entry of $E(i,j)$ is $\delta_{ri}\delta_{sj}$. Hence show that $E(i,j)E(k,l) = \delta_{jk}E(i,l)$ for all i, j, k, l. Describe in words the effect of multiplying a given matrix A by $E(i,j)$.

For applications of the matrix units, see the following problem, and also Chapter 3, Problem 27.

[a]18. True or false? 'If an n by n matrix A commutes with every other n by n matrix, then A must be a multiple of the unit matrix.'

(Hint: use Problem 17.)

19. Stochastic matrices were defined in Example 2C-4.

(a) Show that a square matrix A is stochastic if and only if its entries are all non-negative and it satisfies $uA = u$ where $u = (1, 1, \ldots, 1)$. (Hint: see Problem 8.)

(b) Is the product of two stochastic matrices always stochastic?

(c) Is the inverse of a stochastic matrix always stochastic?

20. Let

$$M = \begin{pmatrix} a & 1 \\ 0 & a \end{pmatrix}$$

Calculate M^2, M^3, guess the answer for M^k, and prove it. Hence show that

$$p(M) = \begin{pmatrix} p(a) & p'(a) \\ 0 & p(a) \end{pmatrix}$$

for any polynomial p.

(Continued in Problem 27 below.)

21. The **commutator** of two n by n matrices A and B is defined to be $AB - BA$, denoted by $[A, B]$. Obviously $[A, B] = -[B, A]$. Prove the following.

(i) $[xA + yB, C] = x[A, C] + y[B, C]$ for any numbers x and y and n by n matrices A, B and C. Similarly for $[A, xB + yC]$.

(ii) $[A, [B, C]]$ is not in general equal to $[[A, B], C]$.

(iii) $[A, [B, C]] + [B, [C, A]] + [C, [A, B]] = 0$ for any A, B, C (this relation is called the 'Jacobi identity').

(Part (i) shows that $[A, B]$ behaves like a kind of product of A and B; (ii) shows that it does not satisfy the associative law. Any set of objects with a 'multiplication' rule $[a, b]$ satisfying $[a, b] = -[b, a]$ and the Jacobi identity is called a 'Lie algebra'. The physics of the fundamental particles of nature is built on Lie algebras.)

Section 2E

ª22. True or false?
(a) If A is invertible then so is A^T.

(b) If $AB = I$ then A and B are invertible.

(c) If $AB = BA = I$ then A and B are square.

(d) If $ABC = BCA = I$ then A is invertible.

ª23. Work out the inverse of the 2 by 2 matrix

$$M = \begin{pmatrix} a & b \\ c & d \end{pmatrix}$$

by taking

$$N = \begin{pmatrix} x & y \\ z & u \end{pmatrix}$$

setting $MN = NM = I$, and solving the resulting equations for x, y, z, u. Show that M is invertible if and only if $ad \neq bc$.

(A deeper understanding of the relation $ad \neq bc$ will emerge in later chapters.)

24. If A is a square matrix such that $A^k = 0$ for some k, show that $I - A$ is invertible, and $(I - A)^{-1} = I + A + A^2 + \ldots + A^{k-1}$.

(This is rather like the binomial expansion $(1 - a)^{-1} = 1 + a + a^2 + \ldots$.)

25. If M and N are invertible and commute, show that the following pairs commute: M and N^{-1}; M^{-1} and N; M^{-1} and N^{-1}; M^T and N^T. Generalise.

Section 2F

ª26. True or false?
(a) If

$$A = \begin{pmatrix} P & Q \\ R & S \end{pmatrix}$$

where P, Q, R, S are square blocks, then

$$A^T = \begin{pmatrix} P & R \\ Q & S \end{pmatrix}$$

(b) If

$$A = \begin{pmatrix} 0 & B \\ 0 & 0 \end{pmatrix}$$

where B is a square block, then $A^2 = 0$ by the block multiplication rule.

27. Write $J_2 =$ the matrix M of Problem 20. For each $n > 2$, define an n by n matrix J_n by

$$J_n = \begin{pmatrix} a & e \\ 0 & J_{n-1} \end{pmatrix} \quad \text{where } e = (1, 0, 0, \ldots, 0) \in \mathbb{R}^{n-1}$$

Evaluate $[J_3]^k$ for $k = 2$ and 3. Guess the answer for general k, and prove it. Hence show that for any polynomial p,

$$p(J_3) = \begin{pmatrix} p(a) & p'(a) & p''(a)/2 \\ 0 & p(a) & p'(a) \\ 0 & 0 & p(a) \end{pmatrix}$$

What about $p(J_n)$ for $n > 3$?

[a]28. Let

$$M = \begin{pmatrix} A & B \\ 0 & C \end{pmatrix}$$

where A, B, C are invertible n by n matrices. Show that

$$M^{-1} = \begin{pmatrix} A^{-1} & Q \\ R & C^{-1} \end{pmatrix}$$

for some matrices Q and R, which you should find in terms of A, B, C.

29. Verify that if A, B, C, H, G, F are invertible n by n matrices, then

$$\begin{pmatrix} A & H & G \\ 0 & B & F \\ 0 & 0 & C \end{pmatrix}^{-1} = \begin{pmatrix} A^{-1} & -A^{-1}HB^{-1} & A^{-1}HB^{-1}FC^{-1} - A^{-1}GC^{-1} \\ 0 & B^{-1} & -B^{-1}FC^{-1} \\ 0 & 0 & C^{-1} \end{pmatrix}$$

Generalise.

Part II

Vector spaces and linear equations

Linear algebra is based on generalising the ideas behind vectors and matrices. Chapter 3 introduces a greatly generalised version of vectors, and Chapter 4 applies the theory to the solution of sets of linear equations. Part III will discuss the generalised version of matrices.

Some sections of these chapters are marked with a star; this means that they can be omitted. The unstarred sections form a basic course in linear algebra. The optional starred sections branch off in many different directions; some give applications of the theory, some discuss computational aspects, and some contain more advanced mathematical theory.

Even if you have no time to read much beyond the bare minimum, I hope that you will at least skim over some of the starred material. The unstarred sections form a main road through the subject; the most enjoyable scenery is often found along side roads.

3

Vector spaces

In this chapter we get to grips with the central ideas of linear algebra. The first two sections introduce abstract vector spaces, and sections 3D–F develop their geometrical properties. Sections 3G and H discuss the dimension of a space, and the next section generalises the idea of the components of a vector. All this material is essential reading.

The rest of the chapter is optional. Sections 3J and 3K develop the mathematical theory further. Section 3L shows how beautifully the mathematical theory describes the way that we see colours. The last section contains proofs of some theorems in the earlier sections.

3A INTRODUCTION

This chapter sets up the basic abstract framework of linear algebra. Some readers will enjoy the abstract theory for its own sake, but others will ask why such generality is needed. There are at least two answers to this question.

The first is that it makes the subject easier to understand. One of the main ideas in linear algebra is that one vector space can be embedded in another. In ordinary three-dimensional space, for example, vectors lying in the xy plane form a self-contained set; we have a two-dimensional set of vectors embedded in three-dimensional space. Are the vectors in the xy plane two-dimensional (because they are confined to the plane) or three-dimensional (because they exist in three-dimensional space)? This is a confusing question. It becomes much clearer when the general concept of a subspace is available.

Dealing with vector spaces in the abstract also saves effort. The general theory of vector spaces includes not only the vectors and matrices discussed in Chapters 1 and 2, but also sets of real- and complex-valued functions of a real variable, and other more exotic mathematical objects. A fact which has been proved once and for all in the general theory applies to a wide range of particular cases. That is why it is worth investing some intellectual effort in understanding abstract linear algebra.

3B   VECTOR SPACES

We shall now distil the essence of elementary vector algebra into an abstract definition. The main thing about vectors is that they can be added together, giving another vector, and multiplied by numbers, giving another vector. Any set of objects with this property is called a 'vector space'. The word 'space' here simply means 'set'; it is used for sets which have a geometrical structure similar to that of ordinary three-dimensional space. We shall see that many of the familiar properties of ordinary space apply to vector spaces in general.

   We shall approach the idea of a vector space in two stages. We begin with a rough definition, which contains the essential idea but misses out some of the details. The full definition which comes later will make more sense when the general idea is clear.

Rough definition 1A. A **vector space** is a set of objects which can be added together and multiplied by numbers, giving in each case another member of the set.   □

The operating of multiplying by a number is sometimes called **scaling**; the object is scaled up or scaled down when it is multiplied by a number greater or less than 1 respectively.

Examples 1. (a) The set of all three-dimensional vectors is a vector space, of course; Definition 1A was modelled on this set.

   For any fixed n, the set of all n-vectors is a vector space; the sum of two n-vectors is an n-vector, and a multiple of an n-vector is an n-vector. Different values of n give different spaces, because one cannot add an n-vector to a k-vector if $n \neq k$.

   The set of all n-vectors is called $\mathbb{R}^n$ ($\mathbb{R}$ for real numbers, n because there are n numbers in each n-vector). Thus the sets of two-dimensional and three-dimensional vectors are written $\mathbb{R}^2$ and $\mathbb{R}^3$ respectively.

   (b) Consider the set of all 3-vectors of the form $(x, y, 0)$. This set is the xy plane; call it P. The sum of two vectors in P is another vector in P, because $(x, y, 0) + (x', y', 0) = (x + x', y + y', 0)$. Scaling a vector in P gives a vector in P, because $k(x, y, 0) = (kx, ky, 0)$. Thus P is a vector space.

   (c) Let Q be the set of all 3-vectors of the form $(x, y, 2x + 3y)$; in other words, $Q = \{(x, y, z): z = 2x + 3y\}$ (if you are unfamiliar with this notation for sets, see Appendix C). The set Q is a plane through the origin. The sum of two elements of Q belongs to Q because $(x, y, 2x + 3y) + (X, Y, 2X + 3Y) = (x + X, y + Y, 2[x + X] + 3[y + Y])$, which belongs to Q because the third entry is twice the first plus three times the second. A multiple of a member of Q belongs to Q because $k(x, y, 2x + 3y) = (kx, ky, 2kx + 3ky)$ for all k. Hence Q is a vector space.

   (d) Now consider the set of all vectors of the form $(x, y, 1)$. The sum of two such vectors does not belong to the set: for example, $(0, 0, 1) + (0, 0, 1) = (0, 0, 2)$. Therefore it is not a vector space.

   (e) Let S be the set of 3-vectors with positive components. The sum of two such vectors also has positive components, therefore belongs to S. But if v belongs to S, then $(-1)v$ has negative components and does not belong to S. This set does not contain all multiples of its members, and is not a vector space.

(f) The set of all 2 by 3 matrices is a vector space, because any two matrices of the same dimensions can be added, giving another matrix of the same dimensions, and a matrix can be multiplied by a number, giving another matrix of the same dimensions. $\square$

Example 1(f) raises the question of what exactly is meant by the 'sum' of two elements of a vector space. When we introduced matrices, we gave definitions of addition and multiplication; we then studied their properties in detail to find out which of the familiar rules of algebraic manipulation apply. It cannot be taken for granted that an operation called 'multiplication' or 'addition' must behave as in elementary algebra. Our rough definition of a vector space is therefore incomplete; we should spell out just what is meant by addition and multiplication.

The answer is that by addition we mean any operation with all the properties of ordinary addition, such as $a + b = b + a$ for example. Similarly for multiplication. There are very many such properties. Fortunately, they can be boiled down to a fairly short list, carefully chosen so that everything else follows logically from the properties on the list.

AXIOMS

Definition 1B. A **real vector space** is a set V such that:

1. There is a rule which, given any x and y in V, determines an element of V called $x + y$ satisfying

(a) $x + y = y + x$ for all x and y in V; *comm.*

(b) $x + (y + z) = (x + y) + z$ for all x, y and z in V; *assoc.*

(c) there is an element of V, called 0, such that $0 + x = x$ for all x in V;

(d) given any x in V there is an element y of V such that $x + y = 0$. *neg.*

2. There is a rule which, given any x in V and any real number k, determines an element of V called kx satisfying

(e) $k(mx) = (km)x$ for all x in V and all real numbers k and m;

(f) $1x = x$ for all x in V;

(g) $(k + m)x = kx + mx$ for all x in V and all real k and m; *dist.*

(h) $k(x + y) = kx + ky$ for all x and y in V and all real k. $\square$

These rules are called the **axioms** for a vector space. Some of them have names. Condition (a) is called the **commutative** law (or axiom) for addition, (b) is the **associative** law for addition, and (g) and (h) are **distributive** laws. The vector y in (d) is called a **negative** of x; notice that as far as this definition goes, there might be more than one negative for a given x (though in the next section we shall prove that in fact there is only one).

Although these laws take up so much space in Definition 1B, the essential idea of a vector space is contained in the rough version 1A above. In Examples 1, it is easy to see that the algebraic rules (a) to (h) are satisfied in every case. The difficulty in 1(d) and (e) was that the operations gave vectors outside the set. This is typical. When deciding whether or not a set is a vector space, it is usually easy to check the algebraic rules. The difficulty, if any, is usually in showing that the sum of two members, and the product of a member with a number, belong to the set.

Example 2. Consider the set V of all real-valued functions of a real variable. We define the sum of two functions f and g to be the function $(f + g)$ whose value at any number x is given by

$$(f + g)(x) = f(x) + g(x) \quad \text{for all real } x$$

Similarly, the product of a function f and a real number k is the function (kf) defined by

$$(kf)(x) = kf(x) \quad \text{for all real } x$$

The rules (a) to (h) follow at once from the corresponding properties of real numbers. For example, (a) is verified as follows. To show that $f + g = g + f$ for any functions f and g, we must have $(f + g)(x) = (g + f)(x)$ for all x. For any x,

$$(f + g)(x) = f(x) + g(x) = g(x) + f(x) \quad \text{(since this is the sum of real numbers, which}$$
$$\text{commute)}$$

$$= (g + f)(x)$$

showing that condition (a) of Definition 1B is satisfied. Condition (b) is proved in much the same way. The zero element of condition (c) is the function O defined by $O(x) = 0$ for all x. Similarly all the other conditions can be verified, showing that V is a vector space.

This example, as developed in Example 3, is the key to many applications of linear algebra. It relates the extremely useful fields of calculus and differential equations to the purely algebraic theory of vector spaces. Linear algebra gives the basic language for modern research on differential equations.

Examples 3. (a) Given an interval $[a, b] = \{x \in \mathbb{R} : a \leq x \leq b\}$, consider the set of all real-valued functions f which are continuous for x in $[a, b]$ (see Appendix D for a discussion of continuous functions). This set is called $C[a, b]$. If f and g are continuous in $[a, b]$, then so are $f + g$ and kf for any constant k; the rules (a) to (h) hold in $C[a, b]$, just as they do in Example 2. Hence $C[a, b]$ is a vector space.

(b) Consider now the subset of $C[a, b]$ consisting of differentiable functions. The sum of two differentiable functions is differentiable, and so is any constant multiple of a differentiable function. The axioms (a) to (h) hold for differentiable functions because they all belong to the vector space $C[a, b]$. Therefore the set of differentiable functions is a vector space in its own right, as well as being a subset of the space $C[a, b]$.

$\square$

You may have noticed that Definition 1B defines a 'real' vector space. The word 'real' refers to the real numbers used to scale the vectors. But we sometimes need to multiply vectors by complex numbers. A set of objects which can be added together and multiplied by complex numbers, giving in each case another member of the set, is called a complex vector space. The algebraic properties (a) to (h) are exactly the same in the complex and real cases.

A real or a complex vector space is often called a vector space 'over' the real or complex numbers respectively. The idea is that the real or complex numbers form a foundation over which the vector space structure is erected.

We now have two kinds of vector space. It often does not matter whether the space is real or complex. The algebraic properties are the same, though the complex case is harder to visualise. It is therefore useful to have a neutral description which refers to both at the same time. We use the word **scalar** to mean the numbers, either real or complex, that are used to scale the vectors. The set of all scalars of a given type is called a **field**. The general concept of a field is discussed in Appendix G, but it is needed only for one section in Volume 2. Everywhere else, a field means either the set of real numbers, denoted by $\mathbb{R}$, or the set of complex numbers, denoted by $\mathbb{C}$. We shall often use $\mathbb{K}$ to denote a field which can be either $\mathbb{R}$ or $\mathbb{C}$.

Definition 2. A **vector space over the field** $\mathbb{K}$ is a set V such that:

1. There is a rule which, given any two elements x and y of V, determines an element $x + y$ satisfying (a) to (d) of Definition 1B.

2. There is a rule which, given any x in V and any scalar k in $\mathbb{K}$, determines an element kx of V satisfying (e) to (h) of Definition 1B.   □

This is to be understood as defining two kinds of vector space: real spaces, in which the scalars are real; and complex spaces, in which the scalars are complex.

Examples 4. (a) The complex analogue of $\mathbb{R}^n$ is the vector space $\mathbb{C}^n$ of n-tuples of complex numbers. A typical element of $\mathbb{C}^n$ is $(z_1, \ldots, z_n)$, where $z_1, z_2, \ldots$ are complex numbers. This vector is often denoted by z. We define the sum of two complex n-vectors z and w to be the vector $(z_1 + w_1, \ldots, z_n + w_n)$. If k is a complex number, then kz is defined as $(kz_1, \ldots, kz_n)$. It is easy to verify that $\mathbb{C}^n$, with these definitions of addition and scaling, is a complex vector space.

(b) Consider the set of all complex-valued functions of a real variable, that is, all functions $f:\mathbb{R} \to \mathbb{C}$. We define the sum of two such functions, and the product of a function with a complex constant, in the obvious way, as in Example 2. Then we have a complex vector space.

(c) Our final example is, perhaps, a little odd. Consider the set $\{x\}$ consisting of a single element. Define addition and scalar multiplication by $x + x = x$ and $kx = x$ for all scalars k (the only possible definitions!). All the vector space axioms are satisfied. The zero vector is x, so we may as well use the notation 0 instead of x; the the addition and scaling rules read $0 + 0 = 0$ and $k0 = 0$, the standard properties of the zero vector. The negative of 0 is 0, and similarly all the other axioms are satisfied. This vector space is called, for obvious reasons, the **trivial** vector space. Although trivial, it must be included in any serious consideration of linear algebra, just as the number zero must be included in any serious discussion of arithmetic.   □

We have now completed the task of defining what is meant by a vector space. The following sections should help to put some flesh on the dry bones of the abstract definition.

3C ALGEBRAIC PROPERTIES OF VECTOR SPACES

This section is devoted to showing that the rules of elementary algebra apply in vector spaces. If you are not interest in logical details, read the statements of the theorems, and take the proofs on trust.

Theorem 1 (uniqueness of zero). In any vector space there can be only one zero vector.

Remark. This statement does not mean that there is a single zero vector applying to all vector spaces. It means that no space can contain more than one zero vector.

Proof. Let 0 and $0'$ be zero vectors; we shall show that $0 = 0'$. We have

$$0 = 0 + 0' \quad \text{because } 0' \text{ is a zero vector}$$

$$= 0' \qquad \text{because } 0 \text{ is a zero vector}$$

Thus $0 = 0'$, so there cannot be two different zero vectors. □

We next show that for any x there is only one negative, that is, there cannot be two vectors y such that $x + y = 0$. If you think that this is obvious, note how similar it is to the statement that for any a there cannot be more than one b such that $ab = 1$. This is true for numbers but false for matrices, as we saw in section 2D, so it is not a trivial statement.

Theorem 2 (uniqueness of negative). For any element x of a vector space, there is only one element y satisfying $x + y = 0$.

Proof. Suppose that $x + y = 0$ and $x + z = 0$; we shall show that $y = z$. We have

$$y = 0 + y = z + x + y \qquad (\text{since } z + x = x + z = 0)$$

$$= z + 0 = z$$

Thus $y = z$, and we cannot have two different negatives of a given x. □

Definition 1. Write $-x$ for the negative of a vector x. **Subtraction** is defined as follows: given two elements x and y of a vector space, we define $x - y$ to be the vector $x + (-y)$.

□

Note that if Theorem 2 had not been proved, we could not have defined subtraction like this, because there might have been several negatives for a given y, hence several possible values for $x - y$: the subtraction operation would not have been well-defined.

Now that we have introduced subtraction, other useful operations become possible, such as cancelling terms which appear on both sides of a vector equation.

Corollary (cancellation law). If $x + a = x + b$, then $a = b$.

Proof. Subtract x from both sides. □

You may thing that this is so obvious that it is not worth proving. If so, observe its similarity to the cancellation law for multiplication, which fails for vector dot products and for matrices as we saw in Chapters 1 and 2. Cancellation laws cannot be take for granted. The proof of the corollary looks trivial, but that is because the essential work has been done in the preceding theorem.

The next result shows how scaling interacts with the additive structure set up in clause 1 of the definition of a vector space.

Theorem 3. Let x be any element of any vector space. Then
(a) $0x = 0$, where the 0 on the left is the scalar 0, and the 0 on the right is the zero vector;
(b) for any scalar a, $a0 = 0$ where 0 is the zero vector;
(c) $(-1)x = -x$ where $-x$ is the negative of x;
(d) for any positive integer n, $nx = x + x + \ldots + x$ (n terms).

Proof. (a) In this proof we write the zero vector as $\mathbf{0}$, to avoid confusion with the number zero. The number 0 satisfies $0 = 0 + 0$, so

$$0x = (0+0)x = 0x + 0x \quad \text{by the distributive law}$$

Hence, by the definition of $\mathbf{0}$,

$$\mathbf{0} + 0x = \mathbf{0} + 0x + 0x = 0x + 0x$$

Cancelling $0x$ from the far left and right sides of this equation (using the above corollary) gives $\mathbf{0} = 0x$, as required.
(b) We have

$$a0 + a0 = a(\mathbf{0} + \mathbf{0}) \qquad \text{by the distributive law}$$

$$= a0 = a0 + \mathbf{0} \quad \text{using properties of } \mathbf{0}$$

Now cancel $a0$ as in (a) above, leaving $a0 = \mathbf{0}$.
(c) $\qquad\qquad x + (-1)x = 1x + (-1)x \quad \text{by axiom (f)}$

$$= [1 + (-1)]x \quad \text{by the distributive law}$$

$$= 0x = \mathbf{0} \qquad\qquad \text{by part (a) of this theorem}$$

Hence $(-1)x$ is the negative of x.
(d) $\qquad\qquad x + x = 1x + 1x \quad \text{by axiom (f)}$

$$= (1+1)x \quad \text{by the distributive law}$$

$$= 2x$$

The general case follows easily by mathematical induction (Appendix B). □

We have now proved most of the essential facts about manipulation in vector spaces. A complete treatment would include some fairly obvious extensions of the algebraic laws. For example, from the commutative and associative laws it follows that vectors can be added together in any order, and the result does not depend on the order; for example, $(a+b)+(c+d) = [c+(a+b)]+d$. To prove this and similar results from the axioms is quite straightforward (unlike some of the proofs above), and is left as an exercise.

The next section introduces a simple yet powerful new idea.

3D   SUBSPACES

The phrase 'vector space' suggests some relation to the space of everyday experience. The connection may not be obvious at this stage; the last section had little to do with everyday experience. But we shall now start to build up a geometrical structure which should eventually convince you that general vector spaces really are like ordinary three-dimensional space.

The idea behind this section is as follows. A plane in three-space has two aspects: it is a subset of the vector space $\mathbb{R}^3$, but at the same time it is a vector space in its own right – see Example 3B-1(b). We often find one vector space embedded in another like this; the first space is said to be a 'subspace' of the second.

Definition 1. A **subspace** of a vector space V is a subset of V which is itself a vector space, over the same field, and with the same operations as V.

Examples 1. (a) Example 3B-1(c) considered a subset of $\mathbb{R}^3$ consisting of all vectors of the form $(x, y, 2x + 3y)$ – a plane through the origin. This subset is itself a real vector space, so it is a subspace of $\mathbb{R}^3$. A similar argument shows that for any fixed real a and b, the plane $\{(x, y, ax + by)\}$ is a subspace.

(b) $\mathbb{R}^3$ is a subset of $\mathbb{C}^3$, the complex space of three-vectors with complex entries. We know that $\mathbb{R}^3$ is itself a vector space. Is it a subspace of $\mathbb{C}^3$?

It is not, because $\mathbb{R}^3$ is a vector space over $\mathbb{R}$, while $\mathbb{C}^3$ is a vector space over $\mathbb{C}$. A subspace of V, according to Definition 1, must have the same field of scalars as V.

(c) Let $F[a, b]$ be the space of real functions on $[a, b]$ and $C[a, b]$ and $D[a, b]$ the spaces of continuous functions and differentiable functions on $[a, b]$ respectively (see Example 3B-3). Then $D[a, b]$ is a subspace of $C[a, b]$, which is itself a subspace of $F[a, b]$. So $D[a, b]$ can be viewed in three ways: as a space in its own right, or as a subspace of $C[a, b]$, or as a subspace of $F[a, b]$. In the same way one can construct spaces of functions which can be differentiated many times, giving large families of subspaces, nested inside each other like the layers of an onion.

(d) Example 3B-1(d) considered the set $\{(x, y, 1)\}$, a plane parallel to the xy plane but not passing through the origin. It is not a subspace of $\mathbb{R}^3$; all subspaces must include the zero vector – this is one of the basic vector space axioms. Planes which do pass through the origin are subspaces.   □

The following theorem gives a quick way of identifying subspaces. For any subset of a vector space V, most of the vector space axioms hold automatically because the vectors belong to V. To show that a subset is a subspace, all we need do is show that the sum of two members belongs to the set and any scalar multiple of a member belongs to the set.

Surely $\mathbb{R} \subset \mathbb{C}$

So $\mathbb{R}^3 \otimes \mathbb{C}^3$ are both over $\mathbb{R}$

?

Theorem 1 (subspace criterion). Let S be a nonempty subset of a vector space V. Then S is a subspace of V if

(i) $x + y \in S$ for all x and y in S, and

(ii) $ax \in S$ for all x in S and all scalars a.

Remark. The word 'nonempty' must appear in this theorem, or else it is not strictly true. You may feel (with some justification) that no sensible person is interested in empty sets, so there is no point in going to the trouble of excluding them. If you agree with this, then ignore the word 'nonempty' whenever it appears. But I prefer true statements to false statements, so qualifications such as 'nonempty' will appear when necessary.

Proof. The rules $x + y = y + x$, etc. all hold in S because all the vectors belong to the vector space V. We only need prove the existence of a zero vector, and of negatives.

Let x be any member of S (note that if S were empty, this step would be impossible). Then $0x \in S$ by condition (ii) of the theorem, and $0x$ is the zero vector (by Theorem 3C-3). Hence S contains the zero vector.

If $x \in S$ then $(-1)x \in S$ by condition (ii) of the theorem, hence (by Theorem 3C-3 again) the negative of x belongs to S for all x. This completes the proof. □

Examples 2. (a) For any fixed real a, b, c, the subset $\{(at, bt, ct) : t \in \mathbb{R}\}$ of $\mathbb{R}^3$ is a subspace (an easy application of Theorem 1). It is a straight line through the origin. Lines and planes through the origin are standard examples of subspaces in $\mathbb{R}^3$; we shall see that they are typical of subspaces in general.

(b) Subspaces of $\mathbb{R}^3$ can be interpreted algebraically as well as geometrically. Let A be any 3 by 3 matrix, and consider the set of 3-vectors x satisfying $Ax = 0$ (where x is written as a column vector). As we saw in Chapter 2, this is a neat way of writing a set of three linear equations for x_1, x_2, x_3. The rules of matrix algebra show that if $Ax = 0$ and $Ay = 0$, then $A(x + y) = 0$ and $A(kx) = 0$ for any real number k. Hence the set of all solutions of $Ax = 0$ is a subspace of $\mathbb{R}^3$. There is a close relation between vector spaces and sets of linear equations, as we shall see in Chapter 4.

Example 3. Consider the space V of all real functions (Example 3B-2). Let S be the set of all real functions f satisfying $f(0) = 0$. The subspace criterion easily shows that S is a subspace of V.

Exercise 1. $\{f : f(0) = k\}$ is a subspace of the space of all real functions if and only if $k = 0$. (Reminder: an exercise like this, in the form of a statement, is a challenge: prove it!)

Exercise 2. The set of all real polynomials (see Appendix F) is a subspace of the space of all real functions. □

We now consider two examples of a different kind.

Examples 4. (a) In any vector space the subset consisting of the single vector 0 is a subspace (see Example 3B-4(c)). It is called, for obvious reasons, the **trivial** subspace. A **nontrivial** subspace means any subspace other than this.

(b) The term 'subset' is defined so that every set is a subset of itself (see Appendix C). Hence every vector space V is a subset of itself. Since V is a vector space as well as a subset of V, it is a subspace of itself. This subspace is not very interesting; we use the term **proper subspace** to mean any subspace of V apart from V itself.

Example 5. Let V be the set of all complex p by q matrices. It is easily shown to be a complex vector space, with addition and multiplication by scalars defined as in Chapter 2. Let D be the set of all diagonal p by q matrices (that is, matrices A with $A_{ij} = 0$ if $i \neq j$). The sum of two diagonal matrices is diagonal; so is a scalar multiple; hence D is a subspace of V by Theorem 1. $\square$

In real 3-space we have seen that lines and planes through the origin O are subspaces. Now, the intersection of two planes through O is usually a line (unless the planes coincide). And the intersection of two lines through O is generally the point O. Thus the intersection of two subspaces in 3-space is itself a subspace ($\{O\}$ is the trivial subspace). We shall now prove that this holds in all vector spaces. (If you are unfamiliar with the set-theoretic notion of intersection, see Appendix C.)

Theorem 2 (intersection of subspaces). In any vector space, the intersection of two subspaces is itself a subspace.

Proof. We use Theorem 1. Let S and T be subspaces of a vector space V. Then $S \cap T$ is nonempty because it contains the zero vector (which belongs to both subspaces S and T, hence to $S \cap T$).

Suppose x and y belong to $S \cap T$; we must show that $x + y \in S \cap T$. We have $x + y \in S$ because x and y both belong to the subspace S. Similarly $x + y \in T$. Since $x + y$ belongs to both S and T, it belongs to $S \cap T$.

Similarly, for any scalar a, $ax \in S$ because x is in the subspace S, and $ax \in T$ because x is in the subspace T, hence $ax \in S \cap T$. The conditions of Theorem 1 are now satisfied, so $S \cap T$ is a subspace. $\square$

Example 6. A matrix is called **upper triangular** if all its nonzero entries lie in the triangular region on and above the main diagonal. In other words, A is upper triangular if $A_{ij} = 0$ whenever $i > j$. A matrix is called **lower triangular** if $A_{ij} = 0$ whenever $i < j$; all its nonzero entries are in the triangle below and on the main diagonal. Let U and L be the sets of upper and lower triangular n by n matrices respectively. Using the subspace criterion it is easy to see that U and L are subspaces of the space V of Example 5, in the case $p = q = n$. Theorem 2 shows that $U \cap V$ is a subspace. This is the set of matrices which are both upper triangular and lower triangular; thus all entries below and above the main diagonal are zero, which means that $S \cap T$ is the set of diagonal matrices. Theorem 2 implies that this is a subspace, in agreement with the result of Example 5. $\square$

Our typical example of a subspace is a line or a plane in $\mathbb{R}^3$. It is clear that $\mathbb{R}^3$ is three-dimensional, while these subspaces are of dimension one and two. Dimension is a basic geometrical idea, which is needed in order to think geometrically about vector spaces. But for the space of polynomials, say, it is not at all clear what 'dimension' means. This question is tackled in the next section.

3E   SPANNING SETS

Planes are two-dimensional, ordinary space is three-dimensional, the space–time of relativity is four-dimensional, and so on. Our aim now is to generalise this concept of dimension. We shall have to express the idea precisely enough to apply in abstract spaces, where intuition is not a reliable guide. This will take some time and effort. As you work through the chain of definitions and theorems below, remember that your labours will eventually be rewarded by a clear concept of dimension in general vector spaces.

The essence of three-dimensionality was expressed in section 1D. There can be three, but no more than three, 'independent' vectors in 3-space, where 'independent' means that none of them can be formed from the others by addition and scaling – geometrically speaking, they do not all lie in the same plane. Similarly, the statement that planes are two-dimensional means that in a plane one can have two, but no more than two, independent vectors. We shall define an n-dimensional space as one in which there can be at most n independent vectors.

This idea of independence is crucial. We shall now develop a language for discussing it in a precise and general way.

Definition 1. Let S be a nonempty set of elements of a vector space. A **linear combination** of S is a vector of the form

$$c_1 x_1 + c_2 x_2 + \ldots + c_n x_n$$

where $x_1, \ldots, x_n$ belong to S, and $c_1, \ldots, c_n$ are scalars.   □

Note that Definition 1 says that a linear combination of S is a sum of a finite number of terms, even if S is an infinite set. Linear algebra deals with finite sums only. Infinite sums are excluded, because they involve the idea of convergence, an enormous extra complication.

Example 1. In the space of all continuous functions, consider the set of all powers: $S = \{1, t, t^2, t^3, \ldots\}$. Here t^r denotes the function defined by $f(t) = t^r$ for all t.

Every linear combination of S is a polynomial. On the other hand, the function $e^t = 1 + t + t^2/2 + \ldots$ is not a linear combination of S, because it is not a finite sum of powers.

Definition 2. For any nonempty set S, the **span** of S is the set of all linear combinations of S. It is written Sp(S). For reasons that will appear below, we define the span of the empty set to be $\{0\}$, the set consisting of the zero vector.

Warning. Do not confuse the empty set, which has no elements, with the set $\{0\}$, which has one element, namely, the zero vector.

Example 2. (a) In $\mathbb{R}^3$ a linear combination of the set $S = \{(1, 0, 0), (0, 1, 0)\}$ is a vector of the form $c(1, 0, 0) + d(1, 0, 0)$ where c and d are real numbers. This equals $(c, d, 0)$, so the span of S is the set of all vectors of the form $(c, d, 0)$, that is, the xy plane.

  (b) In $\mathbb{R}^3$ a linear combination of the set $T = \{(1, 3, -2), (-2, -6, 4)\}$ is a vector of the form $c(1, 3, -2) + d(-2, -6, 4) = a(1, 3, -2)$ where $a = c - 2d$. The span of T is therefore the set of all multiples of the vector $(1, 3, -2)$, which is a line through the origin.

  (c) In Example 1, $\mathrm{Sp}(S)$ is the set of all polynomials. Exponential and trigonometric functions, which are infinite series of powers, do not belong to $\mathrm{Sp}(S)$.

Exercise 1. For every set S, $S \subset \mathrm{Sp}(S)$. $\qquad\square$

In Examples 2, the span was a subspace in each case. There is a general phenomenon here.

Theorem 1. For any set X in a vector space V, the span of X is a subspace of V.

Proof. We use the subspace criterion of section 3D. Suppose first that X is a nonempty set. Then $\mathrm{Sp}(X)$ is nonempty by Exercise 1. Multiplying linear combinations by scalars, and adding them, gives another linear combination. Hence $\mathrm{Sp}(X)$ is a subspace.

  If X is the empty set, then by Definition 2 its span is the trivial subspace $\{0\}$. (In fact, we defined the span of the empty set so that the simple statement of this theorem would hold without exceptions.) $\qquad\square$

  Theorem 1 starts with a set of vectors, and constructs a subspace from their linear combinations. It is called the subspace spanned or **generated** by X. We now turn the idea upside down: starting from a given subspace, we ask whether it can be generated by linear combinations of some set of vectors.

Definition 3. If S is a subspace of a vector space V, a **spanning set** or **generating set** for S is a set X such that $S = \mathrm{Sp}(X)$.

Exercise 2. $\{(1, 0, 0), (0, 1, 0)\}$ is a spanning set for the xy plane in $\mathbb{R}^3$. $\qquad\square$

This exercise shows that a spanning set for S can be finite while S is infinite. Thus a spanning set X can be much smaller than S. Yet X contains all the information about S, since S can be obtained from X by taking linear combinations. Thus X can give a very economical description of S.

Examples 3. (a) $\mathrm{Sp}\{(1, 0, 0), (1, 1, 0)\}$ is the set of all vectors of the form $c(1, 0, 0) + d(1, 1, 0) = (b, d, 0)$ where $b = c + d$. As c and d take all real values, $(b, d, 0)$ ranges over the xy plane. Hence $\{(1, 0, 0), (1, 1, 0)\}$ is another spanning set for the xy plane.

  (b) $\{(1, 0, 0), (1, 1, 0), (0, 1, 0)\}$ is yet another spanning set for the xy plane. Its span is the set of all vectors of the form $c(1, 0, 0) + d(1, 1, 0) + e(0, 1, 0) = (a, b, 0)$ where $a = c + d$ and $b = d + e$. This ranges over the xy plane as c, d, e take all real values.

Exercise 3. Every subspace is a spanning set for itself.   ☐

Examples 3 show that a spanning set for a given nontrivial subspace S can have various sizes. A spanning set can be enlarged by adding other vectors. There is no upper limit to the size of a spanning set; Exercise 3 shows that spanning sets can be infinite. On the other hand, there is a lower limit to the size of a spanning set for a given S. The xy plane, for example, cannot be generated by less than two vectors. This is the essence of the idea that the plane is two-dimensional. Exploring it in detail in the next section will lead to a general definition of the dimension of a vector space.

3F   LINEAR DEPENDENCE

The last section showed how a vector space can be generated from just a few vectors. A spanning set for S contains all the information on S, and the smaller the spanning set, the easier it is to work with, and the more economically it describes S.

In Example 3E-3(b) we took

$$\{u, v, w\} = \{(1, 0, 0), (1, 1, 0), (0, 1, 0)\} \tag{1}$$

as a spanning set for the xy plane in $\mathbb{R}^3$. This spanning set is larger than necessary. The smaller set $\{u, w\}$ also spans the xy plane. This is because $v = u + w$, so that any linear combination $au + bv + cw$ can be rewritten as $(a + b)u + (b + c)w$, a combination of u and w. It follows that $\mathrm{Sp}\{u, v, w\} = \mathrm{Sp}\{u, w\}$, and the spanning set $\{u, v, w\}$ can be replaced by $\{u, w\}$.

We shall now express this simple idea in general terms. A set which is needlessly large, in the sense explained above, is called 'linearly dependent'. The idea is that one of the vectors depends on the others; in the above example, $v = u + w$.

Now, the choice of v as depending on u and w was arbitrary; we could just as well have used the equation $u = v - w$ to reduce the spanning set to $\{v, w\}$. It is better not to make an arbitrary distinction between the vectors; we can treat them on an equal footing by saying that u, v and w are related by the equation

$$u - v + w = 0 \tag{2}$$

A set of vectors satisfying such an equation is called 'linearly dependent'.

There is a minor subtlety here. Consider the set $\{u, v\}$ where $v = u$. Then $u - v = 0$, so u and v are related by an equation similar to (2), and $\{u, v\}$ is therefore linearly dependent.

Yet if $v = u$ then $\{u, v\} = \{u, u\} = \{u\}$, and for this set there is no relation similar to (2) between the members. We have a contradiction: for the set $\{u, u\} = \{u\}$, there is and there is not a relation between the elements.

The difficulty is resolved by regarding the sets as ordered sets (discussed in Appendix C). Then $\{u, u\}$ is a two-element ordered set in which the first and second elements happen to be the same; it is different from the set $\{u\}$. Linear dependence is a property of ordered sets.

Definition 1. An ordered set S of vectors is called **linearly dependent** if there is a nontrivial linear combination of S which equals the zero vector. A **nontrivial linear**

combination is a linear combination in which not all the coefficients are zero. Thus S is linearly dependent if there are elements x_i of S and scalars c_i, not all zero, such that $c_1 x_1 + \ldots + c_n x_n = 0$.   □

    The restriction to nontrivial linear combinations in Definition 1 is vital. If it were omitted, then the definition would be useless because every ordered set in every vector space would satisfy the definition: a linear combination with all coefficients zero will always equal the zero vector. Such a combination is excluded by Definition 1.

    For the sake of brevity we usually omit the word 'ordered'. The order in which the elements are written is not important; rearranging a set does not affect its linear dependence. The crucial distinction here between ordered and unordered sets is that ordered sets can have repeated elements.

Examples 1. (a) Any (ordered) set including a repeated element is linearly dependent: if the vector x appears twice, then $x - x$ is a linear combination which equals zero.

    (b) The set $\{u, v, w\}$ in equation (1) is linearly dependent because $u - v + w = 0$.

Exercises 1. (a) If a set S is linearly dependent, and T is a set containing S, then T is linearly dependent.

no ?   (b) Every set which includes the zero vector is linearly dependent.

Definition 2. A vector x is **linearly dependent on** a set S if x equals a linear combination of S.

Example 2. In Example 1(b), u is linearly dependent on $\{v, w\}$ because $u = v - w$. Also v is linearly dependent on $\{u, w\}$ because $v = u + w$, and w is linearly dependent on $\{u, v\}$ because $w = v - u$.   □

    We have introduced two different uses of the phrase 'linearly dependent'; a *set* can be linearly dependent, or a *vector* can be linearly dependent *on* a set. These are different ideas, though they are very closely related, as the following theorem shows.

Theorem 1 (linear dependence). A set is linearly dependent if and only if one of its members is linearly dependent on the others.

Proof. (a) Let S be a linearly dependent set. Then there are elements $x_1, \ldots, x_n$ of S, and scalars $c_1, \ldots, c_n$, not all zero, such that

$$c_1 x_1 + \ldots + c_n x_n = 0 \tag{3}$$

Since not all the c's are zero, there must be at least one k such that $c_k \neq 0$. Rearranging (3) gives

$$x_k = (-1/c_k)\sum c_r x_r \quad \text{summed over all } r \neq k$$

This shows that x_k is a linear combination of the other elements of S, as required.

    (b) Now suppose that S is a set in which one member, x say, is linearly dependent on the others. Then $x = c_1 x_1 + \ldots + c_n x_n$ for some scalars c_i and elements x_i of S

other than x. Therefore $x - \sum c_r x_r = 0$, and this is a nontrivial linear combination of S because the coefficient of x is nonzero. Hence S is linearly dependent.   □

This theorem is useful in the following way. We are interested in spanning sets which are as small as possible. Given a spanning set which is too big, because it is linearly dependent, Theorem 2 says that we can find a member of the set which depends on the others and can therefore be eliminated. If the original set was finite, then repeating this operation will eventually lead to a set which is not linearly dependent.

Definition 3. A set is **linearly independent** if it is not linearly dependent.   □

This very obvious definition can be expressed in another form, which is more helpful in practice. To prove that a set S is linearly independent, one must show that no nontrivial linear combination of S can equal the zero vector; in other words, if any linear combination of S equals zero, then it must be the trivial one with all coefficients zero. Hence Definition 3 is equivalent to the following.

Definition 3'. A set S is **linearly independent** if the equation $c_1 x_1 + \ldots + c_n x_n = 0$ (where $x_1, \ldots$ belong to S) implies that all the scalars c_i are zero.

Example 3. The set $\{u, v\}$, in the notation of equation (1), is linearly independent. To prove this, we must show that if $cu + dv = 0$ then the scalar coefficients c and d are zero. So suppose that

$$c(1, 0, 0) + d(1, 1, 0) = (c + d, d, 0) = (0, 0, 0)$$

It follows that $c + d = 0$ and $d = 0$, so that $d = c = 0$. Therefore the set is linearly independent.

Exercise 2. $\{u, w\}$ and $\{v, w\}$, in the above notation, are linearly independent.

Exercise 3. Every subset of a linearly independent set is linearly independent.   □

Our general strategy is to reduce the size of spanning sets by removing elements which are linearly dependent on the others. The following theorem says that this does not change the subspace spanned by the set.

Theorem 2. Let S' be obtained from a set S by removing a vector which is linearly dependent on S'. Then $\mathrm{Sp}(S') = \mathrm{Sp}(S)$.   □

A moment's thought should make this obvious. A proof will be found in section 3M.

Now, suppose we have a finite spanning set X for a space S. If X is linearly dependent, we can remove an element which depends on the others, leaving a smaller spanning set for S (by Theorem 2). If this set is linearly dependent, we can repeat the process, and stop when we are left with a linearly independent set. This is the 'smallest possible' spanning set. It is by far the most useful kind of spanning set, and it is called a 'basis' for the space. Before discussing it in detail, however, we shall pause for breath.

3G   BASIS AND DIMENSION

We have now reached the heart of linear algebra. We have seen how a vector space can be described by means of a spanning set, and how a finite spanning set can be reduced in size until it is linearly independent. The resulting set is called a 'basis'; it is the central idea of vector space theory.

Definition 1. A **basis** for a vector space is a linearly independent set which spans it.

Example 1. The set $\{(1,0,0),(0,1,0)\}$ is a basis for the xy plane in $\mathbb{R}^3$ – we saw in the last section that it is linearly independent, and it spans the xy plane because $\mathrm{Sp}\{(1,0,0),(0,1,0)\} = \{x(1,0,0) + y(0,1,0):x, y\in\mathbb{R}\} = \{(x,y,0):x, y\in\mathbb{R}\}$, which is the whole of the xy plane.

The set $\{(1,0,0),(1,1,0)\}$ is another basis for the xy plane. It is linearly independent (Example 3F-3) and spans the plane (see the beginning of section 3F).

Exercise 1. The set $\{(1,0,0),(0,1,0),(0,0,1)\}$ is a basis for $\mathbb{R}^3$. $\square$

A linearly independent spanning set is called a basis because it gives a set of basic elements from which the whole space can be constructed. In $\mathbb{R}^3$, for example, any vector (x,y,z) can be expressed as $x(1,0,0) + y(0,1,0) + z(0,0,1)$.

Example 2. Let P be the space of all real polynomials, a subspace of the space of all real functions (Exercise 3D-2). In Example 3E-1 we showed that the set of all powers spans P. We now show that it is linearly independent, and therefore is a basis.

Suppose some linear combination of powers equals zero:

$$a + bt^i + ct^j + \ldots + ft^m = 0 \tag{1}$$

for some real numbers $a, b, \ldots, f$ and positive integers $i < j < \ldots < m$. We must show that $a = b = \ldots = f = 0$.

Equation (1) means that the polynomial on the left, regarded as an element of the space P, equals the zero element of P. This is the function which equals 0 everywhere. Hence the polynomial on the left-hand side vanishes for all t. By the corollary in Appendix F, it is the zero polynomial, with all coefficients zero. Hence (1) implies that $a = \ldots = f = 0$, showing that the powers are linearly independent and therefore give a basis for P. $\square$

We are now close to our goal, a general definition of the dimension of a vector space. We know that the three-dimensional space $\mathbb{R}^3$ has a basis consisting of three vectors. This suggests that in general we can take the dimension of a space to be the size of a basis for it. However, there is a potential difficulty: if a space has two bases of different sizes, there are two conflicting values for the dimension. Fortunately, this cannot happen. The following theorem says that for a given space, the size of a basis for that space is uniquely defined.

Theorem 1 (the dimension theorem). If a vector space V has a basis consisting of n elements, then:

(a) every basis for V contains n elements;
(b) every linearly independent set of n vectors is a basis; and
(c) every set of more than n vectors is linearly dependent.   □

The proof of this theorem involves no new ideas, but it is rather long. You will find it at the end of the chapter, in section 3M. Understanding the meaning of the theorem is more important than the proof at this stage.

The theorem says that if a space V has a basis, the number of vectors in a basis for V is the same for all bases; this number is a fixed property of the space V. The examples above show that for the xy plane the number is two, and for $\mathbb{R}^3$ the number is three. In general the number n gives the dimension of the space.

Definition 2. An **n-dimensional space** is a space V with a basis consisting of n elements. We call n the **dimension** of V, and write $n = \dim(V)$.   □

We now have a precise and general version of the intuitive idea of dimension. It is not the last word on the subject, however. There are other ways of expressing the idea of dimension in mathematical terms, some of which lead to the striking concept of an object which is $(2/3)$-dimensional, or even d-dimensional where d is an irrational number. Beautiful pictures, and beautiful mathematics, will be found in Mandelbrot (1982). However, in linear algebra all dimensions are integers.

Example 3. (a) Let e_r denote the n-vector with rth entry 1 and all other entries zero. Then $\{e_1, \ldots, e_n\}$ is a basis for $\mathbb{R}^n$. Proof: for any n-vector x we have

$$x = (x_1, x_2, \ldots, x_n) = x_1 e_1 + x_2 e_2 + \ldots + x_n e_n \tag{2}$$

Hence every x is a linear combination of the e's: the e's span $\mathbb{R}^n$. We must now show that they are linearly independent. Suppose that the linear combination $x_1 e_1 + \ldots + x_n e_n$ is zero. Then (2) shows that the n-vector $x = (x_1, \ldots, x_n)$ is zero. Hence all the entries x_i are zero. Thus a linear combination of the e's is zero only if all coefficients are zero, so the e's are linearly independent, and form a basis for $\mathbb{R}^n$. Thus $\mathbb{R}^n$ is n-dimensional.

The basis $\{e_1, \ldots, e_n\}$ is called the **standard basis** or **usual basis** for $\mathbb{R}^n$.

(b) Exactly the same argument works in the complex space $\mathbb{C}^n$, showing that the e's are a basis for $\mathbb{C}^n$, called the standard or usual basis for $\mathbb{C}^n$. The real and complex spaces of n-vectors have the same dimension.   □

This example shows that it makes little difference whether a space is real or complex as far as the theory of this chapter goes. The essential differences between the real and complex cases will not appear until Chapter 5.

Exercise 2. (a) The set of all real cubic polynomials is not a subspace of the space P of all real polynomials.

(b) The set P_3 of real polynomials of degree ≤ 3 is a subspace of P.

(c) $\{1, t, t^2, t^3\}$ is a basis, so P_3 is a 4-dimensional space.

(d) The set of all complex polynomials of degree ≤ 3 is a complex vector space. $\{1, t, t^2, t^3\}$ is a basis, so it too is 4-dimensional.

Exercise 3. The space of polynomials of degree $\leq n$ has dimension $n + 1$ in both the real and the complex cases.

3H   THE DIMENSION OF A SUBSPACE

This section develops the theory of dimension introduced in section 3G. We start from the observation that subspaces of ordinary three-dimensional space have dimension less than three – except of course for the subspace which is $\mathbb{R}^3$ itself. This case is excluded by the term 'proper subspace' introduced in Example 3D-4. Then common sense says that any proper subspace of a vector space V should be smaller than V, in some sense. It would be reasonable to interpret this as saying that a proper subspace has smaller dimension. That is the content of the following result.

Theorem 1 (dimension of subspace). If V is an n-dimensional vector space, every proper subspace of V has dimension less than n.

Proof. Let S be a proper subspace of V. There cannot be a linearly independent set of more than n elements in V, by the dimension theorem. So a basis for S must have n or fewer elements. If it had n elements, then it would be a basis for V by the dimension theorem. Since its span is not V but the proper subspace S, it must have fewer than n elements.   □

There is a trivial exception to this theorem. Every vector space has the trivial subspace S_0 consisting of the single vector 0 (see Example 3D-4). To find its dimension, we look for a basis. The only possible candidate is $\{0\}$, the only subset of S_0. But a set containing the zero vector is linearly dependent (Exercise 3F-1). Hence $\{0\}$ cannot be a basis, and the subspace S_0 has no basis. Therefore it has no dimension according to Definition 3G-2. This contradicts Theorem 1.

Now, since there is no basis, it is reasonable to think of the number of elements in a basis as zero. Definition 3E-2 defined the span of the empty set to be $\{0\}$. We can therefore regard the space S_0 as spanned by the empty set, and say that its dimension is the number of elements in the empty set, namely, zero.

Definition 1. The trivial vector space is said to have dimension 0.   □

This convention removes the exception to Theorem 1.

If S is a proper subspace of an n-dimensional space V, then a basis B for S contains less than n vectors. It is reasonable to expect that adding more vectors to B will give a basis for V. In fact, the following theorem shows that any linearly independent set of less than n elements in an n-dimensional space can be turned into a basis for V by adding more vectors.

Theorem 2 (extension to a basis). Let V be an n-dimensional vector space. If $k < n$, then every linearly independent set of k elements $\{u_1, \ldots, u_k\}$ can be extended to give a basis $\{u_1, \ldots, u_k, u_{k+1}, \ldots, u_n\}$ for V.   □

The proof is not as simple as one might expect; see section 3M.

The following example illustrates other aspects of the idea of dimension.

Example 1. Let P be the space of all polynomials, and P_n the space of all polynomials of degree $\leq n$. All polynomials of degree $\leq n-1$ belong to P_n, so P_{n-1} is a subset of P_n; it is a vector space, so it is a subspace of P_n. This holds for all n; we have infinitely many subspaces of P, each one lying inside the next:

$$P_0 \subset P_1 \subset P_2 \subset \ldots \subset P_{n-1} \subset P_n \subset \ldots$$

$\mathrm{Dim}(P_n) = n+1$ (Exercise 3G-3), so these spaces have dimensions $1, 2, 3, \ldots, n$, $n+1, \ldots$, illustrating Theorem 1.

All the spaces P_n are subspaces of the space P of all polynomials. Hence P cannot be finite-dimensional (proof: if it were k-dimensional for some k, then the $(k+1)$-dimensional space P_k could not be a subspace – contradiction). It would be sensible to call P infinite-dimensional. The following definition allows us to do this.

Definition 2. A vector space is **infinite-dimensional** if for *every* positive integer n it contains a linearly independent set of n elements.

Exercise 1. The space of all real functions is infinite-dimensional.

Example 2. Consider the set S of all infinite sequences of complex numbers (see Appendix C for an introduction to sequences). This set is turned into a vector space as follows. For any sequence $x = (x_1, x_2, \ldots)$ and any complex number k, define kx to be the sequence $(kx_1, kx_2, \ldots)$. For any two sequences x and y, define $x + y$ to be the sequence $(x_1 + y_1, x_2 + y_2, \ldots)$. These are the obvious extensions of the rules for scaling and addition in $\mathbb{C}^n$, and with these operations, S becomes a vector space.

The space S is rather like $\mathbb{C}^n$ with '$n = \infty$', roughly speaking. This suggests that it is infinite-dimensional. To prove this, we must, according to Definition 2, produce a linearly independent set of n elements for each n.

For each positive integer r, let e_r be the sequence whose rth entry is 1 and all other entries are 0 (it will avoid confusion later to refer to the 'entries' rather than the 'terms' of a sequence). Thus $e_1 = (1, 0, 0, 0, \ldots), e_2 = (0, 1, 0, 0, \ldots)$, etc. Then e_r belongs to S for each r. We shall show that for each n, the set $\{e_1, \ldots, e_n\}$ is linearly independent. Suppose that for some scalars c_r we have

$$c_1 e_1 + \ldots + c_n e_n = 0 \tag{1}$$

Here 0 denotes the zero element of the space S, which is the sequence $(0, 0, 0, \ldots)$. Now, all terms but the first in (1) have 0 as the first entry of the sequence. Hence equating the first entries of the left and right sides of (1) gives $c_1 = 0$. Considering the kth entries similarly shows that $c_k = 0$ for all k. Hence for any n the set $\{e_1, \ldots, e_n\}$ is linearly independent. Thus S contains linearly independent sets of any given size, so it is infinite-dimensional according to Definition 2. $\square$

An n-dimensional space has, by definition, a basis of n elements. It is reasonable to expect that an infinite-dimensional space has an infinite basis – in Example 1, the infinite set $\{1, t, t^2, \ldots\}$ is a basis for P. However, it is often very difficult to decide whether an infinite-dimensional space has a basis or not (for example, try to write down a basis for the space of all real functions). Definition 2 has been carefully framed to avoid the idea of an infinite basis.

Infinite bases are (like dragons) subtle and dangerous to deal with. But (like dragons) they guard a great treasure, the theory of infinite-dimensional vector spaces, which is not only very beautiful but very useful – see section 5G, for example. This book deals mainly with finite-dimensional spaces, where there are no real difficulties.

3I   COMPONENTS

The last section used basis sets to define the dimension of a space. They are useful in other ways. A basis is used in linear algebra rather like axes are used in geometry. Points in 3-dimensional space can be described by their coordinates with respect to a set of three axes. An n-dimensional vector space has a basis of n elements, and every vector in the space can be expressed as a linear combination of the basis vectors. The n scalar coefficients play the same role as the three coordinates of a point in 3-space.

Definition 1. Let $\{b_1,\ldots,b_n\}$ be a basis for a vector space V. If $x \in V$, and $c_1,\ldots,c_n$ are scalars such that

$$x = c_1 b_1 + \ldots + c_n b_n$$

then the scalars c_r are called **components** of the vector x with respect to the basis $\{b_1,\ldots,b_n\}$.   ☐

Because the b's span V, every x in V can be expressed as a linear combination of them, so there always exist numbers c_r satisfying this equation. But are they unique?

Example 1. The set $\{u, v, w\}$, as defined in equation (1) of section 3F, spans the xy plane; $(1, 2, 0) = au + (1 - a)v + (1 + a)w$ for any real a.   ☐

This example shows that there can be infinitely many possible values for the coefficients in the expression of a vector in terms of a spanning set. We now prove that this cannot happen if the spanning set is linearly independent.

Theorem 1 (uniqueness of components). In an n-dimensional space, each vector has a unique set of n components with respect to a given basis.

Proof. Let $\{b_1,\ldots,b_n\}$ be a basis for V. Suppose that for some x in V we have

$$x = c_1 b_1 + \ldots + c_n b_n$$

and

$$x = d_1 b_1 + \ldots + d_n b_n$$

Subtracting these equations gives

$$0 = (c_1 - d_1)b_1 + \ldots + (c_n - d_n)b_n$$

Since the b's form a basis, they are linearly independent, so this linear combination of them can equal 0 only if all the coefficients $(c_r - d_r)$ are zero. Hence $c_r = d_r$ for all r, and there cannot be two different sets of components of x with respect to the basis $\{b_1, \ldots, b_n\}$. $\square$

Example 2. Let $S = \{(1, 0, 0), (1, 1, 0), (1, 1, 1)\}$. We shall now show that S is a basis for $\mathbb{R}^3$, and at the same time find the components of an arbitrary vector with respect to S.

Let $v = (x, y, z)$ be a linear combination of S, with coefficients a, b, c say:

$$v = (x, y, z) = a(1, 0, 0) + b(1, 1, 0) + c(1, 1, 1) \tag{1}$$

S is a basis if (i) $v = 0$ implies $a = b = c = 0$ (this means S is linearly independent), and (ii) every v in $\mathbb{R}^3$ can be expressed in this form for some a, b, c (this means S spans $\mathbb{R}^3$).

Equation (1) is equivalent to the three equations

$$x = a + b + c \qquad y = b + c \qquad z = c$$

The last equation gives $c = z$; substituting into the second gives $b = y - c = y - z$, and now the first equation gives $a = x - b - c = x - y$. Hence (1) is equivalent to

$$a = x - y \qquad b = y - z \qquad c = z \tag{2}$$

It follows that if $(x, y, z) = 0$ then $a = b = c = 0$. Hence S is linearly independent. We also see that for every x, y, z there are values of a, b, c such that (1) is satisfied. Therefore S spans $\mathbb{R}^3$, and we have proved that it is a basis. Finally, the numbers a, b, c given in (2) are the components of the vector (x, y, z) with respect to the basis S.

Example 3. Let P_2 be the space of real polynomials of degree ≤ 2. Consider the set $Q = \{1, x, 2x^2 + x + 1\}$. Is it a basis? And if so, how do we find components with respect to it?

Take an arbitrary polynomial $Ax^2 + Bx + C$. We wish to know whether it can be expressed as

$$Ax^2 + Bx + C = a + bx + c(2x^2 + x + 1)$$

for some a, b, c. Equating the coefficients of the powers on the two sides of this equation gives

$$C = a + c \qquad B = b + c \qquad A = 2c$$

Solving these equations gives

$$c = A/2 \qquad b = B - A/2 \qquad a = C - A/2 \tag{3}$$

Since there is a solution for all A, B, C, and it is zero when $A = B = C = 0$, we deduce in the same way as in Example 2 that the set Q is a basis. The components of the polynomial $Ax^2 + Bx + C$ with respect to this basis are the numbers a, b, c given in (3).

Exercise 1. With respect to a basis $\{b_1, \ldots, b_n\}$, the components of b_j are all zero except the jth, which is 1. Thus the i-component of b_j is δ_{ij}, the Kronecker delta (Definition 2D-3). $\square$

We now know that every vector can be represented by a set of n components with respect to a given basis. What happens to the components when vectors are scaled and added? The answer is simple and obvious.

Theorem 2 (algebra of components).

    Version 1. For any vectors, the components of the sum are the sums of their components, and the components of a scalar multiple are the multiples of the components.

    Version 2. Let V be a vector space with basis $B = \{b_1, \ldots, b_n\}$. If vectors v and w in V have components v_r and w_r with respect to the basis B, then the components of $v + w$ are $v_r + w_r$, and for any scalar k the components of kv are kv_r.

Note. These are two different wordings of the same theorem. Version 1 is simple, clear, and memorable. But in order to prove it one needs the more explicit and precise version 2. You don't really understand a precise statement like version 2 until you can phrase it informally; you don't really understand the informal version 1 until you can express it in precise terms. You will not often see both versions printed, as here; but I suggest that you make a habit of formulating the 'other version' of the results you meet. (Example: the 'precise version' of Theorem 1 is: if $\{c_r\}$ and $\{d_r\}$ are two sets of components of a vector x with respect to the same basis $\{b_r\}$, then $c_r = d_r$ for each r.)

Exercise 2. Prove Theorem 2.   □

It follows from Theorem 2 that any set of operations on vectors in a finite-dimensional space is exactly mirrored by corresponding operations on the components. This useful fact allows us to work with components, with confidence that the results will apply to the original vectors. It can also be viewed from a more general point of view, as explained in the next section.

*3J   ISOMORPHISM OF VECTOR SPACES

This optional section deals with one of the key ideas of abstract algebra; it gives a new perspective on vector spaces. But it is not needed for the unstarred sections of this book (the stars are explained in the introduction to Part II).

    The results of section 3I can be viewed in the following way. We have an n-dimensional vector space V, with a basis $B = \{b_1, \ldots, b_n\}$. To each x in V there corresponds a set of n scalars, the components of x with respect to B. If V is a real space, these n numbers are real and can be regarded as entries of a real n-vector, X say. Thus to each x in V there corresponds an X in $\mathbb{R}^n$. Conversely, given any n-vector $X = (X_1, \ldots, X_n)$ in $\mathbb{R}^n$, there is a corresponding element x of V, defined by $x = \sum X_i b_i$. We have a one-to-one correspondence between the elements of V and the elements of $\mathbb{R}^n$.

    The same applies to an n-dimensional space V over any field $\mathbb{K}$. If a basis for V is given, then to each x in V there correspond n scalars, giving an n-vector in $\mathbb{K}^n$. We have a one-to-one correspondence between the n-dimensional space V and the space $\mathbb{K}^n$.

From this point of view, Theorem 3I-2 says that algebraic operations in the space V are exactly mirrored in $\mathbb{K}^n$. The one-to-one correspondence between the two spaces fits in neatly with the algebraic structure (in a sense which is made precise in Definition 1). Such a correspondence is called an 'isomorphism' (meaning 'the same shape': the two spaces look the same in all essential respects).

Definition 1. Let V and V' be two vector spaces over the same field $\mathbb{K}$. An **isomorphism** between V and V' is a rule which for each element v of V gives a corresponding element v' of V' such that

(i)   each v' in V' corresponds to exactly one v in V;

(ii) $v' + w'$ corresponds to $v + w$;

(iii) kv' corresponds to kv for every scalar k.

Two spaces are said to be **isomorphic** if there is an isomorphism between them. We say that v' corresponds to v **under** the isomorphism.

Exercise 1. If V is isomorphic to U and U is isomorphic to W, then V is isomorphic to W.
□

If V is an n-dimensional space over $\mathbb{K}$, and a basis for V is given, the rule explained at the beginning of this section gives an isomorphism between V and $\mathbb{K}^n$. This means that the algebraic properties of V and $\mathbb{K}^n$ are the same; they can be regarded as the 'same space' in different notation.

Proposition 1. Every n-dimensional space over $\mathbb{K}$ is isomorphic to $\mathbb{K}^n$.

Proof. Let B be a basis for an n-dimensional space V over $\mathbb{K}$. For each v in V there is an n-vector v' in $\mathbb{K}^n$ whose entries are the components of v with respect to B; and for each $u' \in \mathbb{K}^n$ there is a corresponding vector in V whose components are the entries of u'. Thus condition (i) of Definition 1 is satisfied; the others hold by Theorem 3I-2.
□

Corollary 1. All n-dimensional spaces over $\mathbb{K}$ are isomorphic to each other.

Proof. They are all isomorphic to $\mathbb{K}^n$, hence they are isomorphic to each other by Exercise 1.
□

This is a remarkable and beautiful fact. It means that there can be no real surprises in finite-dimensional vector spaces; every n-dimensional space over $\mathbb{K}$ behaves just like $\mathbb{K}^n$. For infinite-dimensional spaces, however, things are different; see Problem 37.

We shall now consider the relation between isomorphic spaces in more detail. The rest of this section states and proves some rather obvious facts; it can be skipped without great loss.

Exercise 2. If V and V' are isomorphic, then the zero vector in V corresponds to the zero vector in V' under the isomorphism.

Proposition 2. Suppose V and V' are isomorphic. If $\{v_1, \ldots, v_r\}$ is a linearly dependent set in V, then the corresponding set $\{v'_1, \ldots, v'_r\}$ in V' is linearly dependent.

Proof. Since $\{v_1, \ldots, v_r\}$ is linearly dependent, there are scalars c_i, not all zero, such that

$$c_1 v_1 + \ldots + c_r v_r = 0$$

The corresponding vectors v'_i in V' satisfy

$$c_1 v'_1 + \ldots + c_r v'_r = 0' \tag{1}$$

where $0'$ is the element of V' corresponding to $0 \in V$. Exercise 2 shows that $0'$ is the zero element of V', so (1) shows that the set $\{v'_1, \ldots, v'_r\}$ is linearly dependent.

$\square$

Corollary 2. If V and V' are isomorphic, then linearly independent sets in one space correspond to linearly independent sets in the other.

Proof. Let S' be the set of vectors in V' corresponding to the set S in V under the isomorphism. If S' is linearly dependent, then so is S by Proposition 2. Hence if S is linearly independent, so is S'.

$\square$

Theorem 1. Isomorphic spaces have the same dimension.

Proof. Suppose V is n-dimensional and V' is isomorphic to V. There is a linearly independent set of n vectors in V. By Corollary 2, the corresponding set of n vectors in V' is linearly independent, so $\dim(V') \geq \dim(V)$. The same argument starting from V' shows that $\dim(V) \geq \dim(V')$. So they are equal.

If V is infinite-dimensional, then so is V', for if V' were n-dimensional for some integer n, then V would be n-dimensional by the result of the first paragraph of this proof.

$\square$

Theorem 2. If V and V' are isomorphic, then a subspace of V corresponds to a subspace of V', and the dimensions of these subspaces are the same.

Proof. Let S be a subspace of V, and let S' be the set of all elements of V' corresponding to elements of S under the isomorphism. We must show that S' is a subspace. We use the subspace criterion.

The set S' is nonempty because it contains the zero vector (see Exercise 2). Let s', t' be elements of S'; we must show that for all scalars a and b, the vector $as' + bt'$ belongs to S'.

If s and t are the corresponding elements of S, we have $as + bt = u$ where $u \in S$, because S is a subspace of V. Hence, by Definition 1(ii) and (iii), $as' + bt' = u'$ where u' is the vector in V' corresponding to u. But $u' \in S'$ by definition of S'. Hence S' is a subspace.

We now know that S and S' are vector spaces. If we restrict our attention to S and S' rather than V and V', the isomorphism between V and V' gives an isomorphism between S and S'. Since the vector spaces S and S' are isomorphic, they have the same dimension by Theorem 1 above.

$\square$

The results of this section can be summed up briefly by saying that if two spaces or subspaces are isomorphic, they have the same algebraic properties.

*3K   SUMS AND COMPLEMENTS OF SUBSPACES

This optional section discusses the geometry of subspaces, developing ideas which bring the subject into sharper focus. It is essential to the more advanced theoretical sections of later chapters.

In section 3D we found that the intersection of two subspaces of a vector space is always a subspace. But the same does not apply to unions.

Exercise 1. The union of the subspaces $\{(x,0):x\in\mathbb{R}\}$ and $\{(0,y):y\in\mathbb{R}\}$ in $\mathbb{R}^2$ is not a subspace of $\mathbb{R}^2$. $\square$

One often needs to combine two subspaces to form a larger one. Since it cannot be done by taking their union, we need another way of putting two subspaces together.

Definition 1. If S and T are nonempty subsets of a vector space, their **sum** $S+T$ is defined by $S+T=\{s+t:s\in S \text{ and } t\in T\}$. $\square$

In other words, $S+T$ is the set of all vectors obtained by adding a member of S to a member of T.

Examples 1. (a) if S and T each consist of a single vector, then $S+T$ consists of a single vector, their sum.

(b) If S is a subspace, and T consists of a single vector t, then $S+T=\{s+t:s\in S\}$. This set can be visualised as the subspace S displaced through the vector t. For example, if S is a plane through O in 3-space, then $S+\{t\}$ is a plane passing through the point with position vector t. A moment's thought will show that if $t\in S$ then $S+\{t\}=S$, otherwise $S+\{t\}$ is a plane not passing through the origin.

(c) If S is the x-axis in 3-space and T is the line $\{(t,t,0):t\in\mathbb{R}\}$, then $S+T=\{(x,0,0)+(t,t,0):x,t\in\mathbb{R}\}=\{(x+t,t,0)\}$ which is the xy plane.

(d) If S is the x-axis and T is the xy plane, then $S+T=\{(x,0,0)+(x',y',0):x,x',y'\in\mathbb{R}\}=\{(x+x',y',0)\}$ which is the xy plane again.

(e) If S is the xy plane in $\mathbb{R}^3$ and T is the yz plane, then $S+T=\{(x,y,0)+(0,y',z')\}=\{(x,y+y',z'):x,y,y',z'\in\mathbb{R}\}$ which is the whole of $\mathbb{R}^3$.

Theorem 1 (sum of subspaces). If S and T are subspaces of a vector space V, then $S+T$ is a subspace of V, and S and T are subspaces of $S+T$.

Proof. The set $S+T$ is nonempty because it contains $0+0=0$. If x_1 and x_2 belong to $S+T$, then

$$x_1=s_1+t_1 \quad \text{and} \quad x_2=s_2+t_2$$

for some s_1, s_2 in S and t_1, t_2 in T. Hence

$$x_1+x_2=s_1+t_1+s_2+t_2=(s_1+s_2)+(t_1+t_2)$$

which is a member of S plus a member of T and therefore belongs to $S + T$. It is easy to verify the other part of the subspace criterion (Theorem 3D-1), showing that $S + T$ is a subspace.

For each s in S we have $s = s + 0$, and $0 \in T$, so $s \in S + T$. Hence $S \subset S + T$. We are given that S is a vector space, so it is a subspace of $S + T$. Similarly for T.   ☐

Examples 1(c) and (d) illustrate Theorem 1. Example 1(d) illustrates the following general fact.

Exercise 2. If S is a subspace of V, and X a nonempty subset of S, then $S + X = S$.
                               ☐

A member of $S + T$ can, by definition, be expressed as a member of S plus a member of T. It is natural to ask if this expression is unique.

Examples 2. (a) In Example 1(e) we have $(1, 1, 1) = (1, a, 0) + (0, 1 - a, 1)$ for any real a. Hence there are infinitely many ways of expressing $(1, 1, 1)$ as a sum of members of S and T.

(b) In Example 1(c) we showed that the xy plane in $\mathbb{R}^3$ is the sum of the line $S = \{(x, 0, 0)\}$ and the line $T = \{(t, t, 0)\}$. Expressing a vector $(x, y, 0)$ in the xy plane as a sum of members of S and T gives $(x, y, 0) = (a, 0, 0) + (b, b, 0)$ for some real a, b. Thus $(x, y, 0) = (a + b, b, 0)$, so $x = a + b$ and $y = b$. Hence $a = a - y$ and $b = y$; the numbers a and b are uniquely determined by x and y. Hence each vector in $S + T$ has a unique expression as a sum of members of S and T.   ☐

Examples 2 illustrate two different types of sum. The type with the uniqueness property is particularly important.

Definition 2. If $S + T = U$ and for each u in U there is only one s in S and t in T such that $u = s + t$, then U is called the **direct sum** of S and T, and we write $U = S \oplus T$.
                               ☐

The sum in Example 1(e) is not direct, but that in Example 1(c) is. Now, in Example 1(e), S and T intersect in a line, whereas in Example 1(c), $S \cap T$ is a single point, $\{0\}$. This illustrates a general property of direct sums – the subspaces are as nearly disjoint as subspaces can be, intersecting in 0. The following theorem shows that this is in fact a necessary and sufficient condition for directness.

Theorem 2 (direct sums). The sum $S + T$ is direct if and only if $S \cap T = \{0\}$.

Proof. (a) Suppose $S \cap T = \{0\}$. If $s, s' \in S$ and $t, t' \in T$ with $s + t = s' + t'$, then $s - s' = t' - t$. But $s - s' \in S$ and $t' - t \in T$, and since they are equal, they belong to both S and T and hence to $S \cap T$. But $S \cap T = \{0\}$, hence $s - s' = t' - t = 0$, so $s = s'$ and $t = t'$, showing that there cannot be two different expressions for a vector in the form $s + t$.

(b) If the sum is direct, take any x in $S \cap T$. Then $x = x + 0$ where $x \in S$ and $0 \in T$, and $x = 0 + x$ where $0 \in S$ and $x \in T$. By uniqueness it follows that $x = 0$, so the only element of $S \cap T$ is 0. □

Direct sums $S \oplus T$ are particularly useful because their dimension can be written down immediately in terms of the dimensions of S and T.

Theorem 3 (dimension of direct sum). If S and T are finite-dimensional subspaces of V, and their sum is direct, then

$$\dim(S \oplus T) = \dim(S) + \dim(T)$$

Proof. Let $\{s_1, \ldots, s_p\}$ and $\{t_1, \ldots, t_r\}$ be bases for S and T respectively. We shall show that the set $B = \{s_1, \ldots, s_p, t_1, \ldots, t_r\}$ is a basis for $S + T$.

B spans $S + T$ because every vector in $S + T$ is the sum of a member of S and a member of T and therefore can be expressed as a combination of the s's and t's. To show that B is linearly independent, suppose that some linear combination of B equals zero:

$$\sum c_i s_i + \sum d_j t_j = 0$$

Then

$$\sum c_i s_i = -\sum d_j t_j \tag{1}$$

Now, the left-hand side of (1) belongs to S and the right-hand side belongs to T; since they are equal, they both belong to S and to T, hence to $S \cap T$. Therefore

$$\sum c_i s_i = 0 \tag{2}$$

by Theorem 2. But $\{s_1, \ldots, s_p\}$ is a basis for S, so (2) implies that all the c's are zero. A similar argument shows that the d's are zero. Therefore B is linearly independent.

Hence B is a basis for $S \oplus T$. The number of vectors in B is $p + r = \dim(S) + \dim(T)$. This completes the proof. □

Definition 3. If $S \oplus T = U$, each of the subspaces S and T is said to be a **complementary subspace in** U of the other. □

Note that in this definition, U may or may not be a subspace of some larger space. Given a subspace S of a vector space V, we can consider a complementary subspace of S in V, or a complementary subspace of S in U for any subspace U of V such that $S \subset U$.

In Example 2(b), S is a complementary subspace of T in the xy plane, and vice versa. A complementary subspace of S in $\mathbb{R}^3$ must be 2-dimensional by Theorem 3. Indeed, it is easy to show that $\mathbb{R}^3 = S \oplus P$ where P is the yz plane; hence the yz plane is a complementary subspace of S in $\mathbb{R}^3$.

Warning. The phrase 'complementary subspace' is sometimes abbreviated to **complement**; but it must not be confused with the following.

Definition 4. For any subset X of a set U we define the **set-theoretic complement** of X in U to be $\{u \in U : u \notin X\}$. It is denoted by $U \setminus X$.

Exercise 3. If S is a proper subspace of V, then every complementary subspace of S in V is a proper subset of $\{0\} \cup V \setminus S$.   □

The word **complement** in linear algebra always means the complementary subspace unless the set-theoretic complement is specified.

Exercise 4. Let S be the x-axis in $\mathbb{R}^3$. Then for any $c \neq 0$ the subspace $\mathrm{Sp}\{(1, c, 0)\}$ is a complement of S in the xy plane. The yz plane is a complement of S in $\mathbb{R}^3$; so is $\{(y, y, z) : y, z \in \mathbb{R}\}$.

Theorem 4 (complements). Every subspace of an n-dimensional vector space V has a complement in V. If T is a complement of S in V, then

$$\dim(T) = n - \dim(S)$$

Proof. Let $\{b_1, \ldots, b_k\}$ be a basis for a subspace S. It can be extended to a basis $\{b_1, \ldots, b_n\}$ for V (Theorem 3H-2). Set $T = \mathrm{Sp}\{b_{k+1}, \ldots, b_n\}$, then $S + T = \mathrm{Sp}\{b_1, \ldots, b_k\} + \mathrm{Sp}\{b_{k+1}, \ldots, b_n\} = \mathrm{Sp}\{b_1, \ldots, b_n\} = V$. And $S \cap T = \{0\}$ because the b's are linearly independent, so that no linear combination of the first k of them can equal a combination of the others (except for the trivial combination 0). Hence $V = S \oplus T$ by Theorem 2, so T is a complement of S.

The dimension formula follows from Theorem 3.   □

Direct sums are particularly useful because of the dimension formula of Theorem 3. The following lemma shows how to replace ordinary sums by direct sums.

Lemma 1. For any subspaces S and T of a vector space V, $S + T = S \oplus T'$ where T' is a complement of $S \cap T$ in T.

Proof. By definition of T' we have $T = S \cap T \oplus T'$. Hence

$$S + T = S + S \cap T + T' = S + T'$$

by Exercise 2 (clearly $S \cap T \subset S$). To prove that the sum $S + T'$ is direct, suppose $x \in S \cap T'$, then $x \in S$ and $x \in T$ (because $T' \subset T$). Hence $x \in S \cap T$. But in the space T, the subspaces T' and $S \cap T$ are complementary; x belongs to them both, hence $x = 0$ by Theorem 2. Thus $S \cap T' = \{0\}$, so the sum $S + T'$ is direct.   □

We can now derive an elegant formula for the dimension of the sum of two subspaces.

Theorem 5 (dimension of sum). For any two subspaces S and T of a finite-dimensional vector space,

$$\dim(S + T) = \dim(S) + \dim(T) - \dim(S \cap T)$$

Proof. In the notation of Lemma 1, $\dim(S + T) = \dim(S \oplus T') = \dim(S) + \dim(T')$ by Theorem 3, $= \dim(S) + \dim(T) - \dim(S \cap T)$ by Theorem 4.   □

We have seen that a vector space can often be expressed as a sum of two subspaces.

These subspaces can themselves often be split. Suppose that $V = S \oplus T$ and $T = X \oplus Y$ where X and Y are subspaces of T and therefore of V. We write $V = S \oplus (X \oplus Y)$. It is natural to wonder whether $S \oplus (X \oplus Y) = (S \oplus X) \oplus Y$.

Theorem 6 (associativity of sums). Sums and direct sums of subspaces obey the associative law.

Proof. $S + (X + Y) = (S + X) + Y$, since both equal $\{s + x + y : s \in S, x \in X, y \in Y\}$.

Now suppose $V = S \oplus (X \oplus Y)$. Then $X \cap Y = \{0\}$, and $S \cap (X + Y) = \{0\}$. It follows that $S \cap X = S \cap Y = \{0\}$. Therefore the sum $S + X$ is direct. Finally we must show that $(S + X) \cap Y = \{0\}$. If $u \in (S + X) \cap Y$ then $u = s + x$ and $u = y$ (where $s \in S, x \in X, y \in Y$). Hence $s = (-x) + y$ and $-x \in X$, therefore $s = 0$ because $S \cap (X + Y) = \{0\}$. This completes the proof that if $V = S \oplus (X \oplus Y)$ then $V = (S \oplus X) \oplus Y$. $\square$

We can now write direct sums of three and more subspaces without brackets. Since the dimension of a direct sum of subspaces is the sum of their dimensions, there cannot be more than n terms in a direct sum in an n-dimensional space (not counting the trivial subspace $\{0\}$). If there are n terms then they are one-dimensional. Expressing a space as a direct sum of one-dimensional subspaces is very much like expressing a vector as a combination of basis vectors. Each basis vector spans a one-dimensional subspace, and every vector is expressible as a unique sum of multiples of the basis vectors, that is, of members of these subspaces. Conversely, if a space is a direct sum of one-dimensional subspaces, then a basis can be obtained by taking one nonzero vector in each subspace. Direct sums of one-dimensional subspaces give another way of expressing the concept of a basis. This idea will be developed further in Chapters 11 and 12.

*3L   THE THEORY OF COLOUR VISION

This section is optional; it does not depend on any previous optional section. It deals with a subject which has attracted many mathematicians and mathematical physicists (Newton, Grassman, Maxwell, Helmholtz, Rayleigh, and Schrödinger, amongst others).

It is well known that there are three primary colours, and all other colours can be obtained by combining them. This sounds rather like the idea of a three-dimensional space in which all vectors are combinations of three basis vectors.

There are two ways of combining colours. We can mix coloured paints to give new colours, or we can combine coloured lights, as in stage lighting for example. We shall consider the mixing of coloured lights; it is simpler to think about than mixing paints. So we begin with an outline of the physics of light.

The physics of coloured light

The classic experiments of Newton showed that when white light passes through a prism, it splits into a band of different colours, each of which is pure and cannot be split any further. This pure coloured light is called 'monochromatic'; it consists of electromagnetic waves of a definitive wavelength. Red monochromatic light has a wavelength of about 700 mμ, and blue has wavelength about 400 mμ (1 mμ = 10^{-9} m).

Monochromatic light is an idealisation. Most light is a mixture of wavelengths, and can be described by a function $f(\lambda)$ giving the distribution of light energy over the possible wavelengths λ. (Formal definition: the energy in wavelengths between a and b is $\int_a^b f(\lambda)\,d\lambda$). The function f is called the 'spectral density' of the light. Every continuous function f corresponds to a possible beam of light, and different density functions f correspond to physically different lights. If f is large for λ near 700 and very small elsewhere, it corresponds to nearly monochromatic red light; if f is roughly constant, it corresponds to a whitish light, and so on.

Given a beam of light, one can change its intensity without changing its spectral composition (the relative amounts of different wavelengths). This corresponds to multiplying the function f by a constant. Given two beams of light, one can superimpose them. The combined beam has a spectral density equal to the sum of the densities of the original beams. We thus have a physical interpretation of scaling a light beam, and adding two light beams. The set of all light beams, or equivalently the set of all spectral density functions, is a vector space.

There is one awkward point here. There is no physical interpretation for the negative of a light beam. Negative light intensity makes no sense. We should therefore say that the set of all continuous functions is a vector space, and that lights correspond to the subset of this space consisting of positive-valued functions. This restriction is to be understood whenever we say that lights form a vector space.

Now, the set of all continuous functions is infinite-dimensional. Yet we began with the remark that three basic coloured lights suffice to construct all others by their combinations. How is this possible if the space is infinite-dimensional?

The answer is that the three-dimensionality is psychological, not physical. It is a feature of our minds, not of the physical properties of light beams.

The psychology of colour perception

When light enters our eyes, it produces sensations in our minds which we call colour. The sensations are related to the light that produces them, but not in a simple way. The same colour sensation can be produced by many different lights. We shall see that colour sensations can be represented as points in a three-dimensional vector space.

Consider the set of all coloured lights, that is, the set of all positive-valued spectral density functions. Given two lights A and B, we write $A = B$ if they look the same. The equation $A = B$ represents psychological equality, not physical equality; A and B can be physically different, yet look the same.

'Look the same to whom?', you may ask. It is not obvious that two lights which look the same to me will look the same to you. However, careful experiments have shown that most people agree with each other on judgements of whether two colours match. There is a sizeable minority whose colour judgements differ from the majority view; they are called colour-blind, and we shall return to them below.

We use the term 'colour space' for the set of all the different coloured-light sensations that we can perceive. If two different light beams look the same, they correspond to the same element of colour space. Colour sensations will be denoted by capital letters; thus G may refer to a particular shade and brightness of green. We shall label coloured lights by the colour sensations that they evoke; thus 'the light G' will mean any one of the many lights (that is, many spectral distribution functions) which evoke the sensation G.

Colour space as a vector space

We now define vector space operations on colour space. Suppose that combining two lights R and G gives an effect identical to that of a light Y. For example, mixing red and green light produces yellow. Then we write $R + G = Y$. Write $2R$ for the light obtained by combining R with a second identical light R. That is, $2R$ is the same colour as R but twice as intense. Similarly, kR is the light of the same colour as R but k times as intense.

These definitions need justification. The sensation R can be produced by many different lights, and similarly for G. Will combining R and G give the same sensation regardless of which lights are used for R and G? The answer is not obvious; it must be investigated experimentally. The answer turns out to be 'yes, under normal conditions'. It does not hold for very dim or very strong light. 'All cats are grey in the dark' – in moonlight we cannot discern colours at all. The vector space theory of colour vision applies over a very wide range of conditions, but fails at the extremes of the brightness range.

We have now defined addition and scaling; but it is not obvious that they satisfy the vector space axioms. Consider, for example, the equation $k(A + B) = kA + kB$. According to our definitions it means that if $X = A + B$ is a colour which matches the combination of colours A and B, then they still match if all the intensities are increased by a factor k. It is quite conceivable that changing the intensities of the lights might disturb the matching of colours, in which case $k(A + B)$ would not equal $kA + kB$. However, experiments show that colour matches are independent of brightness over a very wide range of brightnesses. At the extreme ends of the brightness range the law $k(A + B) = kA + kB$ breaks down; the vector space theory does not apply to extremely dim or bright lights.

The other vector space axioms can be experimentally tested in the same sort of way; the upshot is that under normal conditions the set of colour perceptions is indeed a vector space. Or, more properly, the set of colour perceptions is in some sense the positive region of a vector space; negative colours do not make sense.

Even though 'minus red' does not exist, we can make sense of equations with negative terms. The equation $A - B = C$ is equivalent to $A = B + C$, which is directly verifiable by experiment. In a similar way we can interpret equations containing any number of terms with positive or negative coefficients; transferring all negative terms to the other side gives an equation between positive combinations.

The dimension of colour space

Experiments show that given three different colours P, Q, R, say, all possible colours can be obtained by combining them in suitable proportions (including negative proportions interpreted as in the above paragraph). But starting from only two, it is not possible to obtain all colours. This means that $\{P, Q, R\}$ spans the set of all colours, but there is no spanning set of two elements. In other words, colour space is three-dimensional.

We know that the choice of basis in a vector space is arbitrary: any three linearly independent vectors can be used as a basis. This means that the question 'which are the three primary colours?' is meaningless: any three different colours can be used to generate all the others.

However, not all bases are equally convenient. Choosing three different shades of red as a basis would be like taking three almost parallel vectors as a basis for 3-space. It would be inconvenient because many vectors would have very large negative coordinates. It is better to choose a basis such that as many colours as possible have positive coordinates.

Establishing a basis is important from the practical point of view. A large industry is based on dyes and pigments of one sort or another. A systematic way of specifying colours is essential, and the industry uses components with respect to a standard basis, laid down by an international commission.

Why is colour space three-dimensional? Is it because our minds are conditioned to three dimensions because the space in which we live is three-dimensional?

No. The answer is physiological, not philosophical. Our eyes contain three different kinds of colour-sensitive cell, sensitive to different parts of the spectrum. The information picked up by the nervous system thus consists of three numbers, leading to the three-dimensional structure described above.

Subspaces and colour-blindness

How does the phenomenon of colour-blindness fit into our theory? A natural guess is that one or more of the three kinds of colour-sensitive cell in the eye of a colour-blind person is either defective or missing. If all three kinds work, but not perfectly, then the person will have a three-dimensional colour space with different characteristics from the normal. This description fits the majority of people with defective colour vision; they are called anomalous trichromats (trichromatism means having a three-dimensional colour space). They perceive colours differently from others, that is, their judgements of when two colours match disagree with the majority; but there is no colour to which they are strictly blind.

There is another theoretical possibility. One or more of the three kinds of colour-sensitive cell may be missing. This would lead to a two-dimensional or one-dimensional colour space, a subspace of the three-dimensional space described above. Someone with a two-dimensional colour space is called a dichromat, and someone with a one-dimensional colour space is called a monochromat. One expects three different kinds of dichromats, corresponding to the three possible missing pigments. Each type would have a characteristic pattern of colour matching. Similarly there are three kinds of monochromats, but this time they cannot be distinguished by the usual colour-matching experiments because all colours look alike to a monochromat.

This theoretical account of colour-blindness agrees well with the facts. There are indeed three kinds of dichromats, and monochromatism does exist although it is very rare. Generally speaking, the three-dimensional theory outlined here is remarkably satisfactory over a wide range of conditions.

Beyond the three-dimensional theory

The theory sketched above is used a great deal in practice, in the dye and paint industries. See, for example, Wyszecki and Stiles (1967), where a thorough mathematical development will be found, as well as the technical, experimental, and physiological aspects of the subject.

However, it is not the whole truth. It is a good description of the way we see isolated patches of colour; but in everyday life we look at complex scenes with many colours mixed together. About 30 years ago, Edwin Land (the inventor of the Polaroid polarising material, and developer of instant photography) set the cat among the pigeons by demonstrating that for complex scenes the three-dimensional theory does not describe the way that we see colour. The brain processes complex visual information in subtle ways that are still not understood. For an outline of Land's work, see Gregory (1977); for more details, see Land (1977).

3M   PROOFS OF THEOREMS ON DIMENSION

Theorem 3F-2. Let S' be obtained from a set S by removing a vector which is linearly dependent on S'. Then $\mathrm{Sp}(S') = \mathrm{Sp}(S)$.

Proof. We have $S = S' \cup \{x\}$ where x is a linear combination of S'. We must prove that every linear combination of S' belongs to $\mathrm{Sp}(S)$, and vice versa.

Every linear combination of S' belongs to $\mathrm{Sp}(S)$ because every element of S' belongs to S.

Every linear combination of S is the sum of a linear combination of S' and a multiple of x, which is itself a linear combination of S'. Thus every linear combination of S is the sum of two members of $\mathrm{Sp}(S')$, and belongs to $\mathrm{Sp}(S')$ because $\mathrm{Sp}(S')$ is a vector space. This completes the proof that $\mathrm{Sp}(S) = \mathrm{Sp}(S')$.   □

Theorem 3G-1 (the dimension theorem). If a vector space V has a basis consisting of n elements, then:
   (a) every basis for V contains n elements;
   (b) every linearly independent set of n vectors is a basis; and
   (c) every set of more than n vectors is linearly dependent.

Proof of (b). Let $\{b_1, \ldots, b_n\}$ be a basis for V. To prove that a linearly independent set $\{u_1, \ldots, u_n\}$ is a basis, we shall start with the set $\{b_1, \ldots, b_n\}$, and replace the b's by u's one at a time, showing that it remains a basis at each stage.

Since the b's form a basis, the vector u_1 can be expressed as a linear combination of them:

$$u_1 = c_1 b_1 + \ldots + c_n b_n \tag{1}$$

Since u_1 belongs to a linearly independent set, it is nonzero, so at least one of the coefficients c_i must be nonzero. We can number the b's so that $c_1 \neq 0$. Then (1) can be rearranged to give b_1 as a linear combination of $\{u_1, b_2, \ldots, b_n\}$. Now,

$V = \text{Sp}\{b_1,\ldots,b_n\} = \text{Sp}\{u_1,b_1,b_2,\ldots,b_n\} = \text{Sp}\{u_1,b_2,\ldots,b_n\}$ by Theorem 3F-2 above. Therefore

$$\text{Sp}\{u_1,b_2,\ldots,b_n\} = V$$

Now consider u_2. It equals a linear combination of the spanning set $\{u_1,b_2,\ldots,b_n\}$. At least one of the b's must have a nonzero coefficient, otherwise u_2 would equal a linear combination of $\{u_1\}$ which is impossible because the u's are linearly independent. Again, we can number the b's so that b_2 has a nonzero coefficient; it follows that b_2 is a linear combination of $u_1,u_2,b_3,\ldots,b_n$. Hence $\text{Sp}\{u_1,b_2,\ldots,b_n\} = \text{Sp}\{u_1,u_2,b_2,\ldots,b_n\} = \text{Sp}\{u_1,u_2,b_3,\ldots,b_n\}$. Thus

$$\text{Sp}\{u_1,u_2,b_3,\ldots,b_n\} = V$$

Proceeding in this way, we can replace all the b's by u's, and deduce that $\{u_1,\ldots,u_n\}$ spans V. Since it is given to be linearly independent, it is a basis.

Proof of (c). Suppose $\{u_1,u_2,\ldots\}$ is a linearly independent set of more than n elements. Then $\{u_1,\ldots,u_n\}$ is linearly independent, therefore it is a basis by (b). Hence u_{n+1} is a linear combination of $\{u_1,\ldots,u_n\}$, which is impossible because all the u's are linearly independent. Therefore there cannot be more than n linearly independent vectors.

Proof of (a). Suppose that B is a basis for V with n elements, and B' is another basis with n' elements. Since B' is a linearly independent set of n' elements in a space with a basis of n elements, it follows from (c) that $n' \leq n$. Since B is a linearly independent set of n elements in a space with a basis of n' elements, it follows from (c) that $n \leq n'$. Hence $n' = n$. $\square$

Theorem 3H-2 (extension to a basis). Let V be an n-dimensional space. If $k < n$, then every linearly independent set of k elements $\{u_1,\ldots,u_k\}$ can be extended to give a basis $\{u_1,\ldots,u_k,u_{k+1},\ldots,u_n\}$ for V.

Proof. Let $\{b_1,\ldots,b_n\}$ be a basis for V. Consider the set $T_1 = \{b_1,u_1,u_2,\ldots,u_k\}$. If it is linearly dependent, remove b_1 leaving the linearly independent set $\{u_1,\ldots,u_k\}$. If T_1 is linearly independent, then give b_1 the new name u_{k+1}. In both cases the result is a linearly independent set of u's with either k or $k+1$ members.

Now consider $T_2 = \{b_2,u_1,u_2,\ldots\}$. If it is linearly dependent, discard b_2 leaving the linearly independent set of u's constructed in the paragraph above. If T_2 is linearly independent, rename b_2 as the next u (either u_{k+1} or u_{k+2} depending on what happened in the previous step). In both cases the result is a linearly independent set of u's.

Now consider $T_3 = \{b_3,u_1,u_2,\ldots\}$, and proceed in the same way; similarly with $T_4,\ldots,T_n$. The result is a linearly independent set of vectors, which includes the original set $\{u_1,\ldots,u_k\}$. It is obtained from the set $A = \{u_1,\ldots,u_k,b_1,\ldots,b_n\}$ by removing vectors which are linearly dependent on the remaining vectors. By Theorem 3F-2, its span equals $\text{Sp}(A) = V$. Hence it is a basis for V. This completes the proof. $\square$

PROBLEMS FOR CHAPTER 3

Hints and answers for problems marked [a] will be found on page 279.

Sections 3A, 3B, 3C

[a]1. Which of the following are real vector spaces, with addition and scaling defined in the obvious way?
   (a) the set of all real numbers;
   (b) the set of all positive numbers;
   (c) the set of all complex numbers;
   (d) the set of all integers $\ldots, -2, -1, 0, 1, 2, \ldots$;
   (e) the set of all rational numbers (that is, fractions of the form p/q, where p and q are integers and $q \neq 0$).

[a]2. True or false?
   (a) Every real vector space has infinitely many elements.
   (b) The product of a vector and a scalar is a vector.
   (c) The product of two vectors is a scalar.
   (d) The elements of a complex vector space are sets of complex numbers.
   (e) The vectors in a real vector space cannot be multiplied by complex numbers.
   (f) The vectors in a complex vector space cannot be multiplied by real numbers.
   (g) if $k\dot{v} = 0$, where k is a scalar and v belongs to a vector space, then either $k = 0$ or v is the zero vector.

3. Show that the set of all infinite sequences of real numbers (see Appendix C) is a real vector space, where addition and scaling are defined in the obvious way, analogous to $\mathbb{R}^n$.

[a]4. Which of the following sets of real functions are vector spaces, if addition and scaling are defined as in Example 3B-2?
   (a) all even functions (i.e., such that $f(-x) = f(x)$ for all x);
   (b) all odd functions (i.e., such that $f(-x) = -f(x)$ for all x);
   (c) all functions which are either even or odd.

5. A magic square of order n is an n by n matrix such that each row adds up to the same number k, and each column adds up to k, and the main diagonal and the opposite diagonal also add up to k. We do not assume here that the entries are integers (but see Chapter 5, Problem 27).
   Show that for any fixed n, the set of all magic squares of order n is a vector space.

6. $\mathbb{R}^2$ is the set of all pairs (x, y) where x and y belong to the vector space $\mathbb{R}$. It is a space constructed from $\mathbb{R}$ by combining it with itself. This problem shows how new vector spaces can be constructed in a similar way by combining two given spaces.
   Given two vector spaces V and W over the same field $\mathbb{K}$, form pairs (v, w) by taking an element v of V with an element w of W. Let U be the set of all these pairs. In symbols, $U = \{(v, w) : v \in V \text{ and } w \in W\}$.

Show that U is a vector space under the operations $(v, w) + (v', w') = (v + v', w + w')$ and $k(v, w) = (kv, kw)$.

(U is called the 'external direct sum' of V and W; see also Problems 33, 38 and 42.)

Section 3D

[a]7. Which of the following are subspaces of the space of Problem 3?
(a) the set of all sequences with first term 0;
(b) the set of all sequences with first term 1;
(c) the set of all arithmetic progressions, that is, sequences (x_i) such that $x_2 - x_1 = x_3 - x_2 = x_4 - x_3 = \ldots$;
(d) the set of all geometric progressions, that is, sequences (x_i) with $x_2/x_1 = x_3/x_2 = x_4/x_3 = \ldots$.

[a]8. Pat says: '$\mathbb{R}^2$ is a subspace of $\mathbb{R}^3$; the xy plane is obviously part of 3-dimensional space.' Chris argues: 'No it isn't! The elements of $\mathbb{R}^2$ are 2-vectors, not 3-vectors, so how can they belong to $\mathbb{R}^3$?' Whose side are you on?

9. Let V be the space of all n by n matrices (Example 3D-5). Which of the following are subspaces of V?
(a) the set of all symmetric n by n matrices;
(b) the set of all skew-symmetric n by n matrices (Chapter 2, Problem 4);
(c) the set of invertible n by n matrices;
(d) the set of circulants (defined in Chapter 2, Problem 5).

[a]10. For what values of a and b is $C[a, b]$ a subspace of $C[0, 1]$?

11. Find two non-trivial subspaces of the space of magic squares of order 3 (Problem 5).

12. M and N are subspaces of a space V. Under what conditions is $M \cup N$ a subspace of V?

Section 3E

[a]13. In $\mathbb{R}^3$, set $p = (2, 1, 0)$, $q = (1, 0, 0)$, $r = (-1, 1, 1)$, $s = (0, -1, -1)$, $U = \text{Sp}\{p, q\}$, $V = \text{Sp}\{r, s\}$. Give clear geometrical descriptions of U, V, $U \cup V$, $U \cap V$. Are they subspaces of $\mathbb{R}^3$?

[a]14, (For readers acquainted with infinite series.) Let V be the space of all continuous real-valued functions, and let $S = \{1, t, t^2, \ldots\}$. Are the following true or false?
(a) Since $\sin t = t - t^3/6 + \ldots$, it belongs to $\text{Sp}(S)$.
(b) By Taylor's theorem, every continuous function can be expressed as a linear combination of powers, and therefore belongs to $\text{Sp}(S)$.
(c) $\text{Sp}(S)$ contains only those polynomials with finitely many terms; infinite polynomials such as $1 + t + t^2 + \ldots$ belong to V but not to $\text{Sp}(S)$.

^{a}15. Find a spanning set for $U \cap V$ where $U = \{(x, y, 0):x, y \in \mathbb{C}\}$ and $V = \mathrm{Sp}\{(1, 2, 3),$ $(i, -i, 1)\}$.

16. Prove that if X is a subset of a vector space V, then $\mathrm{Sp}(X)$ is the intersection of all the subspaces which contain X. (In this sense, $\mathrm{Sp}(X)$ is the 'smallest' subspace containing X.)

Section 3F

^{a}17. True or false?
    (a) If u and v belong to $\mathrm{Sp}(S)$, then u and v are linearly dependent.
    (b) If $u \in \mathrm{Sp}\{v, w\}$ then $\{u, v, w\}$ is linearly dependent.
    (c) If $\{u, v, w, x\}$ is linearly dependent, then u equals a linear combination of $\{v, w, x\}$.
    (d) Every subset of a linearly dependent set is linearly dependent.
    (e) Every subset of a linearly independent set is linearly independent.

18. Show that the set $\{p, q, r, s\}$ of Problem 13 is linearly dependent. Find a linearly independent subset containing three vectors.

^{a}19. Which of the following sets of functions are linearly independent? They all belong to the complex space of Example 3B-4(b).
    (a) $\sin t$, $\sin 2t$;
    (b) $\sin t$, $\cos t$, e^t, e^{-t};
    (c) $\sin t$, $\cos t$, e^{it} (here i denotes the square root of -1);
    (d) e^t, te^t, e^{2t}.

20. If $\{u, v\}$ is linearly independent, does it follow that $\{u + v, u - v\}$ is? What if $\{u, v\}$ is linearly dependent?

21. Define a PL-function to be a function $f \in C[0, 1]$ defined by $f(x) = a + bx$ for $0 \le x \le 1/2$ and $f(x) = c + dx$ for $1/2 \le x \le b$, where a, b, c, d are numbers satisfying $2a + b = 2c + d$, so that f is continuous at $1/2$. These functions are 'piecewise linear', that is, consist of two linear pieces joined together at $x = 1/2$.
    Show that the set of all PL-functions is a subspace of $C[0, 1]$.
    Functions f_1, f_2, f_3 are given by taking $(a, b, c, d) = (1, -2, 0, 0), (0, 2, 2, -2),$ $(0, 0, -1, 2)$ in the formula for f above. Sketch their graphs. Show that $\{f_1, f_2, f_3\}$ is linearly independent.
    (Continued in Problem 31.)

Sections 3G, 3H

^{a}22. True or false?
    (a) If $\{p, q, r, s\}$ is linearly independent, then every set of four vectors in $\mathrm{Sp}\{p, q, r, s\}$ is linearly independent.
    (b) If $\{p, q, r, s\}$ is linearly independent, then every set of five vectors in $\mathrm{Sp}\{p, q, r, s\}$ is linearly dependent.

(c) For any p, q, r, s in a vector space, every set of five vectors in $\mathrm{Sp}\{p, q, r, s\}$ is linearly dependent.

(d) If $\{u, v, w, x\}$ spans $\mathbb{R}^4$, then it is linearly independent.

(e) There exists a spanning set S of four vectors for $\mathbb{R}^3$ such that every set of three vectors in S is linearly independent.

(f) The set of all powers is a basis for the space of all differentiable functions, since every such function can be expressed by Taylor's series as a linear combination of powers.

ª23. For each of the following, determine whether it is a basis for $\mathbb{C}^3$:
  (a) $\{(1, 2, 3), (3, 2, 1)\}$;
  (b) $\{(1, 2, 3), (0, 2, 3), (0, 0, 3)\}$;
  (c) $\{(1, 0, 0), (0, 1, i), (1, i, -1)\}$.

ª24. Consider the set $\mathbb{C}$ of all complex numbers. Adding two members of $\mathbb{C}$ gives another member, and multiplying a member of $\mathbb{C}$ by a real number gives another member of $\mathbb{C}$. Thus $\mathbb{C}$ is a vector space over the field $\mathbb{R}$. What is the dimension of this space?

ª25. Prove that $\mathrm{Sp}\{v_1, \ldots, v_k\}$ cannot have dimension greater than k.

ª26. Prove or disprove: given linearly independent sets $X = \{x_1, x_2, x_3\}$ and $Y = \{y_1, y_2, y_3\}$, if $\mathrm{Sp}(X) \cap \mathrm{Sp}(Y) = \{0\}$, then $X \cup Y$ is linearly independent. Generalise.

27. Let V be the real vector space of p by q matrices with real entries. For $i \leq p$ and $j \leq q$, define a matrix $E(i, j)$ with all entries zero except the (i, j) entry which is 1. These are p by q versions of the matrix units introduced in Chapter 2, Problem 17.

Show that the matrix units are linearly independent and span V. Hence write down $\dim(V)$.

What is the dimension of the complex vector space of p by q matrices with complex entries?

Consider now the case $p = q = n$. What is the dimension of the subspace consisting of upper triangular matrices (Example 3D-6)?

ª28. Prove that the set $\{e_1, e_2, \ldots\}$ is not a basis for the space S of Example 3H-2.

29. Let E be the n by n matrix with $E_{ij} = 1$ if $j - i = 1$ or $1 - n$, and 0 otherwise. Write down E and E^2 for $n = 4$; note that they are circulants (defined in Chapter 2 Problem 5). Justify the following statements (leading to the construction of a basis for the space of Problem 9(d)).

(a) For any n by p matrix A, EA is the matrix obtained from A by putting the first row at the bottom and shifting all the other rows up one place. For any integer $r < n$, E^r is the circulant with first row consisting of zeros except for 1 in the $(r + 1)$th place. $E^n = I$, and $E^{n+k} = E^k$ for every positive integer k.

(b) The set of all n by n circulants equals $\mathrm{Sp}\{I, E, E^2, \ldots, E^{n-1}\}$.

(c) $\{I, E, E^2, \ldots, E^{n-1}\}$ is linearly independent, and is a basis for the space of all n by n circulants.

30. Use Problem 29 to show that (i) multiplying two circulants gives another circulant, and (ii) circulants commute.

31. Return to Problem 21. Prove that if f and g are PL-functions such that $f(0) = g(0)$, $f(0.5) = g(0.5)$, $f(1) = g(1)$, then $f = g$. Hence prove that $\{f_1, f_2, f_3\}$ is a basis for the space of PL-functions, which is therefore 3-dimensional.

32. Problem 31 can be generalised as follows. Split the interval $[a, b]$ into n parts by giving numbers $x_0, \ldots, x_n$, in increasing order, with $x_0 = a$ and $x_n = b$. A PL (piecewise linear) function in $C[a, b]$ is defined by $f(x) = a_i + b_i x$ if $x_{i-1} \le x \le x_i$, for $i = 1, \ldots, n$. To make f continuous we need $a_i + b_i x_i = a_{i+1} + b_{i+1} x_i$ for $i = 1, \ldots, n$. The x_i's are called the 'knots' for the PL-function. Suppose the values of f are given ('tied down') at the knots; then there is just one PL-function taking the given values at the knots. Its graph is made up of straight lines joining the points corresponding to the knots.

Prove that for a given set of knots, the set of all PL-functions with those knots is a subspace of $C[a, b]$. Find a basis for the subspace (see Problem 31). Hence find its dimension.

(See Chapter 7, Problem 17 for a version of this idea which is very useful in numerical analysis.)

33. If V and W are finite-dimensional vector spaces over $\mathbb{K}$, what is the dimension of their external direct sum (Problem 6)?

Section 3I

[a]34. Find the components of $(i, i, 1)$ with respect to the basis in Problem 23.

[a]35. For each of the following sets in the space P_3 of polynomials of degree ≤ 3, prove that it is a basis, and find the components of the polynomial $1 + t^3$ with respect to it. (a) $1, 1 - t, (1 - t)^2, (1 - t)^3$; (b) $1, 1 + t, t + t^2, t^2 + t^3$.

Section 3J

[a]36. Write down in detail an isomorphism between $\mathbb{R}^6$ and the space of 2 by 3 matrices with real entries.

(If you know about array storage in the computer programming language Fortran, this should be easy.)

[a]37. True or false?
(a) Two n-dimensional spaces over the same field must be isomorphic.
(b) Two infinite-dimensional spaces over the same field must be isomorphic.

38. Return to Problem 33. Show that V is isomorphic to a subspace of $V \oplus W$.

Section 3K

[a]39. True or false?

(a) If $V = S + T$, and B and C are bases for S and T respectively, then $B + C$ spans V.

(b) If $V = S + T$, and B and C are bases for S and T respectively, then $B \cup C$ is a basis for V.

(c) If $V = S + T$, and B and C are bases for S and T respectively, then $B \cup C$ spans V.

(d) If $V = S \oplus T$, and B and C are bases for S and T respectively, then $B \cup C$ is a basis for V.

[a]40. Show that the space of all n by n matrices is the direct sum of the subspaces of symmetric and skew-symmetric matrices (see Problem 9).

41. Given that S and T are 4-dimensional subspaces of $\mathbb{R}^6$, and $S \neq T$, what can you say about the dimensions of $S + T$ and $S \cap T$?

42. If $V = S \oplus T$, prove that V is isomorphic to the external direct sum of S and T (Problem 6).

(This means that the direct sum as defined in section 3K is really the same as the external direct sum. Given two vector spaces over the same field, one can always embed them both in a larger vector space which is their direct sum.)

4

Elementary operations and linear equations

The central theme of this chapter is solving sets of linear algebraic equations. The technique is to eliminate variables one by one to give a single equation for a single unknown. It can be crystallised into three basic operations, which are useful in many parts of linear algebra. In the first two sections these operations are introduced as a way of identifying linearly independent sets of vectors. Sections 4C–4H deal with the solution of sets of linear equations in detail, and introduce the important idea of the rank of a matrix. These sections form the foundation of the general theory of linear transformations in later chapters. In section 4I we discuss the question of when a matrix has an inverse. All this material is essential reading.

The rest of the chapter is optional. Sections 4J and 4K discuss computational techniques for linear equations, and 4L and 4M express the results of this chapter in a new way, as factorisations of matrices.

4A   VECTORS IN ECHELON FORM

Linear independence is one of the leading ideas of Chapter 3; it lies at the heart of the definition of dimension. Yet we have no systematic technique for deciding whether a set is linearly independent. This section introduces such a technique; it consists of juggling the set of vectors into a form which makes its linear dependence properties easy to see.

Example 1. The set $\{(1, 2, 1), (0, 1, -1), (0, 0, 3)\}$ in $\mathbb{R}^3$ is easily shown to be linearly independent, as follows. Suppose some linear combination equals the zero vector, so that

$$a(1, 2, 1) + b(0, 1, -1) + c(0, 0, 3) = (0, 0, 0)$$

for some real a, b, c. The first, second and third components of this equation give

$$a = 0 \qquad 2a + b = 0 \qquad a - b + 3c = 0 \qquad (1)$$

Solving these equations one after the other, we have $a = 0$, hence $b = 0$, hence $c = 0$. Thus if a linear combination of these vectors equals zero, then it is the trivial combination; the set is linearly independent. □

In this example, each vector starts with more zeros then the previous one. This is why the equations (1) can be solved so easily. We now generalise.

Definitions 1. (a) The **leading entry** of an n-vector is the first nonzero entry.

(b) An n-vector has r **leading zeros** if the first r entries are zero and the leading entry is the $(r + 1)$th.

(c) An ordered set of n-vectors is in **echelon form** if the following holds:

(i)   it consists of a finite number of vectors;

(ii) if the zero vector is included, then it occurs at the end (possibly more than once);

(iii) if there is more than one nonzero vector, each one after the first has more leading zeros than the preceding one.

Notes. (1) Being in echelon form is a property of *ordered sets* (see Appendix C), though for brevity we usually omit the word 'ordered'. The odd name 'echelon form' is explained in section 4C.

(2) A set is in echelon form if it does not violate conditions (i), (ii), (iii). A set consisting of a single vector cannot possibly violate these conditions. Therefore the set $\{v\}$ is in echelon form, for any vector v.

Examples 2. The set in Example 1 is in echelon form. The following set of complex 4-vectors is in echelon form: $\{(1, i, 0, 1), (0, 0, 1, 0), (0, 0, 0, i), (0, 0, 0, 0), (0, 0, 0, 0)\}$. But the set $\{(1, i, 0, 1), (0, 0, 1, 0), (0, 0, 2, i)\}$ is not, because the last vector has only two leading zeros. □

If a set is in echelon form, it is extremely easy to see whether it is linearly dependent.

Theorem 1 (linear dependence). A set of vectors in echelon form is linearly dependent if and only if it includes the zero vector.

Proof. If the set includes the zero vector, then it is obviously linearly dependent (see Exercise 3F-1). Suppose it does not include the zero vector; we shall use the method of Example 1 to prove that it is linearly independent.

Let $\{u, v, \ldots, z\}$ be a set of n-vectors in echelon form. Suppose some linear combination equals zero, then

$$au + bv + \ldots + ez = 0$$

for some scalars $a, \ldots, e$. Suppose the leading entry of u is the kth entry. Then the kth entry of all the other vectors in the set is zero because they are in echelon form. So the kth entry of the equation above gives $a = 0$. The equation now reads $bv + \ldots + ez = 0$. The same argument now shows that $b = 0$, and so on for all the other coefficients. This shows that the only linear combination of the set which can equal zero is the trivial linear combination, so the set is linearly independent. □

Exercise 1. The second paragraph of the above proof is based on the assumption that the set does not include 0. At what point of the argument was this used? (It must have been used somewhere, because the conclusion is false if the set includes 0.)
□

We can now see at a glance whether an echelon-form set of vectors is linearly dependent: it is simply a question of whether it includes the zero vector.

Of course, sets in echelon form are a very special case. But the next section shows that this special case is the key to the general problem of determining linear dependence.

4B   ELEMENTARY OPERATIONS

This section shows that every finite set of n-vectors can be transformed into echelon form. It is then easy to see whether it is linearly dependent.

Example 1. The set $\{(0, 1, 1), (1, 0, 1)\}$ is not in echelon form. But changing the order gives $\{(1, 0, 1), (0, 1, 1)\}$, which is in echelon form, showing that the set is linearly independent. □

This is a trivial example, of course; most sets cannot be put into echelon form by simply rearranging the vectors. But by means of this and two other simple operations, any finite set can be put into echelon form.

Definition 1. We define three types of **elementary operation** on ordered sets of vectors, as follows:

Type I: interchange two vectors;

Type II: multiply any one vector by a nonzero scalar;

Type III: add a multiple of one vector to another vector.

Example 2. Let $X = \{u, v, w, \ldots\}$ be an ordered set of elements of any vector space; it may be a finite or an infinite set. Applying a Type I operation might give $\{v, u, w, \ldots\}$ or $\{w, v, u, \ldots\}$, depending on which vectors are interchanged. Applying a Type II operation to X might give, for example, $\{u, 2v, w, \ldots\}$. Applying a Type III operation to X might give, for example, $\{u, v - 4w, w, \ldots\}$ (where -4 times the third vector has been added to the second vector). □

The operations have been carefully chosen so as not to change those properties of the set which determine whether it is a basis. For example, let $S = \{u, v, w\}$; subtract 4 times the last vector from the second, then interchange the first and last elements, then multiply the last by 2; the result is $T = \{w, v - 4w, 2u\}$. Every linear combination of S is a linear combination of T because

$$au + bv + cw = (c + 4b)w + b(v - 4w) + (a/2)2u \tag{1}$$

Furthermore, if S is linearly dependent, then $au + bv + cw = 0$ for some a, b, c not all zero, and (1) shows that there is a nontrivial linear combination of T which vanishes. Generalising from this example gives the following.

Lemma 1. Let T be a set obtained from an ordered set S by applying any number of elementary operations. Then every linear combination of S is a linear combination of T, and every nontrivial linear combination of S is a nontrivial linear combination of T.

Proof. This is a generalised version of the argument above, straightforward but lengthy. See section 4N.   □

We can now prove the main result of this section.

Theorem 1 (elementary operation theorem). Elementary operations do not change the span or the linear dependence properties of a set of vectors.

Proof. Expressed in more detail, the theorem says that if S is an ordered set of elements of a vector space, and T is obtained from S by applying any number of elementary operations, then $\mathrm{Sp}(S) = \mathrm{Sp}(T)$, and S is linearly dependent if and only if T is.

Lemma 1 shows that $\mathrm{Sp}(S) = \mathrm{Sp}(T)$, since the span is just the set of all linear combinations. The set S is linearly dependent if and only if the zero vector is a nontrivial linear combination of S. By Lemma 1, this is so if and only if the zero vector is a nontrivial linear combination of T. Hence S is linearly dependent if and only if T is, which completes the proof.   □

We use this machinery to determine whether a set is linearly dependent, as follows. Given a set S of n-vectors, use elementary operations to transform it into echelon form. Call the resulting set T. It is linearly dependent if and only if it contains the zero vector. By Theorem 1, S is linearly dependent if and only if T is.

In this way we can use elementary operations to determine the linear dependence of a given finite set of n-vectors. The method might seem to be limited to the spaces $\mathbb{K}^n$. But it can be extended to vectors in any finite-dimensional space, by using components with respect to a basis. The abstract vectors are then replaced by vectors in $\mathbb{K}^n$. In section 3I we saw that these n-vectors have the same algebraic properties as the abstract vectors; so we can apply the method above to the n-vectors and then translate the results back to the original vector space. This technique is illustrated in Example 3(b) below.

Examples 3. (a) $S = \{(1, 2, 3, 4), (3, 4, 7, 10), (2, 1, 3, 5)\}$; what is the dimension of the $\mathrm{Sp}(S)$? Writing the vectors underneath each other, we have

$$(1 \quad 2 \quad 3 \quad 4)$$
$$(3 \quad 4 \quad 7 \quad 10)$$
$$(2 \quad 1 \quad 3 \quad 5)$$

The first step is to use Type III operations to make the first entry of the second and third vectors zero. Subtracting 3 times the first vector from the second, and twice the first vector from the third, gives

$$(1 \quad 2 \quad 3 \quad 4)$$
$$(0 \quad -2 \quad -2 \quad -2)$$
$$(0 \quad -3 \quad -3 \quad -3)$$

To tidy things up, multiply the second and third vectors by $-1/2$ and $-1/3$ respectively, giving

$$(1 \quad 2 \quad 3 \quad 4)$$
$$(0 \quad 1 \quad 1 \quad 1)$$
$$(0 \quad 1 \quad 1 \quad 1)$$

Now we must make the last vector have more leading zeros than the second. This is easily done by subtracting the second vector from the third, giving

$$(1 \quad 2 \quad 3 \quad 4)$$
$$(0 \quad 1 \quad 1 \quad 1)$$
$$(0 \quad 0 \quad 0 \quad 0)$$

Call the final set of vectors T. Since it includes the zero vector, it is linearly dependent. But the subset $U = \{(1, 2, 3, 4), (0, 1, 1, 1)\}$ is linearly independent. Since adding the zero vector does not change the span, $\mathrm{Sp}(U) = \mathrm{Sp}(T)$, which equals $\mathrm{Sp}(S)$ by the elementary operation theorem above. Hence U is a basis for $\mathrm{Sp}(S)$, which is therefore 2-dimensional.

(b) Consider the set of all linear combinations of the three polynomials $p(t) = 4t^3 + 3t^2 + 2t + 1$, $q(t) = 10t^3 + 7t^2 + 4t + 3$, and $r(t) = 5t^3 + 3t^2 + t + 2$. They all belong to the space of polynomials of degree ≤ 3, which has the basis $\{1, t, t^2, t^3\}$ (see Exercise 3G-2). The polynomials p, q and r have components $(1, 2, 3, 4)$, $(3, 4, 7, 10)$, $(2, 1, 3, 5)$ with respect to this basis. These are precisely the 4-vectors in (a) above. We have shown that they are linearly dependent, and $\{(1, 2, 3, 4), (0, 1, 1, 1)\}$ is a basis for their span. Translating this result back into the space of polynomials, we see that the set of all linear combinations of p, q and r can be obtained more economically as the set of combinations of the polynomials f and g defined by

$$f(t) = 4t^3 + 3t^2 + 2t + 1 \qquad g(t) = t^3 + t^2 + t$$

Consider, for example, the polynomial $2p - q + r$; it can be expressed as a linear combination of f and g as follows. We want to find numbers a and b such that $2p - q + r = af + bg$, that is,

$$3t^3 + 2t^2 + t + 1 = a(4t^3 + 3t^2 + 2t + 1) + b(t^3 + t^2 + t)$$

Equating coefficients of the powers of t gives $4a + b = 3$, $3a + b = 2$, $2a + b = 1$, $a = 1$. These are satisfied by $a = 1$, $b = -1$. Hence $2p - q + r = f - g$. In the same way, any linear combination of p, q and r can be expressed in terms of f and g. □

We shall now turn to another aspect of the elementary operations. When the three operations were introduced, some of them may have looked familiar. When you first met sets of simultaneous linear equations, you probably learnt the method of eliminating unknowns by multiplying equations by numbers, and adding multiples of one equation to another. These are just the Type II and III operations. In the next section we shall consider this procedure in detail.

4C   SETS OF LINEAR EQUATIONS

We met sets of linear equations in section 2A, and wrote them in matrix notation. We now consider how to solve them. The first question is whether a given set of equations can in fact be solved, that is, whether the equations have a solution.

You may be unfortunate enough to have been taught a 'rule' that given n equations in n unknowns, there is just enough information to find the unknowns. If so, consider the following.

Examples 1. (a) Consider the equations

$$x + 2y = 1$$

$$-2x - 4y = 0$$

We have two perfectly good equations in two unknowns. Yet there is no solution. If you try to find numbers x and y satisfying these equations, you will not succeed.

(b) Now consider the equations

$$x + 2y = 1$$

$$2x + y = 0$$

$$3y = 2$$

We have three equations in two unknowns. Yet there is a solution, $x = -1/3, y = 2/3$.

(c) Now consider the equations

$$x + 2y = 1$$

$$-2x - 4y = -2$$

We have two equations in two unknowns, and there are infinitely many solutions: for any value of x whatsoever, the equations will be satisfied if we take $y = (1 - x)/2$. □

It is clear from the above examples that the statement that you can always find n unknowns from n equations is simply WRONG. It sometimes works, but not always. The theory of linear equations is too subtle to be captured by such a simple rule.

This section and the next give a reliable method of finding the solutions of sets of equations. We shall return later to the question of when solutions exist, and how many there are.

We shall call a set of linear equations a **system**, to avoid confusion with other kinds of set. A system of p equations in q unknowns is written in matrix notation as $Ax = b$, where A is a p by q matrix, x is a q-vector and b is a p-vector, the vectors being written as column matrices. We shall work with real numbers, so that x and b belong to $\mathbb{R}^q$ and $\mathbb{R}^p$ respectively. But everything works in just the same way with complex numbers (or elements of any other field).

Expressed in this language, our problem is to find vectors x in $\mathbb{R}^q$ satisfying $Ax = b$. Examples 1 show that for a given A and b there may be just one solution vector x, or many, or none. So we cannot speak of finding 'the solution'; it may not exist, or it may not be unique. We need a different phrase.

Definition 1. The **solution set** of the system of equations $Ax = b$ (where A is a p by q real matrix and $b \in \mathbb{R}^p$) is the set of all x in $\mathbb{R}^q$ satisfying $Ax = b$.   □

Our problem, then, is to find the solution set of a given system of equations. This set always exists, though it may, as in Example 1(a), be the empty set, and it may, as in Example 1(c), contain infinitely many elements. The best way of finding it is to transform the system of equations into a simpler form, by means of the following elementary operations:

(I)   changing the order in which the equations are written down;
(II)   multiplying an equation by a nonzero number; and
(III) adding a multiple of one equation to another.

These are the same as the operations of section 4B, applied here to equations rather than vectors. It is fairly clear that the operations do not change the solutions of the equations.

Theorem 1 (elementary operations on equations). Applying elementary operations to a system of linear algebraic equations does not change the solution set.

Proof. We shall show that any single elementary operation leaves the solution set unchanged. It will then follow for any sequence of operations.

(a) For Type I operations it is obvious: changing the order in which the equations are written makes no difference.

(b) Suppose the ith equation is multiplied by a nonzero number k. The equation

$$a_{i1}x_1 + \ldots + a_{iq}x_q = b_i$$

is satisfied if and only if the equation

$$ka_{i1}x_1 + \ldots + ka_{iq}x_q = kb_i$$

is satisfied. Hence the operation does not change the solution set. Notice that this would not be true if k were allowed to be zero; in that case the second of these equations might be true even if the first was not.

(c) To write out a general proof for the Type III operation is not hard, but the notation is messy. We shall therefore consider a simple (but typical) case. Operating on

$$\left. \begin{array}{l} ax + by = c \\ dx + ey = f \end{array} \right\} \tag{1}$$

by adding k times the first equation to the second gives

$$\left. \begin{array}{l} ax + by = c \\ (ka + d)x + (kb + e)y = kc + f \end{array} \right\} \tag{2}$$

If the equations (1) are satisfied, then (2) obviously follow. Conversely, if (2) hold, then subtracting k times the first equation from the second gives (1). Hence (1) and (2) have the same solution set. Exactly the same argument applies in the general case.

□

When solving large systems of equations, efficient notation is very helpful. We have packaged the coefficients into a matrix A. We now take the further step of including the right-hand sides of the equation in the same matrix.

Definition 2. The **augmented matrix** for the system of equations $Ax = b$ or

$$a_{11}x_1 + \ldots + a_{1q}x_q = b_1$$

$$\ldots\ldots\ldots\ldots\ldots\ldots\ldots\ldots\ldots$$

$$\ldots\ldots\ldots\ldots\ldots\ldots\ldots\ldots\ldots$$

$$a_{p1}x_1 + \ldots + a_{pq}x_q = b_p$$

is the p by $(q+1)$ matrix $(A|B)$ or

$$\begin{pmatrix} a_{11} \ldots a_{1q} & b_1 \\ a_{21} \ldots a_{2q} & b_2 \\ \ldots\ldots\ldots & \ldots \\ a_{p1} \ldots a_{pq} & b_p \end{pmatrix}$$

□

The vertical line here indicates partitioning as explained in section 2F; it separates the coefficients on the left of the equations from the terms b_i on the right. The word 'augmented' means increased; obviously we have increased the size of the coefficient matrix A by adding the b's.

Example 2. For the system of equations

$$2x + y - z + 2w = 1$$

$$3y + 2z \qquad = 2$$

$$5w = 4$$

the augmented matrix is

$$\begin{pmatrix} 2 & 1 & -1 & 2 & 1 \\ 0 & 3 & 2 & 0 & 2 \\ 0 & 0 & 0 & 5 & 4 \end{pmatrix}$$

The original equations can be read off immediately from the augmented matrix; it is merely a shorthand notation.

These equations are easy to solve. Starting at the bottom we have $w = 4/5$; in the second equation either y or z can be given any value and then the other is determined in terms of it; and the first equation then determines x. The solutions are thus $w = 4/5$, $z = t$, $y = 2(1 - t)/3$, $x = (25t - 19)/30$ where t can take any value.

□

Whenever a set of equations has the structure in Example 2, where each equation has more unknowns than the next, the system can easily be solved by starting at the bottom and working up. Such a system is said to be in **echelon form**. The augmented matrix of an echelon-form system is easily recognised: each row has more leading zeros than the one above.

Definition 3. A matrix is in **echelon form** if its rows, regarded as vectors, are in echelon form as defined in section 4A. In other words, each row has more leading zeros than the one above, with the possible exception of rows consisting entirely of zeros, which are at the bottom. A matrix in echelon form is sometimes called an **echelon matrix**.

□

The matrix in Example 2 is in echelon form. The word comes from the French '*échelle*', essentially the same word as 'scale', meaning ladder or staircase. (The musical scale is a series of steps; other meanings of the word are derived ultimately from the same image.) The connection with steps can be seen by drawing a line round the zeros in the lower left part of the matrix, thus for example:

$$\begin{pmatrix} * & ? & ? & ? & ? & ? \\ 0 & * & ? & ? & ? & ? \\ 0 & 0 & 0 & * & ? & ? \\ 0 & 0 & 0 & 0 & * & ? \\ 0 & 0 & 0 & 0 & 0 & 0 \end{pmatrix}$$

Here * denotes a nonzero entry, and ? an entry which may or may not be zero. The line is shaped like a staircase (an irregular one). Now consider the following:

$$\begin{pmatrix} * & ? & ? & ? & ? & ? \\ 0 & * & ? & ? & ? & ? \\ 0 & 0 & 0 & * & ? & ? \\ 0 & 0 & * & ? & ? & ? \\ 0 & 0 & 0 & 0 & 0 & 0 \end{pmatrix}$$

The line bounding the zeros here has an overhang, and cannot be regarded as an acceptable staircase. The first matrix is in echelon form, the second is not; the staircase test distinguishes them.

Example 2 shows that when the augmented matrix of a system is in echelon form, the system is very easy to solve. The next section gives a method for solving linear equations by turning the augmented matrix into an echelon matrix, and then solving the resulting easy system.

4D   GAUSSIAN ELIMINATION

This section shows how to put any system into echelon form, using elementary operations. In section 4C we proved that applying these operations to a system of equations does not change the solution set. In terms of the augmented matrix, operations on the equations correspond to operations on the rows of the matrix. It follows that elementary operations on the rows of the augmented matrix do not change the solution set. The operations give a new matrix equivalent to the original one, in the sense that it corresponds to a set of equations with exactly the same solutions.

Definition 1. A matrix A is **row-equivalent** to a matrix B if A can be obtained from B by a sequence of elementary row operations. ☐

The argument of the first paragraph can now be summed up as follows.

Proposition 1. Let S and S' be two systems of equations with augmented matrices M and M'. If M is row-equivalent to M', then S and S' have the same solution set.

Proof. This follows at once from Theorem 4C-1, as explained above. ☐

Our strategy for solving sets of equations is to reduce the augmented matrix to echelon form by elementary operations, and solve the resulting simpler system. The method is called **Gaussian elimination**. It is a systematic version of the elementary process of eliminating all but one of the unknowns in a system of equations.

The technique of Gaussian elimination is based on the fact that in an echelon matrix, every leading entry (that is, every entry which is the first nonzero entry in its row) has zeros below it. This can be seen in the examples in the last section. To put a given matrix A into echelon form, the basic step is as follows. Choose one of the leading entries (call it P), and perform a Type III elementary operation on all the rows below it so as to reduce all the entries vertically below P to zero. The leading entry P on which this operation hinges is called the **pivot**. The operation is repeated until the matrix is in echelon form. An example should make the method clear.

Example 1. Consider the matrix

$$\begin{pmatrix} 1 & 2 & 0 & -1 \\ 2 & 0 & -1 & 2 \\ -2 & 1 & 3 & 0 \end{pmatrix}$$

We start with the first row, and take its leading entry 1 as the pivot. Perform the operation $R_2 \to R_2 - 2R_1$ (new 2nd row equals old 2nd row minus twice 1st row). Then the first entry of the second row becomes 0. The operation $R_3 \to R_3 + 2R_1$ gives 0 as the first entry of the third row. We now have

$$\begin{pmatrix} 1 & 2 & 0 & -1 \\ 0 & -4 & -1 & 4 \\ 0 & 5 & 3 & -2 \end{pmatrix}$$

The first step of the process is now complete. We proceed to the second row, and takes its leading entry -4 as the pivot for the next step; we operate on the third row so as to reduce the entry below this new pivot to zero. The operation is $R_3 \to R_3 + (5/4)R_2$, giving

$$\begin{pmatrix} 1 & 2 & 0 & -1 \\ 0 & -4 & -1 & 4 \\ 0 & 0 & 7/4 & 3 \end{pmatrix}$$

This is an echelon matrix, and we have finished.

If the original matrix was the augmented matrix of the equations $x + 2y = -1$, $2x - z = 2$, $-2x + y + 3z = 0$, then the echelon form corresponds to the equations

$$\left. \begin{array}{r} x + 2y \quad\quad = -1 \\ -4y - z = 4 \\ (7/4)z = 3 \end{array} \right\} \tag{1}$$

Solving the last equation gives $z = 12/7$. The equation above then gives $-4y = 4 + 12/7$, so $y = -10/7$. The first equation now gives $x = 20/7 - 1 = 13/7$, which completes the solution. □

We solved the echelon-form system (1) by working upwards, solving the last equation, then substituting into the previous equation and solving that, and so on. This process is called **back-substitution** (one works backwards from the end, substituting the values found at each stage into the next equation up). The method of solving a set of equations can be summarised as Gaussian elimination followed by back-substitution.

We shall now express the Gaussian elimination process in general form. It consists of repeating one basic step, which I call a Gauss step.

A **Gauss step** is the following procedure.

Choose a row of the matrix, say the kth row. Call its leading entry, a_{kp} say, the **pivot**.

Do the following operation on the rows below the kth: for each $i > k$ subtract (a_{ip}/a_{kp}) times the kth row from the ith row, and take the result as the new ith row. The result is a matrix in which all entries vertically below the pivot are zero. □

In Example 2 we took the $(1, 1)$ entry as the pivot for the first Gauss step, then the $(2, 2)$ entry of the resulting matrix as the pivot for the next Gauss step. For a larger matrix one proceeds down the main diagonal in the same way. This method will often reduce the matrix to echelon form, as in Example 1. But not always.

Example 2. Consider the matrix

$$\begin{pmatrix} 0 & 1 & 2 \\ 2 & 1 & 1 \\ 1 & 0 & 1 \end{pmatrix}.$$

The $(1, 1)$ entry is zero. It cannot be used as a pivot because the Gauss step involves dividing by the pivot. The leading entry of the first row cannot be used as a pivot, because it would leave nonzero entries at the beginning of the second and third rows. The method of Example 1 breaks down here.

There is an easy way out of this difficulty: using a Type I elementary operation, interchange the first two rows so as to bring a nonzero entry to the $(1, 1)$ position

Exercise 1. Reduce this matrix to echelon form.   ☐

A combination of row interchanges and Gauss steps, as in this example, will reduce
any matrix to echelon form. We now set out the method in detail. We express it in
the form of an 'algorithm', that is, a clear-cut procedure expressed as a set of
instructions.

The **Gaussian elimination algorithm** for reducing a matrix to echelon form is the
following procedure.
    Do the following with $k = 1$, then repeat with $k = 2$, and so on.
    (a) If the kth row has more leading zeros than one of the rows below, then exchange
the kth row with a row beneath so that no leading zero of the new kth row has a
nonzero entry vertically below it.
    (b) Take a Gauss step with the leading entry of the kth row as pivot.

Example 3. Consider the matrix

$$\begin{pmatrix} 1 & 2 & -3 & -3 & 10 \\ 2 & 4 & 0 & 0 & 2 \\ 2 & 3 & 2 & 1 & 5 \\ -1 & 1 & 3 & 6 & 5 \end{pmatrix}$$

The first step in the algorithm has $k = 1$; the first row has no leading zeros, so no
row interchange is needed. Taking a Gauss step, with the $(1, 1)$ entry as the pivot, gives

$$\begin{pmatrix} 1 & 2 & -3 & -3 & 10 \\ 0 & 0 & 6 & 6 & -18 \\ 0 & -1 & 8 & 7 & -15 \\ 0 & 3 & 0 & 3 & 15 \end{pmatrix}$$

Now take $k = 2$; then the kth row has more leading zeros than the 3rd row, so
exchange the 2nd and 3rd rows:

$$\begin{pmatrix} 1 & 2 & -3 & -3 & 10 \\ 0 & -1 & 8 & 7 & -15 \\ 0 & 0 & 6 & 6 & -18 \\ 0 & 3 & 0 & 3 & 15 \end{pmatrix}$$

Now take the leading entry of the new 2nd row as pivot. This gives

$$\begin{pmatrix} 1 & 2 & -3 & -3 & 10 \\ 0 & -1 & 8 & 7 & -15 \\ 0 & 0 & 6 & 6 & -18 \\ 0 & 0 & 24 & 24 & -30 \end{pmatrix}$$

Finally take $k = 3$. The third row has no more leading zeros than the row below, so
no row interchange is needed and we take the $(3, 3)$ entry as the pivot, giving

$$\begin{pmatrix} 1 & 2 & -3 & -3 & 10 \\ 0 & -1 & 8 & 7 & -15 \\ 0 & 0 & 6 & 6 & -18 \\ 0 & 0 & 0 & 0 & 42 \end{pmatrix}$$

This is an echelon matrix, and the process is complete.

Exercise 2. Carry out Gaussian elimination on a matrix of your own invention. □

By now the mechanics of Gaussian elimination should be clear. In the next section we shall take a more general point of view, and discuss the different kinds of result that the method can give.

4E   FURTHER ASPECTS OF GAUSSIAN ELIMINATION

The main thing about Gaussian elimination is that it is foolproof. The algorithm can be applied to any matrix, and there is no way that it can break down. The end-product is always an echelon matrix. We therefore have the following.

Theorem 1. Every matrix is row-equivalent to an echelon matrix.

Proof. The Gauss process can always be carried out, giving an echelon matrix related by row operations to the original matrix.   □

Remark. There are many echelon matrices row-equivalent to a given matrix. For example, given an echelon matrix we can add a multiple of the last nonzero row to one of the other rows, giving another echelon matrix row-equivalent to the first. It does not matter which echelon form is used for solving a system of equations; they all give the same solution set.   □

   The strategy for solving systems of linear equations is to apply the Gauss algorithm to the augmented matrix, and then solve the resulting echelon-form system. Proposition 4D-1 guarantees that its solutions are the same as the solutions of the original system.
   We have seen that a system of equations may have no solution, or a unique solution, or infinitely many solutions. These three cases correspond to three different types of echelon matrix, as follows.
   Consider the echelon matrix resulting from Gaussian elimination. Any row which consists entirely of zeros can be discarded, because it corresponds to an equation of the form $0 = 0$, which contains no information whatsoever. There are then three cases which can arise.

Case 1. After removal of zero rows, the augmented matrix may have the echelon form $(P|q)$, where q is a column matrix and P is square with zeros below the main diagonal and nonzero entries everywhere on the main diagonal. The echelon form of Example 4D-1 exemplifies this case.

The last row of a matrix of this type corresponds to an equation of the form $ax_n = c$, where $a \neq 0$. This immediately gives the value of x_n; substituting in the equation above gives x_{n-1} and so on. This procedure, called back-substitution, gives a unique value to each unknown, and in this case the system has a unique solution.

Case 2. One row of the augmented matrix may have zeros everywhere except the last entry. This row corresponds to the equation $0 = c$ where $c \neq 0$. This equation can never be satisfied; in this case, therefore, the equations are inconsistent and have no solution. Example 4C-1(a) illustrates this case.

Case 3. If neither of the above apply, then after the removal of entirely zero rows the augmented matrix has the form $(P|q)$, where P is in echelon form and has more columns than rows. (Proof: if P has more rows than columns and is in echelon form, then it must have a row of zeros, and since $(P|q)$ has no zero rows, we have case 2; if P is square then we have either case 2 or case 1 depending on whether or not P has a row of zeros.)

In this case the echelon form has more unknowns than equations, and there are infinitely many solutions. Example 4C-2 illustrates this case.   □

From the practical point of view our discussion of Gaussian elimination is now complete. Writing down the augmented matrix and applying the Gauss algorithm gives either no solution, a unique solution, or infinitely many solutions, according to which of the three cases above arises.

But it would be very unsatisfactory to stop at this point. The obvious question is, given a system of equations, which of the three cases will occur? In the next section we shall begin to answer this question.

4F   RANK

Given the system of equations $Ax = b$, it is natural to wonder which of the three cases of section 4E will arise. The answer depends mainly on the matrix A, and to a lesser extent on the vector b.

Consider case 2, where the equations have no solution. In this case the echelon-form augmented matrix has a row with all entries zero except the last (which corresponds to the right-hand side of the equation). This means that certain elementary operations applied to A produce a row of zeros. Now, think of the rows of A as vectors; if A is a p by q matrix, its rows are q-vectors. Theorem 4B-1 says that elementary operations do not change the linear dependence properties of a set of vectors; and we know that any set which includes the zero vector is linearly dependent. It follows that in case 2, the rows of A are linearly dependent.

Exercise 1. If the rows of A are linearly independent, then $Ax = b$ has at least one solution.   □

It is clearly useful to regard the rows of A as vectors, and to consider their linear dependence properties. We need some technical terms for discussing these matters.

Definition 1. Let A be a p by q matrix with entries in the field $\mathbb{K}$. The **row space** of A is the subspace of $\mathbb{K}^q$ spanned by the rows of A. Here $\mathbb{K}$ is either $\mathbb{R}$ or $\mathbb{C}$ depending on whether A is a real or a complex matrix.

Definition 2. The **row rank** of A is the dimension of its row space. Alternatively, it is the maximum number of linearly independent rows of A.

Exercise 2. The two versions of Definition 2 are equivalent.   □

The argument at the beginning of this section can now be summed up as follows.

Theorem 1. If the row rank of A equals the number of rows, then for any vector b, the equation $Ax = b$ has at least one solution.

Proof. If the row rank equals the number of rows, then the rows are linearly independent by the second version of Definition 2. Hence any matrix obtained by applying elementary operations has linearly independent rows (Theorem 4B-1). Therefore the echelon form cannot have any zero rows, so case 2 cannot occur. Hence there is always a solution.   □

Now, anything that you can do to the rows of a matrix can equally well be done to the columns; hence the following definitions.

Definition 3. Let A be a p by q matrix with entries in the field $\mathbb{K}$. The **column space** of A is the subspace of $\mathbb{K}^p$ spanned by its columns. The **column rank** of A is the dimension of its column space, that is, the maximum number of linearly independent columns.   □

The two versions of the definition of column rank are equivalent, by the same easy argument as for the row rank.

Theorem 1 gives a practical use for the row rank. The column rank is useful too. If A is a p by q matrix with entries in the field $\mathbb{K}$ ($= \mathbb{R}$ or $\mathbb{C}$), then for any x in $\mathbb{K}^q$, Ax is a column vector in $\mathbb{K}^p$. The column lemma of section 2C shows that Ax is a linear combination of the columns of A, with coefficients equal to the entries of x. Therefore the equation $Ax = b$ says that b is a linear combination of the columns of A, with coefficients x_i. It follows that the equation has a solution if and only if b can be expressed as a linear combination of the columns of A, that is, if b belongs to the column space of A.

Theorem 2. (a) The equation $Ax = b$ has a solution if and only if b belongs to the column space of A.

(b) The equation $Ax = b$ has a solution for every b if and only if the column rank of A equals the number of rows.

Proof. (a) This follows from the argument above.

(b) This follows easily from (a). Suppose A is a p by q matrix with entries in the field $\mathbb{K}$. The equation $Ax = b$ has a solution for all b if and only if all vectors b belong to the column space, that is, if and only if the column space is the whole of $\mathbb{K}^p$, that is, if and only if the dimension of the column space is p. ☐

This theorem answers the question of when a solution exists. The other main question is whether there are many solutions or just one. This too can be answered in terms of the column rank.

Theorem 3. If the column rank of A equals the number of columns, then the equation $Ax = b$ cannot have more than one solution.

Proof. If the column rank equals the number of columns, then the columns are linearly independent and form a basis for the column space. Now, the column lemma shows that if the equation $Ax = b$ has solutions, they are the coefficients in an expression for b as a linear combination of the columns of A. Since the columns here are a basis for the column space, the x's are components with respect to a basis, which are unique by Theorem 3I-1. Hence there can be only one set of x's. ☐

We now have an 'if and only if' condition for existence of a solution, and an 'if' condition for uniqueness. There ought to be an 'if and only if' condition for uniqueness too. A more complete analysis will be given in section 5D, when more theoretical machinery is available.

Meanwhile, we shall look more closely at the idea of rank. Theorems 1 and 3 deal with the important special cases when the row rank equals the number of rows, or the column rank equals the number of columns. There is a special term to describe these cases.

Definition 4. A matrix has **full row rank** if the row rank equals the number of rows. It has **full column rank** if the column rank equals the number of columns. ☐

The term 'full' here means that the rank has its maximum possible value. The row rank is the largest number of linearly independent rows, so it clearly cannot exceed the total number of rows; similarly for the column rank.

Theorems 1, 2(b) and 3 can now be expressed as follows.

Theorem 4. (a) The equation $Ax = b$ has a solution for all b if and only if A has full row rank.

(b) There cannot be more than one solution of $Ax = b$ if A has full column rank. ☐

This needs no proof: it is merely a restatement of results proved above.

Finally, we shall work an example, which will reveal a new aspect of the row and column ranks.

Example 1. Let

$$A = \begin{pmatrix} 1 & 2 & 3 & 4 \\ 3 & 4 & 7 & 10 \\ 2 & 1 & 3 & 5 \end{pmatrix}$$

To find its row space, we reduce it to echelon form by row operations in the usual way. Gaussian elimination gives the echelon matrix

$$\begin{pmatrix} 1 & 2 & 3 & 4 \\ 0 & 1 & 1 & 1 \\ 0 & 0 & 0 & 0 \end{pmatrix}$$

row-equivalent to A. The row space is the subspace of $\mathbb{R}^4$ spanned by $\{(1, 2, 3, 4), (0, 1, 1, 1)\}$. This set is linearly independent because it is in echelon form, hence it is a basis for the row space. So the row space is two-dimensional, and the row rank is 2.

The column space is the set of all linear combinations of the columns. To find a basis, we apply elementary operations to the column vectors so as to put them into echelon form. This can be done by transposing A and applying row operations to A^T; alternatively one can apply elementary operations directly to the columns. Thus

$$\begin{pmatrix} 1 & 2 & 3 & 4 \\ 3 & 4 & 7 & 10 \\ 2 & 1 & 3 & 5 \end{pmatrix} \rightarrow \begin{pmatrix} 1 & 0 & 0 & 0 \\ 3 & -2 & -2 & -2 \\ 2 & -3 & -3 & -3 \end{pmatrix} \rightarrow \begin{pmatrix} 1 & 0 & 0 & 0 \\ 3 & -2 & 0 & 0 \\ 2 & -3 & 0 & 0 \end{pmatrix}$$

where the second matrix is obtained by the operations $C_2 \rightarrow C_2 - 2C_1, C_3 \rightarrow C_3 - 3C_1$, $C_4 \rightarrow C_4 - 4C_1$, and the third is obtained from the second by the operations $C_3 \rightarrow C_3 - C_2, C_4 \rightarrow C_4 - C_2$. It follows that the column space is the subspace of $\mathbb{R}^3$ spanned by $\{(1, 3, 2), (0, -2, -3)\}$. These vectors are in echelon form; therefore they are linearly independent and form a basis for the column space, which is thus two-dimensional. The column rank is therefore 2, which equals the row rank. ☐

The row and the column spaces of this matrix are subspaces of quite different vector spaces, $\mathbb{R}^4$ and $\mathbb{R}^3$ respectively. Yet they have the same dimension. If you work out any other example, you will find that the row and column ranks are always the same. The reason for this is not at all obvious; the proof will eventually emerge from the machinery developed in the next section.

4G   GAUSS–JORDAN ELIMINATION

Our general strategy for solving linear equations is to simplify the matrix by elementary operations so that it becomes easier to solve. The echelon form given by Gaussian elimination is not the simplest possible.

Example 1. Consider a system of equations for unknowns v, w, x, y, z, with the following echelon-form augmented matrix:

$$\left(\begin{array}{ccccc|c} 2 & 1 & 4 & 2 & 1 & 0 \\ 0 & 2 & -6 & 5 & 4 & 3 \\ 0 & 0 & 0 & -1 & 2 & -3 \\ 0 & 0 & 0 & 0 & 3 & -7 \end{array} \right)$$

The last equation is $3z = -7$, which determines z. The third equation is $-y + 2z = -3$, so $y = 2z + 3 = -5/3$. Now w is determined by the equation $2w - 6x + 5y + 4z = 3$, and so on.

The equations would be simpler if the coefficient of the unknown were 1, that is, if the leading entry of each row were 1. This can easily be arranged, by the elementary operation of multiplying each row by the reciprocal of the leading entry (which is nonzero by definition). The result is a new matrix row-equivalent to the old:

$$\begin{pmatrix} 1 & 1/2 & 2 & 1 & 1/2 & 0 \\ 0 & 1 & -3 & 5/2 & 2 & 3/2 \\ 0 & 0 & 0 & 1 & -2 & 3 \\ 0 & 0 & 0 & 0 & 1 & -7/3 \end{pmatrix}$$

We can simplify further by subtracting half the second row from the first. This reduces the $(1, 2)$ entry to zero, thus:

$$\begin{pmatrix} 1 & 0 & 7/2 & -1/4 & -1/2 & -3/4 \\ 0 & 1 & -3 & 5/2 & 2 & 3/2 \\ 0 & 0 & 0 & 1 & -2 & 3 \\ 0 & 0 & 0 & 0 & 1 & -7/3 \end{pmatrix}$$

Now we can add multiples of the third row to the two rows above so as to make the $(1, 4)$ and $(2, 4)$ entries zero:

$$\begin{pmatrix} 1 & 0 & 7/2 & 0 & -1 & 0 \\ 0 & 1 & -3 & 0 & 7 & -6 \\ 0 & 0 & 0 & 1 & -2 & 3 \\ 0 & 0 & 0 & 0 & 1 & -7/3 \end{pmatrix}$$

Note that these operations leave the 0 in the $(1, 2)$ position unchanged, and, like all the operations in this example, they preserve the echelon form.

Finally, add multiples of the last row to the rows above, so as to give zeros everywhere above the $(4, 5)$ entry:

$$\begin{pmatrix} 1 & 0 & 7/2 & 0 & 0 & -7/3 \\ 0 & 1 & -3 & 0 & 0 & 31/3 \\ 0 & 0 & 0 & 1 & 0 & -5/3 \\ 0 & 0 & 0 & 0 & 1 & -7/3 \end{pmatrix}$$

The equations corresponding to this form of the augmented matrix are very simple. The last two rows give $y = -5/3$ and $z = -7/3$. The second equation reads $w - 3x = 31/3$, so x can take any value, and $w = 3x + 31/3$. Now the first equation gives $v = -(7/2)x - 7/3$.   □

The final form of the matrix in Example 1 is called the 'reduced echelon form'; it has zeros everywhere above the leading elements, which are all 1. We shall see that every matrix can be reduced to this form by elementary operations. Once this is done, it is very easy to write down the solution of the corresponding set of equations.

Definition 1. A matrix is in **reduced echelon form** (also known as the **Hermite normal form**) if all the following conditions hold:
  (a) the matrix is in echelon form;
  (b) the leading entry of every nonzero row equals 1;
  (c) each leading entry is the only nonzero entry in its column.   □

The final matrix in Example 1 satisfies these conditions.

Exercise 1. A square matrix in reduced echelon form either has a row of zeros or is the unit matrix.   □

We shall now describe in general terms the procedure for obtaining the reduced echelon form of a matrix. The basic step is as follows.

A **Jordan step** is the following procedure.
    Take a given nonzero row of the matrix, the ith row, say. Let a_{ik} be its leading entry. Multiply the ith row by $1/a_{ik}$; then its leading entry becomes 1. If $i = 1$, then stop. Otherwise, for $r = 1, 2, \ldots, i - 1$ subtract a_{rk} times the ith row from the rth row.
                    □

In Example 1 we took a Jordan step for each of the four rows. The sequence of operations was slightly different: we first reduced all four leading elements to 1, then did the Type III operations. But this makes no difference.

Theorem 1 (Gauss–Jordan reduction). Every matrix is row-equivalent to a matrix in reduced echelon form.

Proof. Section 4D showed that every matrix is row-equivalent to an echelon matrix. Apply a Jordan step to each nonzero row in turn, starting from the top. For each row after the first, the Jordan step gives a matrix with zeros everywhere above the leading entry of that row. Because each row has more leading zeros than the rows above, a Jordan step does not interfere with the zeros produced by previous steps. The process also transforms the leading entry to 1. Hence the final result is in reduced echelon form. Since all the steps are sequences of elementary row operations, the final form is row-equivalent to the original matrix.   □

Exercise 2. Write down a matrix in echelon form, and use Jordan steps to obtain the reduced echelon form.

Example 2. Everything works in the same way for complex matrices. The following calculation illustrates Gaussian elimination for a complex matrix:

$$\begin{pmatrix} 2 & 3i & 1-i \\ -i & 1+i & 0 \\ 2 & 3 & i \end{pmatrix} \rightarrow \begin{pmatrix} 2 & 3i & 1-i \\ 0 & i-1/2 & (i+1)/2 \\ 0 & 3-3i & 2i-1 \end{pmatrix} \rightarrow \begin{pmatrix} 2 & 3i & 1-i \\ 0 & i-1/2 & (1+i)/2 \\ 0 & 0 & (1+22i)/5 \end{pmatrix}$$

It is straightforward to apply Jordan reduction; the result is the identity matrix. □

Our procedure for solving sets of linear equations can now be summed up as Gauss reduction followed by Jordan reduction. The whole process is called Gauss–Jordan elimination or Gauss–Jordan reduction.

The reduced echelon form displays the solutions of the equation very clearly. Case 2 of section 4E, in which there are no solutions, is easily recognised: the augmented matrix has a row in which the only nonzero entry is the last. A row with only one nonzero entry apart from the last corresponds to an unknown whose value is uniquely determined. A row with more than one nonzero entry apart from the last corresponds to an unknown which can take many values. See Example 1 for illustrations. Case 1 of section 4E, in which all unknowns are uniquely determined, corresponds to an augmented matrix in reduced echelon form which is either $(I|b)$ or

$$\left(\begin{array}{c|c} I & b \\ \hline 0 & 0 \end{array} \right)$$

where I is an identity matrix, and b is a column.

We have described Gaussian and Gauss–Jordan elimination in considerable detail. However, as always in mathematics, there is more to be said. The Gauss–Jordan method is useful for theoretical purposes, but Gaussian elimination followed by back-substitution is more efficient for numerical calculation. Furthermore, calculations in practice normally work with a limited number of decimal places. The results are therefore approximations, and the numerical errors must be considered; see section 4J. Analysing the computational aspects of Gaussian elimination leads to new theoretical insights; see section 4M.

Finally, a few words about the meaning of 'normal' in 'Hermite normal form'. A given matrix is row-equivalent to many different echelon matrices, but the reduced echelon form is unique. If you study the final matrix in Example 1, you will see that any row operation will destroy its reduced echelon form; this observation can be turned into a proof that a given matrix has a unique Hermite normal form. The word 'normal' here means 'standard'; the Hermite normal form is a standard form to which any matrix can be uniquely reduced. It has the advantage that one can easily read off the solutions of the corresponding system of equations. In the next section we will see that the ranks of the matrix can also be read off immediately from the Hermite normal form.

4H   THE RANK THEOREM

We shall now use the Hermite normal form to prove the fact announced in section 4F, that the row and column ranks of a matrix are equal. The Hermite form displays the ranks of a matrix very clearly. The nonzero rows are linearly independent, so the row rank is the number of nonzero rows. Each nonzero row contains a leading 1. The columns containing the leading 1's form a basis for the column space; if this is not obvious, look at an example in the last section. Hence the column rank equals the number of leading 1's, which equals the number of nonzero rows, which equals the row rank. This shows that the row and column ranks of a reduced echelon matrix are equal. Since every matrix is equivalent to a reduced echelon matrix, the matter seems to be settled.

But of course it isn't. It is true that every matrix is row-equivalent to a reduced echelon matrix, and has the same row rank. But the reduction uses row operations which change the columns, and therefore change the column space. However, it turns out that though the column space changes, its dimension stays unchanged.

Theorem 1. Row operations do not change the row rank or the column rank of a matrix.

Proof. Theorem 4B-1 shows that elementary operations do not change the span of a set of vectors. Therefore row operations do not change the row space, so the row rank does not change.

We now prove that the column rank is unchanged by row operations. Let A and B be row-equivalent matrices. The column rank is the maximum number of linearly independent columns; we shall show that this number is the same for A and B, by proving that any set of columns of A is linearly dependent if and only if the corresponding set of columns of B is linearly dependent.

Let A' be a matrix consisting of some of the columns of A; that is, A' is obtained from A by deleting certain columns. Let B' be the matrix obtained from B by deleting the corresponding columns. Then A' and B' are row-equivalent, since the row operations which turn A into B will obviously turn A' into B'. Now, we proved in section 4D that row operations on the augmented matrix do not change the solution set of a system of equations. Since the augmented matrices $(A'|0)$, $(B'|0)$ are row-equivalent, it follows that

$$x = 0 \text{ is the only solution of } A'x = 0 \text{ if and only if}$$
$$x = 0 \text{ is the only solution of } B'x = 0 \tag{1}$$

But $A'x$ is a linear combination of the columns of A' (Lemma 2C-1). Hence (1) says that the columns of A' are linearly independent if and only if the columns of B' are. It follows that any subset of the columns of A is linearly independent if and only if the corresponding set of columns of B is.

Therefore the maximum number of independent columns of A equals the maximum number of independent columns of B. Thus they have the same column rank.   □

Corollary (the rank theorem). For every matrix, the row rank equals the column rank.

Proof. Every matrix is row-equivalent to a reduced echelon matrix, for which the ranks are equal as shown in the first paragraph of this section. Theorem 1 shows that the ranks of the original matrix are equal.   □

From now on we need not distinguish between row and column ranks, and shall simply refer to the **rank** of a matrix.

The fusing together of row and column ranks affects the notions of full column rank and full row rank as defined in section 4F. A matrix has full row rank if the rank equals the number of rows, and full column rank if the rank equals the number of columns. Each of these gives a limit on the possible values of the rank. If A is a p by q matrix, then $\operatorname{rank}(A) \le p$ and also $\operatorname{rank}(A) \le q$.

Definition 1. A p by q matrix has **full rank** if $\operatorname{rank}(A)$ equals the smaller of p and q.   □

This means that the rank is as large as it can possibly be; it has either full row rank or full column rank, depending on which of p and q is larger. The case $p = q$ is particularly important, becuase only in this case can a matrix have both full row rank and full column rank. The next section is devoted to this case.

4I   MATRIX INVERSES

Gaussian elimination is a good practical method for solving $Ax = b$. But it is sometimes useful to think in terms of matrix algebra rather than the elimination algorithm. This section expresses the theory in terms of matrix inverses.

In Chapter 2 we defined a right inverse of A to be a matrix M such that $AM = I$. It is closely related to the solution of linear systems.

Exercise 1. If M is a right inverse of A, then $x = Mb$ is a solution of the equation $Ax = b$.

$\square$

It follows that if A has a right inverse, then $Ax = b$ always has a solution. Section 4F shows that there is always a solution when A has full row rank. This suggests a connection between right inverses and row rank.

Theorem 1 (right invertibility). A matrix has a right inverse if and only if it has full row rank.

Proof. (a) Suppose A has full row rank. By Theorem 4F-4, for each i there is a vector m_i such that $Am_i = e_i$ where e_i is the ith column of the unit matrix I. Assembling the column vectors m_i into a matrix M gives $AM = I$, so A has a right inverse.

(b) If A has a right inverse M, then the system $Ax = b$ has a solution for all b, namely, $x = Mb$. Hence A has full row rank by Theorem 4F-4. $\square$

There is a similar relation between column rank and left invertibility.

Theorem 2 (left invertibility). A matrix has a left inverse if and only if it has full column rank.

Proof. A has a left inverse if and only if its transpose A^T has a right inverse, since $MA = I$ if and only if $A^T M^T = I$. By Theorem 1, A^T has a right inverse if and only if it has full row rank. This holds if and only if A has full column rank, since rows and columns are interchanged by transposition. $\square$

This is all that can be said in general for rectangular matrices. For square matrices we can go further. We have proved that row and column ranks are equal. For a square matrix, therefore, full row rank implies full column rank, and in this case there is both a left and a right inverse.

Theorem 3 (invertibility). A matrix is invertible if and only if it is square and full rank.

Proof. Invertible means having both a left and a right inverse. By the previous results, this holds if and only the matrix has full row rank and full column rank. This means that the number of rows equals the number of columns equals the rank.   □

Corollary 1. If a square matrix has either a left or a right inverse, then it is invertible.

Proof. Theorems 1 and 2 imply that it has either full column rank or full row rank; in either case it is a full rank matrix, hence invertible by Theorem 3.   □

The statement of this corollary involves only the simple ideas of matrix algebra introduced in Chapter 2. It could well have been stated in section 2E. However, its proof involves the idea of rank, which is built on the ideas of subspaces, dimension, etc. set out in Chapter 3. It illustrates the power of the general theory of vector spaces in proving concrete and useful results.

Theorem 3 gives a satisfactory answer to the question of when inverses exist. But how are they to be calculated? The most efficient method uses Gaussian elimination, as follows.

The inverse A^{-1} of A satisfies $AA^{-1} = I$. The column lemma of section 2C shows that this is equivalent to $Ac_i = e_i$ for each i, where c_i and e_i are the ith columns of A^{-1} and I respectively. Hence c_i can be found by Gaussian elimination applied to the augmented matrix $(A|e_i)$.

It thus appears that for an n by n matrix, Gaussian elimination must be carried out n times to find the n columns of A^{-1}. But we can be more efficient that this. If you think about Gauss–Jordan reduction, you will see that the elementary operations are chosen by looking at the first n columns of the augmented matrix; the last column plays a purely passive role. Hence the operations are the same for all the augmented matrices $(A|e_1)$, $(A|e_2)$, …. We can therefore save effort by combining them into one super-augmented matrix

$$(A|e_1|e_2|\ldots|e_n) \tag{1}$$

If A is invertible, applying Gauss–Jordan reduction to this matrix will give

$$(I|c_1|c_2|\ldots|c_n) \tag{2}$$

where c_i is the ith column of A^{-1}. But the matrix (1) is just $(A|I)$, and (2) is $(I|A^{-1})$. Hence we have the following.

Method for calculating inverses. To compute A^{-1} where A is an n by n matrix, write down the n by $2n$ matrix $(A|I)$ and apply Gauss–Jordan elimination. If A is invertible, the reduced echelon form of $(A|I)$ will be $(I|A^{-1})$.   □

This beautiful method is the best way of calculating inverses. It works for any invertible matrix, and if A is not invertible, then the method tells you so: Gaussian elimination will not give I in the first n by n block, it will give a matrix with a row of zeros.

Example 1. The inverse of

$$\begin{pmatrix} 2 & 4 & 3 \\ 0 & 1 & -1 \\ 3 & 5 & 7 \end{pmatrix}$$

is computed as follows.

$$\begin{pmatrix} 2 & 4 & 3 & | & 1 & 0 & 0 \\ 0 & 1 & -1 & | & 0 & 1 & 0 \\ 3 & 5 & 7 & | & 0 & 0 & 1 \end{pmatrix} \rightarrow \begin{pmatrix} 2 & 4 & 3 & | & 1 & 0 & 0 \\ 0 & 1 & -1 & | & 0 & 1 & 0 \\ 0 & -1 & 5/2 & | & -3/2 & 0 & 1 \end{pmatrix}$$

$$\rightarrow \begin{pmatrix} 2 & 4 & 3 & | & 1 & 0 & 0 \\ 0 & 1 & -1 & | & 0 & 1 & 0 \\ 0 & 0 & 3/2 & | & -3/2 & 1 & 1 \end{pmatrix}$$

$$\rightarrow \begin{pmatrix} 2 & 4 & 3 & | & 1 & 0 & 0 \\ 0 & 1 & -1 & | & 0 & 1 & 0 \\ 0 & 0 & 1 & | & -1 & 2/3 & 2/3 \end{pmatrix}$$

$$\rightarrow \begin{pmatrix} 2 & 4 & 0 & | & 4 & -2 & -2 \\ 0 & 1 & 0 & | & -1 & 5/3 & 2/3 \\ 0 & 0 & 1 & | & -1 & 2/3 & 2/3 \end{pmatrix}$$

$$\rightarrow \begin{pmatrix} 1 & 0 & 0 & | & 4 & -13/3 & -7/3 \\ 0 & 1 & 0 & | & -1 & 5/3 & 2/3 \\ 0 & 0 & 1 & | & -1 & 2/3 & 2/3 \end{pmatrix}$$

It is easy to check that the matrix in the last three columns here is the inverse of the given matrix.

Exercise 2. Compute the inverse of

$$\begin{pmatrix} 1 & 0 & 0 \\ a & 1 & 0 \\ b & c & 1 \end{pmatrix}$$ ☐

Our treatment of linear equations and matrix inverses is now complete, in the sense that you now know enough to proceed to the next chapter. However, as usual, there is more to be said. The next two sections discuss the solution of linear equations from the point of view of practical computation, taking into account the errors introduced by using calculators or computers. The rest of the chapter discusses other aspects of Gaussian elimination.

*4J PRACTICAL ASPECTS OF GAUSSIAN ELIMINATION

This section and the next are optional; they depend on no other optional section. They introduce a subject of great theoretical and practical interest: numerical linear algebra.

Numerical analysis and algebra start from the fact that calculations are hardly ever done precisely. Most calculations use electronic machines which normally work with 8 to 16 decimal places. The answers are therefore usually wrong. For example, if you ask an 8-decimal-place machine to divide 1 by 3, it gives 0.33333333. This differs by 0.0000000033.. from the right answer; the difference is called **roundoff error**. The name comes from the operation of removing all but the first 8 digits, which is called 'rounding' the number. (In practice one sometimes increases the last remaining digit, as when 0.6666... is rounded to 0.66666667, but we are not concerned with such details here.)

If a calculation works with n decimal places, then the relative error in a single multiplication or addition will be no larger than 10^{1-n}, which is usually negligibly small. But in a long calculation, the errors mount up, and they can become so large that the results are hopelessly wrong. A realistic example would be tedious because computers and calculators work with a large number of decimal places. But the idea can be seen from the following simplified example.

Example 1. Consider the system $0.01x + 10y = 10$, $0.1x - 0.1y = 0$. Its solution is $x = y = 0.999000999$. Let us solve it by Gauss–Jordan elimination, working correct to 3 figures. We have

$$\begin{pmatrix} 0.01 & 10 & 10 \\ 0.1 & -0.1 & 0 \end{pmatrix} \rightarrow \begin{pmatrix} 0.01 & 10 & 10 \\ 0 & -100 & -100 \end{pmatrix} \rightarrow \begin{pmatrix} 1 & 0 & 0 \\ 0 & 1 & 1 \end{pmatrix}$$

where we have rounded 100.1 to 100 in the first step.

Solving the final reduced echelon form here gives $x = 0$. This is hopelessly wrong; the right answer to 3 figures is 1.00, so an inaccuracy of 0.1% in the calculation has led to an error of 100% in the answer.   □

This (admittedly artificial) example shows how a small roundoff error in one step can be magnified in later steps to give a final answer that is seriously wrong.

An analysis of roundoff errors is beyond the scope of this book. But there is a simple (though not infallible) remedy for the bad behaviour of Example 1.

In Gaussian elimination, zero pivots are forbidden. A machine working with a finite number of decimal places cannot distinguish between very small numbers and zero; 10^{-9} equals 0 to a calculator working to 8 figures. Hence very small pivots are likely to give trouble. The standard Gauss algorithm interchanges rows when necessary to avoid zero pivots. We can improve it by using row interchanges to avoid small values as well as zero values for the pivots.

An awkward question now arises: when is a number too small to be used as a pivot? How small is too small? We sidestep this interesting question by the following strategy: always use row interchanges to bring the largest element in the relevant column into the pivotal position. This is a simple policy to carry out, and it will clearly help to avoid small pivots.

Exercise 1. Interchange the rows in Example 1 and then carry out Gaussian elimination, working to three places. How accurate is the solution now?   □

The policy outlined in the paragraph above is called **partial pivoting**. There is a method called 'complete pivoting' which interchanges columns as well as rows. It is theoretically more accurate, but it is more complicated, and partial pivoting is nearly always used in practice. The method is as follows.

The Gaussian elimination algorithm with partial pivoting. Do the following with $k = 1$, then repeat with $k = 2$, and so on.

Consider the submatrix consisting of the kth row and all the rows below it. Look at the first nonzero column of this submatrix, and interchange rows so as to bring the largest element of this column to the top, so that it becomes the leading element of the kth row. Use it as pivot for a Gauss step (described in section 4D).   □

This procedure is used by many large-scale computer programs for solving linear systems. It avoids the difficulties illustrated by Example 1, and gives accurate results in such cases. But it is not infallible. Some systems are incurably sensitive to roundoff error.

Example 2. Consider the system $x + y = 2$, $x + 1.006y = 8$. The solution is $y = 1000$, $x = -998$. Rounding the equations to 3 figures gives $x + y = 2$, $x + 1.01y = 8$, the solution of which is $y = 600$, $x = -598$. The equations are so sensitive that rounding one of the coefficients to 3 figures causes a gross error even if the solution is evaluated exactly.   □

It is easy to see the source of the trouble here. The rows of the 2 by 2 matrix of the system are very nearly identical, so although the rank is 2, it is only just 2, so to speak, and a very small change would reduce the rank to 1. The matrix is almost non-invertible, so the system is very close to being insoluble; that is the root of the difficulty.

Similar troubles can arise in a less obvious way. To a sensitive eye the matrix of Example 2 looks suspicious: its rows are nearly the same. But for a larger matrix it is much harder to see when its rows are nearly linearly dependent. A matrix may look quite innocent, yet suffer from the disease of being almost non-invertible, the symptoms of which are shown in Example 2. Such matrices are called **ill-conditioned**.

Example 3 (the Hilbert matrix). The classic example of an ill-conditioned matrix is the n by n matrix with (i, j) entry $1/(i + j - 1)$ for $i, j = 1, \ldots, n$. It is called the Hilbert matrix H_n.

With $n = 4$, solving the system $H_n x = b$ with 3-figure accuracy gives grossly wrong results. For larger values of n things get rapidly worse. If you have access to a computer with routines for solving equations or inverting matrices, try putting in the Hilbert matrix for increasing values of n. You will find that beyond a certain value of n, depending on the quality of the routine you are using, the answers are rubbish.

How can one tell when the answers from the machine are rubbish? This is one of the great questions of numerical computation. Here there is a simple test. Given a program which inverts matrices, evaluate the inverse of $(H_n)^{-1}$. The answer ought to be H_n. The machine should give the correct answer for $n = 5$ or so, but as n increases the answers deteriorate. For example, the LINPACK routines, some of the best available linear algebra software, give wrong answers for $((H_n)^{-1})^{-1}$ with $n \geq 10$, when run on a machine working to 16 figures. The Hilbert matrix becomes more and more ill-conditioned as n increases.   □

So what can one do about solving an ill-conditioned system? There is no general satisfactory answer, but there are various remedies one can try. See, for example Golub and Van Loan (1983).

*4K   ITERATIVE METHODS FOR LINEAR EQUATIONS

This section is optional; it depends on no other optional section. It describes a technique for solving linear systems which is sometimes better than Gaussian elimination.

An n by n matrix of the following form

$$\begin{pmatrix} a & b & 0 & 0 & 0 & \ldots & 0 & 0 \\ c & d & e & 0 & 0 & \ldots & 0 & 0 \\ 0 & f & g & h & 0 & \ldots & 0 & 0 \\ 0 & 0 & i & j & k & \ldots & 0 & 0 \\ 0 & 0 & 0 & l & m & \ldots & 0 & 0 \\ & \cdot & \cdot & \cdot & \cdot & \ldots & \cdot & \cdot \\ 0 & 0 & 0 & 0 & 0 & \ldots & y & z \end{pmatrix}$$

is called **tridiagonal**; the three central diagonals contain all the nonzero entries. Tridiagonal matrices appear frequently in the numerical solution of differential equations (see Problem 31).

The matrix above has $3n - 2$ nonzero entries. If n is large, then n^2 is much larger than $3n - 2$, hence most of the entries are zero. A matrix in which a large proportion of the entries are zero is called **sparse**. Tridiagonal n by n matrices are sparse when n is large.

Computation with sparse matrices can be carried out very efficiently; because most entries are zero, they need much less storage space, and fewer arithmetical operations, than ordinary matrices of the same dimensions. Unfortunately, Gaussian elimination applied to a sparse matrix destroys the sparsity. It gives zeros everywhere below the main diagonal, by a procedure which puts nonzero entries above the main diagonal. The resulting echelon matrix has about half of its entries nonzero, and is no longer sparse. Gaussian elimination is therefore unsuitable for these systems. This section describes a better method.

We introduce the method by way of an example of a tridiagonal system:

$$\begin{pmatrix} 3 & -1 & 0 & 0 \\ -1 & 3 & -1 & 0 \\ 0 & -1 & 3 & -1 \\ 0 & 0 & -1 & 3 \end{pmatrix} \begin{pmatrix} x_1 \\ x_2 \\ x_3 \\ x_4 \end{pmatrix} = \begin{pmatrix} 1 \\ 2 \\ 0 \\ 1 \end{pmatrix} \tag{1}$$

We can rearrange each equation so that it expresses one of the unknowns in terms of the others:

$$\left.\begin{array}{llll} x_1 = & 0.333x_2 & & +0.333 \\ x_2 = 0.333x_1 & & +0.333x_3 & +0.667 \\ x_3 = & 0.333x_2 & +0.333x_4 & \\ x_4 = & & 0.333x_3 & +0.333 \end{array}\right\} \quad (2)$$

or

$$x = Ax + b \quad (3)$$

where A is the matrix of coefficients in (2) and b is the column with entries 0.333, 0.667, 0.0, 0.333.

The form of (3) suggests a way of trying to solve it approximately, as follows.

Jacobi iteration

Make a rough guess at the solution vector; call it $x^{(0)}$. Then plug it into the right-hand side of (3), and define a new vector $x^{(1)}$ by

$$x^{(1)} = Ax^{(0)} + b$$

Repeat the procedure, defining

$$x^{(i+1)} = Ax^{(i)} + b \quad (4)$$

for $i = 0, 1, 2, \ldots$. This is called the **Jacobi iteration** procedure. Iteration means repetition, and (4) defines a sequence of vectors by repeating the operation 'multiply by A and add b'.

The idea here is that $x^{(i+1)}$ may be a better approximation than $x^{(i)}$ to the exact

Table 4.1 – Jacobi iteration on the system (1).

Iteration number	x_1	x_2	x_3	x_4
0	0.333	0.667	0.000	0.333
1	0.556	0.778	0.333	0.333
2	0.593	0.963	0.370	0.444
3	0.654	0.988	0.469	0.457
4	0.662	1.041	0.481	0.490
5	0.680	1.048	0.510	0.494
6	0.683	1.063	0.514	0.503
7	0.688	1.065	0.522	0.505
8	0.688	1.070	0.523	0.507
9	0.690	1.070	0.526	0.508
10	0.690	1.072	0.526	0.509
11	0.691	1.072	0.527	0.509
12	0.691	1.073	0.527	0.509
13	0.691	1.073	0.527	0.509

solution x, and the sequence of vectors $x^{(i)}$ may approach the exact solution. It may seem naive to hope that this works. But it often does.

If we take $x^{(0)} = b$, the calculations in (4) for $i = 0, 1, 2, \ldots$ give the results shown in Table 4.1. The method works. But it is not very efficient: the approximations in Table 4.1 take a long time to settle down. The following variant is quicker.

Gauss–Seidel iteration

Consider one step of the Jacobi iteration. To simplify the notation, write $y = x^{(i)}$ and $Y = x^{(i+1)}$; thus y is an approximation to the solution vector x, and Y is (we hope) a better approximation. Now (4) becomes $Y = Ay + b$, or, in components,

$$Y_j = A_{j1} y_1 + \ldots + A_{jn} y_n + b_j \qquad \text{for } j = 1, \ldots, n \qquad (5)$$

The Gauss–Seidel method improves the Jacobi procedure as follows. We calculate the components Y_j in turn, starting with Y_1, which is determined in terms of the y's by (5). When we calculate Y_2 from (5), the first term involves y_1. But we have just calculated Y_1, a better approximation than y_1 to the exact solution. So it would be better to use Y_1 instead of y_1 in the equation for Y_2:

$$Y_2 = A_{21} Y_1 + A_{22} y_2 + \ldots + A_{2n} y_n + b_2$$

In the same way we calculate the third component using the improved values for the first and second components:

$$Y_3 = A_{31} Y_1 + A_{32} Y_2 + A_{33} y_3 + \ldots + A_{3n} y_n + b_n$$

and so on. The general scheme is

$$Y_j = \sum_{k=1}^{j-1} A_{jk} Y_k + \sum_{r=j}^{n} A_{jr} y_r + b_j \quad \text{for } j = 1, \ldots, n \qquad (6)$$

This is the basic Gauss–Seidel iteration step. In the jth equation, $Y_1, \ldots, Y_{j-1}$ have been determined from previous calculations, and the equation gives Y_j in terms of them and the y's. In the first equation, with $j = 1$, the first sum runs from 1 to 0, and is to be interpreted as zero, so that Y_1 is calculated from $y_1, \ldots, y_n$.

Gauss–Seidel iteration consists of choosing an initial vector $x^{(0)}$, and then calculating $x^{(1)}, x^{(2)}, \ldots$ by means of (6) with $y = x^{(i)}$ and $Y = x^{(i+1)}$. The amount of calculation at each step is the same as for Jacobi's method, but Gauss–Seidel needs fewer steps to achieve a given accuracy. Table 4.2 shows the results for the same equation as Table 4.1.

Equation (6) may look more complicated than the basic step (4) of the Jacobi method, but from the computational point of view is it simpler. A typical line of a computer program is an instruction of the form 'calculate a certain number, and store it under the name p, replacing the value that was previously stored under that name'. This instruction can be written '$p := \ldots$', where $\ldots$ represents instructions for calculating a number. The symbol ':=' can be pronounced 'becomes'. Thus, $p := 2$ gives p the value 2, and $p := p + 1$ (p becomes $p + 1$) has the effect of increasing p by 1. In this language, the Gauss–Seidel step (6) can be expressed very simply:

$$\text{do the following for } j = 1, \ldots, n: \quad y_j := \sum_{k=1}^{n} A_{jk} y_k + b_j \qquad (7)$$

Table 4.2 – Gauss–Seidel iteration on the system (1).

Iteration number	x_1	x_2	x_3	x_4
0	0.333	0.667	0.000	0.333
1	0.556	0.852	0.395	0.465
2	0.617	1.004	0.490	0.497
3	0.668	1.052	0.516	0.505
4	0.684	1.067	0.524	0.508
5	0.689	1.071	0.526	0.509
6	0.690	1.072	0.527	0.509
7	0.691	1.073	0.527	0.509
8	0.691	1.073	0.527	0.509

If you work through this carefully, starting with $j = 1$, and distinguishing the old values y from the new values assigned by the := operation (called Y in equation (6)), you will see that it agrees with (6). In the Jacobi method the old values and the new values must both be remembered; Gauss–Seidel iteration is simpler in that each new value replaces the old value, and there is no need to store the old and the new values separately.

Convergence

In the examples above, the approximations $x^{(i)}$ approach the exact solution x as i increases. This does not always happen. For example, the solution of the equations

$$\left. \begin{array}{l} x + 2y = 1 \\ 2x + y + 2z = 0 \\ 2y + z = 1 \end{array} \right\} \qquad (8)$$

is $x = z = -1/7$, $y = 4/7$. Iteration gives the results shown in Table 4.3. Both methods fail disastrously.

In Tables 4.1 and 4.2 the approximations are said to **converge**, meaning that they approach the exact solution more and more closely. In Table 4.3 the approximations are said to **diverge**: they recede further and further from the solution.

Table 4.3 – Jacobi and Gauss–Seidel iteration on the system (8).

Iteration number	Jacobi			Gauss–Seidel		
	x	y	z	x	y	z
1	1	0	1	1	0	1
2	1	−4	1	1	−4	9
3	9	−4	9	9	−36	73
4	9	−36	9	73	−292	585
5	73	−36	73	585	−2340	4681
6	73	−292	73			

This example puts iterative methods into a different perspective. It raises the question, can one tell in advance whether or not iteration will work? We will study this question in Chapter 7, when more theoretical tools are available. But meanwhile, the following simple-minded approach gives some idea of what is happening.

Write the system (8) in the form used for iteration:

$$\left. \begin{array}{l} x = -2y + 1 \\ y = -2x - 2z \\ z = -2y + 1 \end{array} \right\} \tag{9}$$

The coefficients of x, y, z on the right-hand sides here have magnitudes greater than 1. That is why iteration diverges: any error in x, y, z at one stage becomes magnified by the iteration, and the situation gets worse at each step. For the system (2) on the other hand, the coefficients are all less than 1 in magnitude, and this means that errors are reduced at each stage and the iteration converges.

Expressed in terms of the original forms (8) and (1) of the equations, the argument above leads us to expect divergence when the diagonal coefficients are less than the others, as in (8), and convergence when they are greater, as in (1). For further information see section 7D.

*4L ELEMENTARY MATRICES

This section is optional; it depends on no other optional section. It looks at the key idea of this chapter, reducing a matrix by elementary row operations, from a different point of view.

Exercise 1. Calculate EA where

$$E = \begin{pmatrix} 1 & 0 & 0 \\ 0 & 0 & 1 \\ 0 & 1 & 0 \end{pmatrix} \qquad A = \begin{pmatrix} a & b \\ c & d \\ e & f \end{pmatrix} \qquad \Box$$

Multiplying any 3 by 2 matrix by E has the effect of interchanging the last two rows. A moment's thought will show that the same applies to a 3 by k matrix for any k. Multiplying by E performs a Type I elementary operation.

How was the matrix E obtained? – by interchanging the last two rows of the unit matrix. We have discovered the beautiful fact that applying a Type I elementary operation to the unit matrix I gives a matrix which, when multiplied by any matrix A, performs the same operation on A.

The obvious next step is to consider other types of elementary operation.

Exercise 2. If a Type II or Type III elementary operation is applied to the 3 by 3 unit matrix, the result is a matrix E such that for any A, EA is the matrix obtained by applying the same operation to A. $\Box$

Everything works in the same way for n by n matrices.

Definition 1. An n by n **elementary matrix** is a matrix obtained by applying an elementary row operation to the n by n unit matrix.

Example 1. The following elementary matrices correspond to Types I, II, and III operations respectively.

$$\begin{pmatrix} 0 & 1 & 0 \\ 1 & 0 & 0 \\ 0 & 0 & 1 \end{pmatrix} \quad \begin{pmatrix} 1 & 0 & 0 \\ 0 & k & 0 \\ 0 & 0 & 1 \end{pmatrix} \quad \begin{pmatrix} 1 & 0 & 0 \\ k & 1 & 0 \\ 0 & 0 & 1 \end{pmatrix}$$

These are only some of the elementary matrices, of course. There are three Type I 3 by 3 matrices: the first matrix here, the matrix in Exercise 1, and a third corresponding to interchanging the first and third rows. Similarly there are other elementary matrices of Types II and III.

Theorem 1 (elementary matrix theorem). If E is an n by n elementary matrix, then for any n by r matrix A, EA is the matrix obtained by applying the corresponding elementary row operation to A.

Proof. Proving the theorem just means carrying out the matrix multiplication. The calculation is clearer if A is partitioned into its rows. For the Type II operation of multiplying the third row by k, we have

$$\begin{pmatrix} 1 & 0 & 0 & \dots & 0 \\ 0 & 1 & 0 & \dots & 0 \\ 0 & 0 & k & \dots & 0 \\ . & . & . & \dots & . \\ 0 & 0 & 0 & \dots & 1 \end{pmatrix} \begin{pmatrix} a_1 \\ a_2 \\ a_3 \\ .. \\ a_n \end{pmatrix} = \begin{pmatrix} a_1 \\ a_2 \\ ka_3 \\ .. \\ a_n \end{pmatrix}$$

where a_r denotes the rth row of A. Similarly for Type II operations on the other rows.

The proof works in the same way for the other operations; the details are left as an exercise.   □

Theorem 2 (inverses of elementary matrices). Every elementary matrix is invertible, and its inverse is another elementary matrix of the same type.

Proof. We consider the three types separately.

(a) If a row interchange is applied to I twice in succession, the result is I again. So if E is a Type I elementary matrix, we have $E(EI) = I$; hence $EE = I$. Therefore E is invertible and its inverse is E.

(b) Now let E be the matrix correspond to multiplying the rth row by a nonzero number k. Let E' be the elementary matrix corresponding to multiplying the rth row by $1/k$. Then by Theorem 1, $E'(EI)$ is the matrix obtained by multiplying the rth row of I first by k and then by $1/k$. Hence $E'E = I$. A similar argument shows that $EE' = I$, so E' is the inverse of E.

(c) If E is the matrix corresponding to the operation $R_i \rightarrow R_i + kR_j$, let E' be the matrix for $R_i \rightarrow R_i - kR_j$. In the same way as in (b), we have $E'E = I$ and $EE' = I$, so E' is the inverse of E.   □

We have introduced elementary matrices as a technical device for doing elementary operations. But they have another aspect. They are building blocks from which other matrices can be constructed. In fact, every square nonsingular matrix can be expressed as a product of elementary matrices. This follows easily from the Gauss–Jordan algorithm.

Theorem 3 (factorisation into elementary matrices). Every nonsingular square matrix equals a product of elementary matrices.

Proof. By the Gauss–Jordan algorithm, every matrix A can be transformed into reduced echelon form by elementary operations. If A is square and nonsingular, the reduced echelon form is I. Let $e_1, \ldots, e_k$ denote the elementary operations which, applied one after the other, turn A into I. If E_j are the corresponding elementary matrices, then

$$E_k E_{k-1} \ldots E_1 A = I$$

This is the elementary-matrix expression of the fact that applying the operation e_1 to A, followed by operation e_2 and so on, gives I.

Each elementary matrix is invertible by Theorem 2. Multiplying the above equation on the left by $(E_k)^{-1}$, then by $(E_{k-1})^{-1}$ and so on, gives

$$A = (E_1)^{-1}(E_2)^{-1} \ldots (E_k)^{-1}$$

Since the inverse of an elementary matrix is another elementary matrix by Theorem 2, this completes the proof.   □

This is a nice result: many of the central theorems of algebra are of a similar type, showing that the most general object of a certain kind can be built up from simpler objects. But Theorem 3 is useful too. In Chapter 6 we shall use it to prove a key fact about determinants. And in the next section we shall use the same ideas to reveal yet another aspect of Gaussian elimination.

*4M TRIANGULAR FACTORISATION

This optional section is a continuation of section 4L, which expressed elementary operations in terms of matrix multiplication. Using this point of view, we shall look again at Gaussian elimination.

Any matrix A (not necessarily square) can be reduced to echelon form by elementary row operations. Hence

$$EA = U \tag{1}$$

where E is a product of elementary matrices and U is an echelon matrix, called U because it is upper triangular, that is, all entries below the main diagonal are zero. Now, elementary matrices are invertible, hence E is invertible. Multiplying (1) by E^{-1} gives

$$A = E^{-1}U \tag{2}$$

We now investigate the structure of the matrix E^{-1}.

Gaussian elimination is based on subtracting a multiple of one row from a row below. The corresponding elementary matrix is of the following type.

Definition 1. A triangular matrix is called **unit triangular** if all entries on the main diagonal are 1.

Remark. If a triangular matrix is not square, then the nonzero entries may form a trapezium, not a triangle – see the second matrix in (3) below. The term 'triangular' is therefore sometimes restricted to square matrices, and the rectangular case is then called 'trapezoidal'. We shall not use this term.

Exercise 1. (a) A unit lower triangular square matrix has an inverse which is unit lower triangular.

  (b) The product of unit lower triangular square matrices is unit lower triangular.

  (c) The elementary operation $R_i \to R_i + kR_j$, with $i > j$, corresponds to a unit lower triangular elementary matrix.   □

It follows that the operations in Gaussian elimination (without row interchanges) correspond to unit lower triangular elementary matrices; subtracting a multiple of a row from a row beneath gives $i > j$ in Exercise 1(c). Exercise 1(b) shows that combining several such operations gives a matrix of the same type. Hence the matrix E in (1) is unit lower triangular. By Exercise 1(c), E^{-1} is also unit lower triangular. We therefore write $E^{-1} = L$ (for lower), so that (2) becomes

$$A = LU$$

Thus a matrix can be factorised into lower and upper triangular factors L and U, provided that no zero pivots occur in Gaussian elimination, forcing row interchanges.

Working through an example reveals some interesting things.

Example 1. Let

$$A = \begin{pmatrix} 2 & 1 & 0 & 4 \\ 4 & 5 & 1 & 7 \\ 2 & -8 & -1 & 12 \end{pmatrix}$$

The first Gauss step, consisting of two elementary operations, is equivalent to multiplying by a 3 by 3 matrix:

$$\begin{pmatrix} 1 & 0 & 0 \\ -2 & 1 & 0 \\ -1 & 0 & 1 \end{pmatrix} A = \begin{pmatrix} 2 & 1 & 0 & 4 \\ 0 & 3 & 1 & -1 \\ 0 & -9 & -1 & 8 \end{pmatrix}$$

The next step gives

$$\begin{pmatrix} 1 & 0 & 0 \\ 0 & 1 & 0 \\ 0 & 3 & 1 \end{pmatrix}\begin{pmatrix} 1 & 0 & 0 \\ -2 & 1 & 0 \\ -1 & 0 & 1 \end{pmatrix} A = \begin{pmatrix} 2 & 1 & 0 & 4 \\ 0 & 3 & 1 & -1 \\ 0 & 0 & 2 & 5 \end{pmatrix}$$

Multiplying the two matrices on the left and inverting the result gives

$$A = \begin{pmatrix} 1 & 0 & 0 \\ 2 & 1 & 0 \\ 1 & -3 & 1 \end{pmatrix} \begin{pmatrix} 2 & 1 & 0 & 4 \\ 0 & 3 & 1 & -1 \\ 0 & 0 & 2 & 5 \end{pmatrix} \tag{3}$$

which is the LU factorisation of A.

Observe that the first two leading entries of U equal the pivots used in the Gauss reduction. This is no coincidence. If you look at the calculation, you will see that each pivot survives into the next stage of the reduction, and is unchanged thereafter. The leading entries of U (except the last) are always the pivots used in Gauss reduction without row interchanges.

Now consider L. The first Gauss step was $R_2 \rightarrow R_2 - 2R_1$ and $R_3 \rightarrow R_3 - 1R_1$. The multipliers 2 and 1 equal the $(2, 1)$ and $(3, 1)$ entries of L. Likewise, the second Gauss step is $R_3 \rightarrow R_3 - (-3)R_2$; and L_{32} equals -3. Again, this is no coincidence. The entries of L always give the multipliers used in Gauss reduction. Thus the LU factorisation contains a complete record of the details of the Gauss reduction. The leading entries of U are the pivots, and the subdiagonal entries of L are the multipliers.

Computational remark. Although L and U have different shapes, the three subdiagonal entries of L in (3) are arranged in the same pattern as the three subdiagonal zeros in U. At each stage of the reduction process, the multipliers appear in L in the same positions as the new zeros in U. Efficient computer codes exploit this fact by storing L and U in the same place. Each Gauss step reduces some entries of A to zero, and their spaces can be used to store the corresponding L_{ij}. ☐

The argument so far has taken no account of row interchanges. The elementary matrix for row interchanges is not triangular, so the simple LU factorisation does not work when there are zero pivots requiring row interchanges. The following theorem gives conditions under which no interchanges are needed; the theorem requires a preliminary lemma and definition.

Lemma 1. A triangular square matrix is invertible if and only if all diagonal entries are nonzero.

Proof. Let A be an n by n upper triangular matrix with all diagonal entries nonzero. Then A is an echelon matrix with no zero rows, hence it has rank n and is invertible.

Now suppose A is n by n upper triangular with not all diagonal entries nonzero; say $A_{kk} = 0$. Applying Gaussian elimination to the rows from the kth onwards gives a new kth row with at least k leading zeros, a new $(k + 1)$th row with at least $k + 1$ leading zeros, and so on. Hence the last row must be entirely zero, so A has rank less than n and is not invertible.

If A is lower triangular, apply the above arguments to A^T which is upper triangular; A is invertible if and only if A^T is. ☐

Definition 2. The rth **leading submatrix** of a matrix A is the r by r matrix consisting of the first r rows and columns of A.

Theorem 1 (*LU* decomposition). If all the leading submatrices of A are invertible, then $A = LU$ where L and U are lower and upper triangular respectively, with nonzero entries on their diagonals.

Proof. The first leading submatrix is just the $(1, 1)$ entry, which is nonzero because the submatrix is invertible. Hence a Gauss step can be taken, giving a matrix A_1, say.

Elementary operations do not change the rank of a matrix; hence the second leading submatrix of A_1, call it A'_1, is invertible. The Gauss step has made A'_1 upper triangular, so its $(2, 2)$ entry is nonzero by Lemma 1. The $(2, 2)$ entry can therefore be used as pivot for another Gauss step, leading to a matrix A_2.

The third leading submatrix of A_2 is upper triangular and invertible, by the same arguments as above. By Lemma 1 its $(3, 3)$ entry is nonzero, and can therefore be used as pivot for another Gauss step.

Proceeding in this way shows that all the pivots are nonzero, and no row interchanges are needed. The final echelon form of U is upper triangular, and its leading submatrices are triangular invertible matrices. Hence all the diagonal entries are nonzero.

The elementary operations are equivalent to multiplying by unit lower triangular elementary matrices. By Exercise 1(b), $QA = U$ where Q is unit lower triangular. Hence $A = LU$ where L is unit lower triangular by Exercise 1(a).   □

This theorem is not very easy to apply in practice; for a large matrix it means considering a large number of submatrices and determining whether they are invertible. But it can be used to identify certain recognisable types of matrix for which LU factorisation works without row interchanges. See section 7C and 12E. See Problem 36 for the question of what happens when the conditions of Theorem 1 are not satisfied.

There is more than one form of the LU decomposition. If L is the product of elementary matrices as above, then it is square and unit lower triangular. If A is rectangular, then U will be rectangular in this scheme. We can arrange for U to be square and unit triangular, as follows. Apply the standard LU decompositon to A^T; then $A^T = LU$ where L is square and unit triangular. Transposing this equation gives $A = L'U'$ where $L' = U^T$ is lower triangular and $U' = L^T$ is square and unit upper triangular.

One can also make both L and U unit triangular, at the cost of introducing a third factor. Suppose $A = LU'$ where L is a square unit lower triangular matrix. Under the conditions of Theorem 1, the diagonal entries of U' are nonzero; call them $d_1, \ldots, d_r$. Let D be the diagonal matrix with entries $d_1, \ldots, d_r$, and set $U = D^{-1}U'$. It is easy to see that the rows of U are the rows of U' divided by their diagonal entries, so U is unit upper triangular. Thus $A = LU' = LDU$ where D is diagonal and L and U are unit triangular.

This factorisation into LDU is not as simple as the LU factorisation for practical purposes. But it has a great theoretical advantage: the factors are uniquely determined.

Theorem 2 (Turing's theorem). Under the conditions of Theorem 1, there is a diagonal matrix D, of the same dimensions as A, and unit lower and upper triangular square matrices L and U, such that $A = LDU$. For a given A this representation is unique.
     □

The proof is not difficult, but is fairly long, and has been relegated to section 4N.

Exercise 2. The entries of D are the pivots used when Gaussian elimination without row exchanges is applied to A.

Question. What's so wonderful about triangular factorisation? Why such a long discussion?

Answer. Linear equations with triangular matrices are very easy to solve, as we saw in section 4C. For upper triangular matrices we called the process back-substitution; the procedure for lower triangular matrices is very similar but proceeds in the other order; it is called forward-substitution. Now, the equation

$$LUx = b \tag{4}$$

can be written as

$$Ly = b \tag{5}$$

where

$$Ux = y \tag{6}$$

If b is given, then y can be found from (5) by forward-substitution and then x can be found from (6) by back-substitution. Thus the equation $Ax = b$ can easily be solved when the LU decomposition of A has been found. The LU decomposition does the same job as the inverse: it gives an efficient solution of $Ax = b$ for many different b's without having to perform Gauss elimination each time. Triangular factorisation is a standard element of computer programs for solving linear systems.

4N   PROOFS

Lemma 4B-1. Let T be a set obtained from an ordered set S by applying any number of elementary operations. Then every linear combination of S is a linear combination of T, and every nontrivial linear combination of S is a nontrivial linear combination of T.

Proof. If we prove this for single elementary operations of each type, then it will follow for any series of operations.

(a) For Type I operations the result is obvious: changing the order in which the elements of a set are written makes no difference to the linear combinations.

(b) Now consider Type II operations. We can change the order of the set S by Type I operations, so as to put the vector which is multiplied by a nonzero scalar in the first position. Then we have $S = \{u, v, w, \ldots\}$ and $T = \{au, v, w, \ldots\}$ where a is a nonzero scalar. Any linear combination which does not include u is clearly unaffected by the operation. A linear combination which includes u can be written

$$cu + dv + ew + \ldots = (a^{-1}c)(au) + dv + ew + \ldots$$

The dots here indicate a finite sum in which the number of terms is not specified.

Thus each linear combination of S can be written as a linear combination of T. Each nontrivial linear combination of S is a nontrivial combination of T, since if not all of $c, d, e, \ldots$ are zero, then not all of $a^{-1}c, d, e, \ldots$ are zero. Notice that this argument would break down if the multiplier a were allowed to be zero.

(c) For Type III operations we can again change the order of the set so that the operation is adding a multiple of the second element to the first. Thus if $S = \{u, v, w, \ldots\}$, then $T = \{u + av, v, w, \ldots\}$ where a is a scalar (which in this case may be zero). A linear combination of S can be written

$$cu + dv + ew + \ldots = c(u + av) + (d - ac)v + ew + \ldots$$

with the dots indicating a finite sum as in part (b) of this proof. This shows that every linear combination of S is a linear combination of T. And if it is a nontrivial linear combination of S, then $c, d, e, \ldots$ are not all zero, therefore $c, (d - ac), e, \ldots$ are not all zero, so it is a nontrivial linear combination of T. This completes the proof. $\square$

Theorem 4M-2. Under the conditions of Theorem 1, there is a diagonal matrix D, of the same dimensions as A, and unit lower and upper triangular square matrices L and U, such that $A = LDU$. For a given A this representation is unique.

Proof. Theorem 4M-1 gives a unit lower triangular square matrix L, and a p by q echelon matrix U' with nonzero diagonal entries, such that $A = LU'$. The argument now depends on whether the rectangular matrix A is tall and thin or short and fat.

(a) Suppose A is a p by q matrix with $p < q$. Let D be the p by q matrix with $D_{ii} = U'_{ii}$ for $i = 1, \ldots, p$ and all other entries 0. Define a q by q matrix U as follows: for $i = 1, \ldots, p$ the ith row of U is the ith row of U' divided by U'_{ii}, and for $i > p$ the ith row is zero except for 1 in the (i, i) position. Then U is unit upper triangular. If you work out an example and then think about the general case, you will see that $U' = DU$, so that $A = LDU$.

(b) Now suppose $p \geq q$. Let D be the p by q matrix with $D_{ii} = U'_{ii}$ for $i = 1, \ldots, q$ and all other entries zero. Let U be the q by q matrix whose ith row is the ith row of U' divided by U'_{ii}. Again U is unit upper triangular and square, and $A = LDU$.

(c) We have now shown that an LDU factorisation always exists. To prove uniqueness, suppose $A = LDU$ and also $A = L'D'U'$. Then $LDU = L'D'U'$, hence $(L')^{-1}LDU = D'U'$, and

$$(L')^{-1}L = D'U'U^{-1}D^{-1} \tag{1}$$

By Exercise 4M-1, the left-hand side of (1) is unit lower triangular. The right-hand side is a product of upper triangular matrices, and is therefore upper triangular. Hence both sides are both upper and lower triangular, and are therefore diagonal. Since the left-hand side is unit triangular, it must equal the unit matrix I. Hence $(L')^{-1}L = I$, so $L = L'$.

Since $LDU = L'D'U'$, we have $DU = D'U'$, and a similar argument gives $U = U'$ and thence $D = D'$. $\square$

PROBLEMS FOR CHAPTER 4

Hints and answers for problems marked [a] will be found on page 280.

Sections 4A, 4B, 4C

[a]1. Which of the following sets are in echelon form? Which are linearly dependent?
  (a) $\{(0, -1, 0, 0), (0, 0, 0, 1)\}$;
  (b) $\{(0, 0, 0, 0), (0, 0, 0, 0)\}$;
  (c) $\{(1, 1, 1, 1), (1, 1, 1, 1)\}$;
  (d) $\{(1, 1, 1, 1), (0, 0, 1, 1), (0, 1, 0, 0), (0, 0, 0, 0)\}$.

[a]2. What is wrong with the following argument? 'In Problem 1(c), the set $\{(1, 1, 1, 1), (1, 1, 1, 1)\}$ is simply $\{(1, 1, 1, 1)\}$, and therefore linearly independent.'

[a]3. True or false?
  (a) If a set of vectors is in echelon form, and a vector is removed, the resulting set will be in echelon form.
  (b) If a set of vectors is in echelon form, and two vectors are interchanged, then the resulting set is not in echelon form.
  (c) If the rows of a matrix are in echelon form, then its columns are also in echelon form.
  (d) If the rows of a matrix are in echelon form, then its columns cannot be in echelon form unless the matrix is zero.
  (e) If a square matrix is upper triangular (defined in Example 3D-6), then it is in echelon form.
  (f) If a square matrix is in echelon form, then it is upper triangular.

4. Write down all possible shapes for a 3 by 3 matrix in echelon form, in the notation used at the end of section 4C.

5. For what values of t is $\{(1, t, 3), (2, 0, 1), (-1, 2, t)\}$ a basis for $\mathbb{C}^3$?

6. Show how to combine elementary operations of Types II and III to give the same result as a Type I operation.

Sections 4D, 4E

[a]7. Use Gaussian elimination to find the solution sets of the following systems:
  (a) $\begin{aligned} x + 3y + z &= 1 \\ 2x + 6y + 9z &= 7 \\ 2x + 8y + 8z &= 6 \end{aligned}$
  (b) $\begin{aligned} x_1 + x_2 + x_4 &= 0 \\ x_1 + 2x_2 + x_3 + x_4 &= 1 \\ 2x_1 + 3x_2 + x_3 + 2x_4 &= -1 \\ x_2 + 3x_3 - x_4 &= 3 \end{aligned}$

  (c) $\begin{aligned} a + b - c + 2d &= 0 \\ 3b - c + 3d &= 0 \\ 2a - b - c + d &= 0 \end{aligned}$

8. For each of the following, (i) determine whether or not S is linearly independent, (ii) find a basis for $Sp(S)$, (iii) write down the dimension of $Sp(S)$:

(a) $S = \{(1, 0, 1, 1), (3, 2, 5, 1), (0, 4, 4, -4)\}$;

(b) $S = \{(1, 3, 0), (0, 2, 4), (1, 5, 4), (1, 1, -4)\}$;

(c) $S = \{(i, -1, 2i), (1, i, 0), (1 + i, i, 2)\}$.

9. Show that the effect of each type of elementary operation can be reversed by applying another operation of the same type. Deduce that if a matrix A is row-equivalent to a matrix B, then B is row-equivalent to A.

[a]10. True or false?

(a) A matrix with nonzero entries cannot be row-equivalent to the zero matrix.

(b) I is row-equivalent to $-I$.

[a]11. Find a condition on a, b, c which will ensure that the system below has a solution. Find the solutions in the case when this condition is satisfied.

$$x_1 + x_2 + x_3 + x_4 = a$$
$$5x_2 + 2x_3 + 4x_4 = b$$
$$3x_1 - 2x_2 + x_3 - x_4 = c$$

12. Using Gaussian elimination, prove that the system $ax + by = p$, $cx + dy = q$ has a unique solution (x, y) for each p, q if and only if $ad \neq bc$.

(Hint: you should treat separately the cases where a and c do or do not vanish.) What can you say about the solution set if $ad = bc$?

13. A **tridiagonal** matrix is one for which the only nonzero entries are on the three central diagonals. In other words, A is tridiagonal if $A_{ij} = 0$ when $|i - j| > 1$. An example is given at the beginning of section 4K.

If Gaussian elimination is applied to a tridiagonal matrix, without row interchanges, is the resulting echelon form still tridiagonal?

Section 4F

[a]14. True or false?

(a) If the rows of a matrix are linearly independent, so are the columns.

(b) For any matrix A, the row ranks of A and $\bar{A}$ are equal, where $\bar{A}$ is the matrix whose entries are the complex conjugates of those of A.

(c) Every matrix of row rank 1 equals xy^T for some column vectors x and y.

(d) Every matrix of column rank 1 equals xy^T for some column vectors x and y.

(e) A matrix of full row rank cannot have more columns than rows.

(f) A matrix of full column rank cannot have more columns than rows.

[a]15. (a) Find the row rank and the column rank of

$$\begin{pmatrix} 1 & 0 & 1 & 1 \\ 3 & 2 & 5 & 1 \\ 0 & 4 & 4 & -4 \end{pmatrix}$$

(see Problems 8(a) and (b).

(b) Write down some other matrices and calculate their row and column ranks. (You should find that the row and the column ranks turn out to be equal.)

ª16. Prove that the system $Ax = b$ has solutions if and only if the column rank of A equals the column rank of the augmented matrix.

17. Prove that row $\text{rank}(A + B) \le$ row $\text{rank}(A) +$ row $\text{rank}(B)$ for all p by q matrices A, B, and similarly for the column rank.

Sections 4G, 4H

18. Return to Problem 7 and, where appropriate, use Gauss–Jordan elimination to find the solutions.

ª19. If a matrix is in reduced echelon form, and a row is deleted, is the resulting matrix necessarily in reduced echelon form? What if a column is deleted?

ª20. '$\text{Rank}(A) = \text{rank}(B)$ implies $\text{rank}(A^2) = \text{rank}(B^2)$.' True or false?

21. Show that every matrix can be reduced to a form in which the first r entries of the main diagonal are 1 and all other entries are zero, by applying elementary operations to the columns as well as the rows. Here r is the rank of the matrix.

22. A and B are matrices with the same number of columns, so that they can be put together to form a matrix

$$C = \begin{pmatrix} A \\ B \end{pmatrix}$$

Prove that $\text{rank}(C) \le \text{rank}(A) + \text{rank}(B)$.

Section 4I

ª23. For each of the following matrices, determine whether it is invertible, and if so, find the inverse:

$$\text{(a)} \begin{pmatrix} 1 & 1 & 0 \\ 0 & 1 & 1 \\ 0 & 0 & 1 \end{pmatrix} \quad \text{(b)} \begin{pmatrix} 1 & 3 & 1 \\ 2 & 6 & 9 \\ 2 & 6 & -5 \end{pmatrix} \quad \text{(c)} \begin{pmatrix} 2 & 4 & 0 & 0 \\ 4 & 1 & 1 & 0 \\ 0 & 3 & 1 & 2 \end{pmatrix}$$

ª24. True or false?
(a) The sum of two invertible matrices is invertible.
(b) Every square matrix with full rank is invertible.
(c) Converse of (b).
(d) If A is row-equivalent to an invertible matrix, then A is invertible.
(e) If a diagonal matrix is invertible, its inverse is diagonal.
(f) The inverse of a matrix with real entries is a matrix with real entries.
(g) Same as (f) with 'real' replaced by 'pure imaginary'.

(h) Same as (f) with 'real' replaced by 'positive'.
(i) A left invertible matrix cannot have more rows than columns.
(j) Converse of (i).
(k) A right invertible matrix cannot have more rows than columns.

25. Find the inverse of

$$\begin{pmatrix} 1 & 1 & 1 \\ 1 & \omega & \omega^2 \\ 1 & \omega^2 & \omega \end{pmatrix}$$

where $\omega = e^{2\pi i/3}$ (so that $\omega^3 = 1$).
(Hint: show that $1 + \omega + \omega^2 = 0$, and then use intelligent guesswork.)

26. Every 3 by 3 skew-symmetric matrix (see Chapter 2, Problem 4) can be written in the form

$$A = \begin{pmatrix} 0 & -s & t \\ s & 0 & -u \\ -t & u & 0 \end{pmatrix}$$

Show that A is not invertible, but $I + A$ is. Find the inverse of $I + A$.
(See also Chapter 6, Problem 8.)

Section 4J

27. Solve the equations $x + y + z = 1$, $x + 0.9999y + 2z = 2$, $x + 2y + 2z = 1$ by Gaussian elimination, with and without partial pivoting. Work to 4 significant figures in both cases. Compare your answers with the exact solution.

28. Divide the third equation in Problem 27 by 2×10^4. Solve the new system using partial pivoting, working to 4 significant figures again. Compare your answer with the exact solution.

(This is an example of a system for which partial pivoting fails, but which can be solved by different techniques. See Burden and Faires (1985) under the heading 'scaled column pivoting'.)

[a]29. Adding and subtracting numbers takes about the same amount of work; multiplications and divisions take about the same amount as each other, but longer than additions or subtractions. Estimate roughly the number of addition/subtraction operations and the number of multiplication/division operations needed to solve an n by n system, assuming that n is so large that n^2 is much larger than n.

Hence show that if the number of unknowns is doubled, the amount of work increases by a factor of about 8.

Section 4K

[a]30. Apply Jacobi and Gauss–Seidel iteration to the system $1.5x - 0.8y = 1.0$, $x - 2.5y = -1.0$, working to 4 figures. Take $(1, 1)$ as the starting point, and calculate

the first 6 iterates in each case. Compare the answers with the exact solution (easily found).

31. (For readers familiar with differential equations.) Consider the equation $y'' = 2y$ for $0 < x < 1$, with boundary conditions $y(0) = y(1) = 1$. Split the interval $[0, 1]$ into $n + 1$ subintervals of length $h = 1/(n + 1)$. Write $Y_1 = y(h)$, $Y_2 = y(2h)$, ..., $Y_n = y(nh) = y(1 - h)$. The n-vector Y contains the values of y at equally spaced points in the interval.

Using the approximation $y''(x) \simeq [y(x + h) - 2y(x) + y(x - h)]/h^2$, show that the given equation and boundary conditions lead to the system $AY = b$, where A is an n by n tridiagonal matrix with constant diagonals, and $b = (-1, 0, \ldots, 0, -1)^\mathsf{T}$.

Take $n = 3$ and solve the system by Gauss–Seidel iteration. Start with $(1, 1, 1)$, and do several iterations. Compare the results with (i) the exact solution of the 3 by 3 system, and (ii) the exact solution of the original differential equation, evaluated at the points $x = 0.25$, 0.5, 0.75.

(Gauss–Seidel iteration converges quite slowly here. There are more efficient iterative methods; see Burden and Faires (1985) or Todd (1978), for example.)

Sections 4L, 4M

[a]32. Express the matrix

$$\begin{pmatrix} 2 & -1 \\ 3 & 2 \end{pmatrix}$$

as a product of elementary matrices.

[a]33. Find the LU and LDU factors of the matrix

$$\begin{pmatrix} 2 & -1 & 0 \\ -1 & 2 & -1 \\ 0 & -1 & 2 \end{pmatrix}$$

34. Problem 13 shows that if a tridiagonal matrix has an LU factorisation, then U is tridiagonal. In Problem 33, both are tridiagonal. Show that in general, both L and U are tridiagonal if A is.

L and U can be found without using elementary operations, as follows. For 4 by 4 tridiagonal matrices, the LU factorisation is

$$A = \begin{pmatrix} a_1 & b_1 & 0 & 0 \\ c_2 & a_2 & b_2 & 0 \\ 0 & c_3 & a_3 & b_3 \\ 0 & 0 & c_4 & a_4 \end{pmatrix} = \begin{pmatrix} 1 & 0 & 0 & 0 \\ p_1 & 1 & 0 & 0 \\ 0 & p_2 & 1 & 0 \\ 0 & 0 & p_3 & 1 \end{pmatrix} \begin{pmatrix} q_1 & r_1 & 0 & 0 \\ 0 & q_2 & r_2 & 0 \\ 0 & 0 & q_3 & r_3 \\ 0 & 0 & 0 & q_4 \end{pmatrix}$$

By taking the entries of this matrix equation one by one, derive a system of equations determining the unknown p's, q's and r's in terms of the a's, b's and c's.

(This technique for finding LU factors is called 'Doolittle's method'. It works for any matrix, not just tridiagonal matrices; see Burden and Faires (1985).

35. The uniqueness part of Theorem 4M-2 does not need the assumption that all leading submatrices are invertible. Prove that if $A = LDU$ where L and U are unit lower and upper triangular, and D is diagonal and nonsingular, then L, D, U are uniquely determined. Deduce that if $A = LDU$ is symmetric, then $U = L^T$.

36. This problem works out what happens to the LU factorisation when row interchanges are needed.

(a) Return to Example 4D-3; call the given matrix A. Rearrange the rows of A so that Gaussian elimination can be applied to the rearranged matrix with no further interchanges. Hence show that $PA = LU$ where P is a 'permutation matrix', that is, a product of Type I elementary matrices.

(b) Show that for any invertible 2 by 2 matrix A there is a permutation matrix P such that $PA = LU$.

(c) Use mathematical induction to show for all k that if A is a k by k invertible matrix, then $PA = LU$.

(d) Extend the result to rectangular matrices: show that if the largest leading submatrix of A is invertible, then $PA = LU$.

(e) Can the invertibility condition be removed? (See Horn and Johnson (1985).)

37. This problem factorises every matrix into two full rank matrices. For an application, see Chapter 13, Problem 32.

(a) Equation (2) of section 4M shows that every matrix A can be expressed as $A = VU$, where V is invertible and U is upper triangular. The matrix U has r nonzero rows, where $r = \text{rank}(A)$.

The factorisation VU can be stripped down. Discard all but the first r columns of V, leaving a matrix M, and all but the first r rows of U, leaving N. Prove that $A = MN$, and that M and N are matrices of full rank.

(b) Prove that if $A = MN$ where M and N have full rank, then every other factorisation of A into full rank factors can be expressed in the form $A = (MY)(Y^{-1}N)$ for some invertible matrix Y.

(c) Find a full rank factorisation of

$$\begin{pmatrix} 1 & 2 & 3 \\ 1 & 1 & 1 \\ 2 & 3 & 4 \end{pmatrix}$$

Part III

Linear transformations

Linear algebra studies vectors and their algebraic properties, in the same way that elementary algebra deals with real numbers. Now, the real numbers themselves are not very exciting; but functions defined on real numbers are much more interesting and useful. In linear algebra too, functions play a leading role; they are the subject of the next few chapters.

Chapter 5 introduces the type of function studied in linear algebra, called linear transformations. Chapters 6 and 7 deal with determinants and their application to linear transformations. Chapter 8 develops the theory and its application to differential equations and to quadratics in several variables, leading to the geometry of conic sections and their higher-dimensional analogues.

The system of starred sections is explained in the introduction to Part II.

5

Linear transformations on vector spaces

The ideas of this chapter develop naturally out of the study of linear algebraic equations in Chapter 4. If A is a p by q matrix, the equation $Ax = y$ can be regarded as transforming a q-vector x into a p-vector y. Section 5A generalises this idea, considering transformations from any vector space to another. The following sections generalise the notions of row and column spaces, rank, and invertibility.

Section 5F introduces an important new theme: eigenvectors and eigenvalues. The optional section 5G describes one of their physical interpretations, and 5H develops their mathematical properties.

The rest of the chapter is optional. Section 5I discusses rotations in the language of linear transformations and matrices. Section 5J returns to eigenvalues, from the point of view of efficient numerical computation. Section 5K reverts to the general theory of linear transformations, and fills in some mathematical details not needed earlier in the chapter; and the last section gives proofs of some theorems stated in earlier sections.

5A   BASIC DEFINITIONS

Linear transformations are generalisations of matrices, in the same way that abstract vectors are generalisations of n-vectors. A p by q matrix A applied to a q-vector x (written as a column, as usual) transforms it into a p-vector Ax. There are two vector spaces here, $\mathbb{R}^q$ and $\mathbb{R}^p$ (assuming that we are working with real numbers). A p by q matrix applied to a member of $\mathbb{R}^q$ gives a member of $\mathbb{R}^p$. We say that the matrix 'maps' $\mathbb{R}^q$ into $\mathbb{R}^p$. A similar notion can be defined for any two spaces.

Definition 1. A mapping f from a vector space V to a vector space W is a rule which, given any v in V, specifies a corresponding w in W. We say that f maps the vector v into w, and also that f maps the space V into W.   □

A mapping is the same kind of thing as a function, as defined in elementary calculus. Indeed, taking $V = W = \mathbb{R}$ in Definition 1 gives precisely a real function of a real argument. This is why the letter f is often used for mappings in general.

Notation. $f: V \to W$ means 'f is a mapping from the space V to the space W'. We write $w = f(v)$, or $w = fv$, to mean 'w is the element of W into which f maps v'.

Examples 1. (a) Take $V = \mathbb{R}^2$ and $W = \mathbb{R}$; a mapping $f: V \to W$ can be defined by the rule: for all (x, y) in $\mathbb{R}^2$, $f(x, y) = x^2 + y^2$.

(b) In the notation of (a), another mapping g can be defined by $g(x, y) = 2x - 3y$. In fact, any real-valued function of two real variables gives a mapping $\mathbb{R}^2 \to \mathbb{R}$.

(c) The function $h(x, y) = (xy)^{1/2}$ does not define a mapping $\mathbb{R}^2 \to \mathbb{R}$, because it does not give a real number for all (x, y) in $\mathbb{R}^2$ (consider $(1, -1)$, for example).

(d) Let P_3 be the space of all polynomials of degree ≤ 3 with complex coefficients. A mapping $K: P_3 \to \mathbb{C}^4$ can be defined by the rule $K(p) = c$, where c is the 4-vector whose entries are the coefficients of the polynomial p.

(e) In any space V we can define a mapping $f: V \to V$ by $f(v) = v$ for all v. This is called the **identity** map, since the image of any vector is identical to the original vector. It is usually called i or I.   □

The idea of a mapping is very simple. But there are many words to be learnt.

Definitions 2. The words **map**, **operator**, and **transformation** mean the same as mapping. The space V is called the **domain** and W is called the **codomain** for the transformation $f: V \to W$. If u is any element of the domain, then the vector $f(u)$ is called the **image** of u under the mapping f. See Fig. 5.1.   □

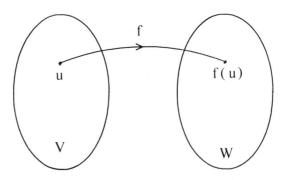

Fig. 5.1 – A map $V \to W$.

Some authors make a distinction between the terms mapping, operator, and transformation; I don't.

Linear algebra deals mainly with maps of a special type, which fit naturally into the algebraic structure of vector spaces. Roughly speaking, they have the property of mapping sums into sums, and scalar multiples into scalar multiples. The following definition expresses this precisely.

Definition 3. Let V and W be vector spaces over the same field $\mathbb{K}$. A **linear transformation** (or linear map, etc.) of V into W is a map f such that

$$f(u+v) = f(u) + f(v) \quad \text{and} \quad f(ku) = kf(u)$$

for all u and v in V and all scalars k in $\mathbb{K}$. Expressed in words: the image of a sum is the sum of the images, and similarly for scaling.   □

Note that the two $+$ signs in this definition mean different things: in $f(u+v)$ the $+$ means addition in the space V, and in $f(u)+f(v)$ it means addition in the space W. Similarly, the two multiplications by k are in different spaces. Since the same scalar k is used in both spaces, it is essential that V and W have the same field of scalars.

Exercise 1. The map f of Example 1(a) is not linear, but the map g of Example 1(b) is.

Examples 2. (a) Let P be the space of all real polynomials, and S the space of all real infinite sequences (the real version of the space of Example 3H-2). Define a map $f: P \to S$ by the rule that a polynomial maps to the sequence of its coefficients; thus

$$f(p) = (c_0, c_1, c_2, \ldots)$$

where c_r is the coefficient of x^r in the polynomial p. If the degree of p is n, then $c_{n+1} = c_{n+2} = \ldots = 0$.

Now, the coefficients of the polynomial $p+q$ are the sums of the coefficients of p and q, and for any real number k the coefficients of the polynomial kp are k times the coefficients of p. Hence $f(p+q) = f(p) + f(q)$ and $f(kp) = kf(p)$, so f is a linear map.

(b) Let S be the space of all smooth real functions (see Appendix D). Define a map $D: S \to S$ by $Df = f'$ for all f in S, where f' is the derivative of f. The rules of calculus show that $D(f+g) = Df + Dg$ for all f and g, and $D(kf) = kDf$ for any constant k. So D is a linear mapping.

More generally, define $L: S \to S$ by $Lf = pf'' + qf' + r$, where p, q, and r are smooth functions chosen once and for all. Then

$$L(f+g) = p(f'' + g'') + q(f' + g') + r(f + g)$$

$$= pf'' + qf' + rf + pg'' + qg' + rg = Lf + Lg$$

Similarly $L(kf) = kLf$ for any constant k. Thus L is a linear operator. It will be discussed further in Example 5D-1.   □

We now work out in detail the relation between matrices and linear maps outlined at the beginning of this section.

Example 3. Let A be a p by q real matrix. We define a map from $\mathbb{R}^q$ to $\mathbb{R}^p$ as follows. For each q-vector u, let u_c be the column matrix obtained by writing u out vertically. Define $a: \mathbb{R}^q \to \mathbb{R}^p$ by

$$a(u) = Au_c \quad \text{for all } u \text{ in } \mathbb{R}^q$$

The product Au_c is a column with p entries, giving a member of $\mathbb{R}^p$. The rules of matrix algebra give $A(u_c + v_c) = Au_c + Av_c$, and $Aku_c = kAu_c$, hence $a(u+v) =$

$a(u) + a(v)$ and $a(ku) = ka(u)$ for all u, v in $\mathbb{R}^q$ and all real numbers k. Therefore a is linear.

Starting from the same matrix A, we can define a different linear map as follows. For any p-vector v, let v_r be the row matrix obtained by writing v out horizontally. Define a map $a': \mathbb{R}^p \to \mathbb{R}^q$ by

$$a'(v) = v_r A \quad \text{for all } v \text{ in } \mathbb{R}^p$$

An argument similar to the above shows that a' is linear.

If A is a complex p by q matrix, we can define two linear maps $\mathbb{C}^q \to \mathbb{C}^p$ and $\mathbb{C}^p \to \mathbb{C}^q$ in the same way.

The two maps corresponding to a given matrix A are not really very different. Since the transpose of a row matrix is a column matrix, $(v_r A)^T = A^T (v_r)^T = A^T v_c$. Hence the q-vector $v_r A$ can be obtained by applying the matrix A^T to the corresponding column vector. Hence the a'-map corresponding to A is the same as the a-map corresponding to A^T. One can work either with rows or with columns; we shall usually work with columns, and use the a-map rather than the a'-map. But the choice is arbitrary; it is easy to translate from the column version to the row version of the theory.

Example 4. If you have read section 3J, you will see that an isomorphism from V to W is a linear map $f: V \to W$ such that for each w in W there is exactly one v in V satisfying $f(v) = w$. $\qquad\qquad\square$

You may have wondered why linear mappings are called linear. The answer will emerge in the next section, where we shall show that linear maps transform straight lines into straight lines, whereas nonlinear maps in general turn straight lines into curves. This special relationship with straight lines is why the whole subject is called linear algebra. A vector space is the most general structure on which linear mappings can be defined: the definitions of addition and scalar multiplication are just what are needed to formulate Definition 3.

But why do linear mappings deserve special attention? Consider elementary calculus again. Here we deal with all kinds of functions, not just linear ones. A useful way of investigating a function is to differentiate it. The idea behind the calculus is to approximate the graph of a function by the tangent; the approximation is useful near the point of tangency. Thus the curved graph is replaced by a straight line, which means that the function is approximated by a linear function. Similarly, linear mappings give good approximations to nonlinear mappings near any given point. So the answer to the question 'why concentrate on linear mappings?' is that they give good local approximations to nonlinear maps, yet they are much easier to deal with. Linear mappings have many simple properties, as will be seen in the next section.

5B THE IMAGE AND RANK OF A LINEAR TRANSFORMATION

The properties of a mapping can often be seen more clearly by looking at its effect on whole sets of points, rather than one point at a time. Hence the following definition.

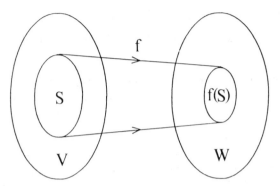

Fig. 5.2 – The image of a set.

Definition 1. Let S be any subset of a vector space V, and f a map from V to W. The **image of S** under f is the set $\{f(v): v \in S\}$. It is written $f(S)$. ☐

In the language of Definition 5A-2, the image of a set is the set of images of its elements (see Fig. 5.2).

We shall now justify the claim, made at the end of section 5A, that the image of a straight line is in general another straight line. In the language of linear algebra, a straight line through the origin is a one-dimensional subspace. We shall prove that linear maps have the beautiful property of mapping subspaces to subspaces. It follows that if $f: \mathbb{R}^3 \to \mathbb{R}^3$ is linear, and L is a straight line through O, then $f(L)$ is either another line or a subspace of some other dimension. A linear map can never map a straight line into a curve.

Theorem 1 (image of a subspace). Let $f: V \to W$ be a linear mapping. If S is a subspace of V, then $f(S)$ is a subspace of W.

Proof. We use the subspace criterion of section 3D.

The set $f(S)$ is nonempty because it contains 0 (proof: $0 = f(0)$ and 0 belongs to the subspace S). We must show that for all x and y in $f(S)$, $x + y$ and kx belong to $f(S)$ for any scalar k. This is not difficult, but it needs care.

Since $f(S) = \{f(v): v \in S\}$, every member of $f(S)$ must be one of the $f(v)$'s; hence, if $x \in f(S)$, then $x = f(s)$ for some vector s in S. Similarly, if $y \in f(S)$, then there is a t in S such that $y = f(t)$.

Hence

$$x + y = f(s) + f(t)$$

$$= f(s + t) \quad \text{using the linearity of } f$$

$$= f(v)$$

where $v = s + t$. Now, $v \in S$ because S is a subspace of V. Hence $x + y = f($a member of S), so $x + y \in f(S)$. Thus the first condition of the subspace criterion holds.

Now consider kx where k is any scalar (remember that V and W have the same field of scalars according to Definition 5A-3). We have

$$kx = kf(s) = f(ks) \quad \text{using the linearity of } f$$
$$= f(u)$$

where $u = ks \in S$ because S is a subspace. Hence $kx = f(\text{a member of } S)$, so $kx \in f(S)$, showing that the second condition of the subspace criterion is satisfied. Hence $f(S)$ is a subspace of W. □

Examples 1. (a) Consider Example 5A-1(b). Let S be the y-axis in $\mathbb{R}^2$; then $g(S) = \{g(0, y)\} = \{-3y : y \in \mathbb{R}\} = \mathbb{R}$. Here the image of S is the whole of the codomain, which is a subspace (though a rather special one – see Example 3D-4).

(b) Consider Example 5A-1(a), and again let S be the y-axis. Then $f(S)$ consists of the positive numbers and zero. This is not a subspace of $\mathbb{R}$. Here the map f is not linear; this example shows that linearity is essential in Theorem 1. □

The main characteristic of a subspace S is its dimension, $\dim(S)$. If f is a linear map, then $f(S)$ is a subspace, and it is natural to ask how $\dim(f(S))$ is related to $\dim(S)$. In Example 1(a) the image has the same dimension as the original subspace. Can this be a general rule?

No. Consider the zero mapping O, defined by $O(x) = 0$ for all x; it is linear (easily verified), and it maps every subspace into the zero-dimensional subspace $\{0\}$. Thus linear maps can reduce dimension. Can a linear mapping increase dimension? The following argument suggests not. Each point of $\text{im}(S)$ is the image of a point of S, so $\text{im}(S)$ cannot contain more points than S. Since $\text{im}(S)$ cannot be larger than S, it cannot have a higher dimension.

The underlying idea here is reasonable, but the argument is unsound. It says that there cannot be more points in $\text{im}(S)$ than in S. But a nontrivial subspace contains infinitely many points, and comparing sizes of infinite sets is a subtle business, fraught with contradictions (see Appendix C). A proof along different lines is needed.

Theorem 2 (dimension of image). If S is a finite-dimensional subspace of the domain of a linear transformation f, then

$$\dim(f(S)) \leq \dim(S)$$

Proof. Suppose S is k-dimensional, with a basis $\{b_1, \ldots, b_k\}$. We shall find a spanning set of k vectors for $f(S)$.

Take any x in $f(S)$; then $x = f(u)$ for some u in S. Let c_r be the components of u with respect to the basis $\{b_1, \ldots, b_k\}$, then

$$x = f(u) = f(c_1 b_1 + \ldots + c_k b_k)$$
$$= c_1 f(b_1) + \ldots + c_k f(b_k)$$

because f is linear. Thus every element of $f(S)$ is a linear combination of $\{f(b_1), \ldots, f(b_k)\}$; this set of k vectors spans $f(S)$. A basis can be obtained by discarding from this set any vectors which are linearly dependent on the others; hence $\dim(f(S)) \leq k$. □

A particularly important case arises when S is the whole of the space.

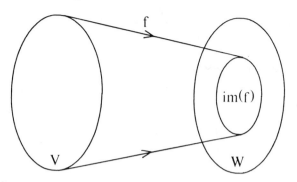

Fig. 5.3 – The image of a map.

Definition 2. If $f: V \to W$ is a linear mapping, the image of V under f is called the **image of the mapping**, and denoted by im (f). Thus im $(f) = f(V) = \{f(v): v \in V\}$ (see Fig. 5.3).

Corollary 1. If $f: V \to W$ is linear, then im (f) is a subspace of W.

Proof. Take $S = V$ in Theorem 1. ☐

Notice that there are now three types of image: the image of an *element*, the image of a *subspace*, and the image of a *mapping*. It should always be clear from the context which of the three is intended (provided that you read the text carefully).

Exercise 1. Consider the maps a, a' of Example 5A-3. Use the column lemma to show that im(a) is the column space of A, and im(a') is its row space. It follows that dim (im (a)) is the column rank of A, and dim (im (a')) is the row rank of A.
  Show also that A has full column rank if and only if im (a) equals the codomain of a. ☐

This exercise suggests how the notion of rank can be extended to any linear map.

Definition 3. The **rank** of a linear transformation is the dimension of its image. If the image is infinite-dimensional, we say the transformation has infinite rank. ☐

Theorem 2 immediately implies the following.

Theorem 3. If f is a linear transformation with a finite-dimensional domain (denoted by dom (f)), then

$$\text{rank} (f) \le \dim (\text{dom} (f))$$

Proof. In Theorem 2, take S to be the whole of the domain. ☐

Examples 2. (a) In Example 5A-3, rank (a) and rank (a') are the column and row ranks of A respectively (see Exercise 1).

(b) Let P_n be the space of all complex polynomials of degree $\leq n$. Define a map $D: P_3 \to P_3$ by $Dp = p'$, the derivative of p. Since p' is at most quadratic, im (D) is a subset of P_2. It is plausible that im (D) is the whole of P_2; to prove it, we must show that (i) every element of im (D) belongs to P_2, and (ii) every element of P_2 belongs to im (D). This is easy: (i) is obvious, and for any a, b, c, we have $ax^2 + bx + c = D(ax^3/3 + bx^2/2 + cx)$ which belongs to im (D), proving (ii). Hence im $(D) = P_2$, and rank $(D) = 3$.

(c) The identity transformation $I: V \to V$, which maps every vector into itself, has rank equal to dim (V), since im $(I) = V$. □

The rank of a map measures how well-behaved it is, in the following sense. We regard a map $f: V \to W$ as well-behaved if it gives a one-to-one correspondence between V and W. In other words, for every y in W there is exactly one x in V such that $f(x) = y$. The equation has a unique solution x for every given y; this is a satisfactory state of affairs. Now, if for each y in W there is a corresponding x, then im $(f) = W$, so rank $(f) = \dim(W)$ which is the largest possible value that the rank can have. If rank $(f) < \dim(W)$, then im (f) is a proper subspace of W, and there are vectors y in W for which the equation $f(x) = y$ has no solution. Thus the smaller the rank, the worse behaved the mapping is.

Now, there are two numbers associated with every linear map: its rank, and another number called its nullity. They tend to go together, like sausages and mash*. In the next section we shall define the nullity and discuss its relation to the rank. These ideas will then be applied to linear equations; they will enable us to improve considerably the results obtained for linear equations in Chapter 4.

5C THE NULL SPACE OF A LINEAR TRANSFORMATION

We have seen that the image of a linear map f is a subspace of its codomain, and dim (im (f)) is the rank of f. For matrix transformations the rank is closely related to the theory of systems of linear equations, as we saw in Chapter 4.

There is another subspace related to linear equations. Consider the set of all solutions of the equation $Ax = b$, where A is a given matrix. If $b \neq 0$, this set is not a vector space; it does not include the zero vector. But if $b = 0$ we have the set of all solutions of $Ax = 0$, that is, the set of all vectors which are mapped to zero. This, as we shall see, is a vector space.

Definition 1. The **kernel** or **null space** of a linear mapping $f: V \to W$ is the set of all vectors in V which are mapped to zero by f. It is written ker (f); thus ker $(f) = \{v \in V: f(v) = 0\}$ (see Fig. 5.4).

Exercise 1. Find the kernel of the map $f: \mathbb{R}^2 \to \mathbb{R}^2$ defined by $f(x, y) = (3x - 2y, 6x - 4y)$. □

In this example the kernel is a line, a subspace of $\mathbb{R}^2$. This is no accident.

---

* There is an algebraic construction called a smash product – but that is a red herring.

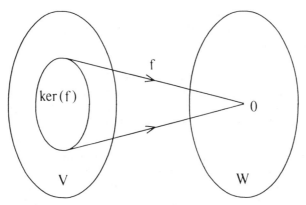

Fig. 5.4 – The kernel of a map.

Theorem 1. The kernel of a linear mapping is a subspace of the domain.

Exercise 2. Prove this theorem, using the subspace criterion.   □

Now that we know that ker (f) is a vector space, it is natural to think about its dimension.

Definition 2. The dimension of the kernel of a linear map is called the **nullity** of the map. A **singular** map is a map with positive nullity; a **nonsingular** map is one with nullity zero.   □

The nullity of a map describes the set of vectors which are annihilated (mapped to zero) by the map. If the nullity of f is zero, then the set of solutions of $f(x)=0$ is zero-dimensional, consisting of the single vector 0. So a nonsingular mapping f is one for which $f(x)=0$ has a unique solution. If f is singular, then $f(x)=0$ has infinitely many solutions.

We regard a mapping as well-behaved if it gives a one-to-one correspondence between the domain and the codomain. A singular map certainly is not one-to-one, since by definition there are many vectors in the domain which are mapped to 0. The nullity measures the degree of badness; the larger the nullity, the larger the set of vectors which are mapped to zero. We shall develop this idea later. Meanwhile, we consider some examples.

Examples 1. (a) Consider the map g of Example 5A-1(b). Its kernel is $\{(x, y): 2x - 3y = 0\}$, a line through the origin in $\mathbb{R}^2$. This is indeed a subspace of $\mathbb{R}^2$, as predicted by Theorem 2, and it is one-dimensional. Hence nullity $(g) = 1$, and g is singular.

(b) Consider Example 5B-2(b). The null space of D consists of polynomials with zero derivative, that is, constant polynomials. This is a one-dimensional space, spanned by $\{1\}$, so nullity $(D) = 1$.

(c) The zero mapping $O: V \to W$ is defined by $O(v) = 0$ for all v in V. The kernel of O is V, so if V is n-dimensional, then nullity $(O) = n$. This mapping is as singular as it can possibly be.

(d) Consider a set of linear equations $Ax = 0$ where A is any matrix. Define a mapping a by $a(x) = Ax$ as usual, then $AX = 0$ becomes the equation $a(x) = 0$. Thus $\ker(a)$ is the solution set of $Ax = 0$.

Example 2. The lens of your eye forms an image on the back surface of the eyeball, called the retina. We can regard the retina as a two-dimensional space, with an origin (zero vector) somewhere in the middle. We ignore details like the curvature of the retina and the fact that one's field of vision is limited.

Then for each point in three-space there is a corresponding image point on the retina, which is idealised as a plane. Thus we have a map $f: \mathbb{R}^3 \to \mathbb{R}^2$. Straight lines in space are perceived as straight lines; this suggests that the mapping f is linear.

What is its kernel? It is the set of points in three-space with images at the zero point of the retina. All points directly in front of you give images on the centre of the retina, so the kernel is the set of all such points, a straight line pointing directly away from you. This is a one-dimensional subspace of $\mathbb{R}^3$, hence the nullity of f is one, and it is singular. The regrettable consequences of this will be mentioned in section 5D.

Of course, this account is oversimplified. For more information, see, for example, Frisby (1980) and Gregory (1970, 1977).   □

The kernel and the image of a mapping f are subspaces of different vector spaces, in general. But there is a remarkable relation between their dimensions: they add up to a number which does not depend on f; it is simply the dimension of the domain. This is the first substantial theorem of the chapter; all the results so far have been more or less simple consequences of the definitions. Theorem 2 is the key to the general theory of linear equations.

Theorem 2 (rank plus nullity theorem). If $f: V \to W$ is a linear map, and V is finite-dimensional, then

$$\text{rank}(f) + \text{nullity}(f) = \dim(\text{domain of } f)$$ □

The proof is rather long; it will be found in section 5L.

We saw at the end of the last section that the larger the rank, the better-behaved the mapping, and we have seen above that the smaller the nullity, the better-behaved the mapping. Theorem 2 fits in with these ideas; if rank plus nullity is constant, then large rank goes with small nullity, both being associated with well-behaved mappings.

Application to matrices

A p by q matrix A, with entries in the field $\mathbb{K}$, gives a linear map $a: \mathbb{K}^q \to \mathbb{K}^p$ defined by $a(x) = Ax$ for all x in $\mathbb{K}^q$. The rank of the matrix equals the rank of the corresponding map, as we saw in Example 5B-2. We can extend other terms defined for linear maps to matrices in a similar way.

Definition 3. The **image**, **kernel** or **null space**, and **nullity** of a matrix A are the image, kernel, and nullity of the linear operator a defined by $a(x) = Ax$. A matrix is **singular** if its nullity is positive.   □

Thus the nullity of a matrix A is the dimension of the space of vectors x satisfying $Ax = 0$. As explained above, the smaller the nullity, the better-behaved the matrix.

Another way of assessing the behaviour of a matrix is to ask if it has an inverse. We showed in section 4I that a p by q matrix has a left inverse if and only if it has rank q. Since the domain of the corresponding map is $\mathbb{K}^q$, it follows from Theorem 2 that A is left invertible if and only if nullity$(A) = 0$. Thus positive nullity means no inverse, and is a Bad Thing.

The basic theory of linear mappings is now complete. In the next section we shall apply it to linear equations.

5D   LINEAR EQUATIONS

Chapter 4 discussed the system of linear equations $Ax = b$. It can be written as $f(x) = b$, where f is the linear map corresponding to the matrix A. We now generalise to linear maps on any vector space; in this way differential equations and many other kinds of equation are brought into the framework of the theory of linear mappings.

It is customary to use the word 'operator' for a mapping in this context.

Linear operator equations are equations of the form $f(x) = c$, where $f: V \to W$ is a linear operator, c is a given element of W, and x is unknown. The **solution set** of an equation is the set of all x which satisfy it. $\square$

In earlier sections we have thought of the formula $f(x) = c$ as mapping a given vector x into a new vector c. Here we are thinking of it differently: we regard c as given and try to find x. The formula is the same, but this difference in attitude is crucial*.

Examples 1. (a) Define $f: \mathbb{K}^q \to \mathbb{K}^p$ by $f(u) = Au$, where A is a p by q matrix with entries in $\mathbb{K}$ ($= \mathbb{R}$ or $\mathbb{C}$). The operator equation $f(u) = c$, where c is a given p-vector, is the familiar system of p linear equations for the q unknown entries of u.

(b) Let S be the space of smooth real functions (see Appendix D). Define $L: S \to S$ by $Lf = f'' + 2f' - f$. This is a linear operator, a special case of Example 5A-2(b). Let g be a given smooth function; then the linear operator equation $Lf = g$ is the equation $f'' + 2f' - f = g$. Equations of this type, involving derivatives of an unknown function, are called **differential equations**; they are fundamental to much of science and engineering (see, for example, section 5G).

(c) With S as in (b), define $M: S \to S$ by $Mf = f'' + ff'$. It is not linear, because $M(2f) = 2f'' + 4ff' \neq 2Mf$. (If you are familiar with differential equations, you will see that linear differential equations correspond to linear operators, while nonlinear equations do not.)

(d) Let V be the space of infinite sequences of complex numbers (Example 3H-2). Write the members of V as $x = (x_r) = (x_1, x_2, \ldots)$. Let $T: V \to V$ be the operator which maps (x_r) into the sequence whose rth term is $x_{r+2} - 3x_{r+1} + x_r$. It is easy to verify that T is linear.

* Contrary to popular belief, mathematics is not primarily about formulas, it is about ideas.

The equation $T(x) = y$ is equivalent to $x_{r+2} - 3x_{r+1} + x_r = y_r$ for $r = 1, 2, \ldots$, where the y's are given and the x's are unknown. This is called a difference equation or recurrence relation; they occur in many parts of applied and computational mathematics.

(e) Let $C[0, \pi]$ be the space of continuous functions on the interval $[0, \pi]$. Let P be the operator on $C[0, \pi]$ which maps any function f into the function g defined by $g(t) = \int_0^\pi \sin(st) f(s) \, ds$. The operator equation $P(u) = f$, where f is a given function and u an unknown function, reads $\int_0^\pi \sin(xy) u(x) \, dx = f(y)$. Integral equations of this kind appear in Fourier analysis and many branches of mathematical physics. ☐

The general linear operator equation includes many different kinds of equation, occurring in many different branches of mathematics. We can make some general statements which apply to them all.

Theorem 1 (the linear operator equation theorem). Consider the equation $f(x) = c$ where f is a linear operator.

1. The equation has a solution if and only if $c \in \text{im}(f)$.
2. If a solution exists, then:
(a) if f is nonsingular, then there is exactly one solution;
(b) if f is singular, then there are infinitely many solutions; if X is any one solution, then the solution set is $\{X + k : k \in \ker(f)\}$.

Proof.

1. This follows immediately from the definition of $\text{im}(f)$.
2. Let X be a solution, and set $S = \{X + k : k \in \ker(f)\}$. We shall show that S is the solution set of the equation.

If x satisfies $f(x) = c$ then $f(x - X) = f(x) - f(X) = c - c = 0$, hence $x - X \in \ker(f)$. Thus $x = X + h$ where $h = x - X \in \ker(f)$, showing that $x \in S$.

Conversely, if $s \in S$ then $s = X + k$ for some $k \in \ker(f)$. Hence $f(s) = f(X + k) = f(X) + f(k) = c + 0 = c$. This completes the proof that S is the solution set.

(a) If f is nonsingular, then its nullity is zero and $\ker(f)$ is the zero-dimensional space $\{0\}$. Hence the solution set $\{X + k : k \in \ker(f)\}$ consists of the single vector X.

(b) We have shown that the solution set is $\{X + k : k \in \ker(f)\}$. If f is singular, then $\ker(f)$ has dimension ≥ 1 and contains infinitely many vectors. ☐

This theorem implies that a linear equation cannot have exactly 2 or 3 solutions: the number of solutions is either 0, 1, or infinity. In the case 2(b), where there are infinitely many solutions, the solution set $\{X + k : k \in \ker(f)\}$ seems to depend upon which solution X is chosen. But if X and Y are any two solutions, then every vector in the set $\{Y + k\}$ belongs to the set $\{X + k\}$, and vice versa. (Proof: $Y + k = X + k'$ where $k' = k + Y - X$; we have $f(k') = f(k) + f(Y) - f(X) = f(k) + c - c = f(k)$, hence $k \in \ker(f)$ if and only if $k' \in \ker(f)$.) So the choice of the particular solution X makes no difference.

If you are acquainted with differential equations, clause 2(b) of the theorem may remind you of the rule 'General Solution = Particular Integral + Complementary

Function' for linear differential equations. Example 1(b) above shows that such equations do indeed fit into the framework of linear operator equations.

Example 2. Inside our eye, there is a two-dimensional picture formed by the lens. Our brain interprets this picture as the image of an object in three dimensions. In the notation of Example 5C-2, the brain solves the equations $f(x) = y$, where y is the image on the retina of a point x in three-dimensional space.

We saw in Example 5C-2 that f is singular. Hence the equation $f(x) = y$ does not have a unique solution; there are many solutions, differing from each other by members of $\ker(f)$. This means that there are many three-dimensional structures consistent with a given retinal image. Only one of these is the true one; there are infinitely many false interpretations of what our eyes see.

Usually our brain does an excellent job of interpreting the world three-dimensionally. But it can go wrong; there are visual illusions where the brain insists on choosing an incorrect member of the solution set of $f(x) = y$. For example, concave surfaces are sometimes perceived as convex, and vice versa. Visual illusions make amusing games; but they are a serious matter in situations where clear vision is essential (for example, for drivers or aircraft pilots).

For more information, and illustrations of visual illusions, see Gregory (1970, 1977).

Example 3. If you have read section 3J and Example 5A-4, you will see that an isomorphism is a linear map $f: V \to W$ such that $\mathrm{im}(f) = W$ and f is nonsingular. These conditions ensure that for each w in W there is just one v in V such that $f(v) = w$.

Linear systems of algebraic equations

Let A be a p by q matrix with entries in $\mathbb{K}$ ($= \mathbb{R}$ or $\mathbb{C}$). Then $Ax = y$ is a set of p equations in q unknowns. We write it in the form $a(x) = y$, where the linear operator a is defined by $a(x) = Ax$ for all x in $\mathbb{K}^q$. We shall use Theorem 1 to give a systematic account of linear equations which includes and extends the results of Chapter 4. There are three cases.

1. If $p > q$, there are more equations than unknowns. Since $\mathrm{rank}(A) \leq q$, we have $\mathrm{rank}(A) < p$. Hence $\mathrm{im}(A)$ is a proper subspace of $\mathbb{K}^p$, and there are vectors y in $\mathbb{K}^p$ for which $Ax = y$ has no solution.

If $\mathrm{rank}(A) < q$, then $\mathrm{nullity}(A) > 0$ by the rank plus nullity theorem, so a is singular and there are either no solutions or infinitely many. If $\mathrm{rank}(A) = q$, then there is either no solution or a unique solution.

2. If $p < q$, there are fewer equations than unknowns. Since $\mathrm{rank}(A) \leq p$, we have $\mathrm{rank}(A) < q$ here. By the rank plus nullity theorem, $\mathrm{nullity}(A) > 0$, so a is singular. It follows from Theorem 1 that if a solution exists, then there are infinitely many solutions. If $\mathrm{rank}(A) = p$, its maximum value, then $\mathrm{im}(A) = \mathbb{K}^p$ and there are solutions for all y; if $\mathrm{rank}(A) < p$, then for some y there are no solutions.

3. If $p = q$, then again there are two possibilities. If $\mathrm{rank}(A) = p$, then $\mathrm{im}(a)$ is the whole of $\mathbb{K}^p$ so there is always a solution. And $\mathrm{rank}(A) = p$ implies $\mathrm{nullity}(a) = 0$, hence the solution is unique.

If $\mathrm{rank}(A) < p$, then $\mathrm{im}(a)$ is a proper subspace of $\mathbb{K}^p$, so for some y there are solutions and for other y there are none. Also, if rank $(A) < p$, then nullity $(A) > 0$; hence if there is a solution, then there are infinitely many by clause 2(b) of Theorem 1.

Summing up the case $p = q$; if A has full rank, then the equations have a unique solution; if rank $(A) < p$, then there may be infinitely many solutions or no solution, depending on y.

Exercise 1. Fit examples 4C-1 into the above classification.

Remark. If y does not belong to im (A), the equations have no solution. Yet there is a sense in which they can be solved; see Chapter 13. □

This completes our discussion of algebraic equations. The next section returns to general linear operator equations, from a different point of view.

5E INVERSES

This section looks at the material of section 5D in a different way, focusing on the mapping f rather than the linear equation $f(x) = c$. We have seen that for some mappings f, the equation has a unique solution for each c. This means that given a vector x, one can determine a vector v such that $f(v) = x$. In other words, one can 'undo' the effect of the mapping f. This operation is called inverting f; if it can be done, then f is called 'invertible'.

Definition 1. A map $f : V \to W$ is **invertible** if for each vector w in W there is one and only one v in V such that $f(v) = w$.

Examples 1. (a) Let P_n be the space of all polynomials of degree up to n. Let $D : P_3 \to P_2$ be the differentiation operator. Then D is not invertible, because for each w in P_2 there are many polynomials v in P_3 such that $w = Dv$; indeed, if $Dv = w$ then $D(v + C) = w$ for any constant C.

(b) Now let V be the subspace of P_3 consisting of all polynomials with zero constant term. Define a map $D_1 : V \to P_2$ by the rule $D_1 v = v'$. This is the same rule as for D, but the map is given a different name because the domain is different. Now, for each w in P_2 there is exactly one v in V such that $w = D_1 v$: if $w(x) = ax^2 + bx + c$ then the corresponding v in V is given by $v(x) = ax^3/3 + bx^2/2 + cx$. Hence D_1 is invertible. □

In Example 1(b), the mapping back from $ax^2 + bx + c$ to $ax^3/3 + bx^2/2 + cx$ is called the inverse of D_1. It corresponds to integrating, whereas D_1 corresponds to differentiating. In general, if Definition 1 is satisfied, then there is a rule which, given any w in W, specifies a unique v in V satisfying $f(v) = w$. This rule specifies a map $W \to V$, called the inverse of f.

Definition 2. If $f : V \to W$ is invertible, its **inverse** is the transformation $W \to V$ which maps each w in W to the unique v in V such that $f(v) = w$. It is denoted by f^{-1}. □

We have already met the notion of inverse, for matrices. There is, of course, a close connection between inverses of matrices and inverses of mappings.

Exercise 1. A real p by p matrix A generates a linear map $a: \mathbb{R}^p \to \mathbb{R}^p$ by the rule $a(x) = Ax$, where the vector x is interpreted as a column matrix. Show that the map a is invertible if the matrix A is invertible.   □

Some mappings are invertible, some are not. It is natural to ask how one can tell whether a map is invertible or not. One answer is that singular matrices are not invertible (where singular means having nullity greater than zero, see Definition 5C-2). If v_1 satisfies $f(v) = w$, then so does $v_1 + k$ for any k in ker (f). If ker (f) is nontrivial, it follows that there are many vectors v satisfying $f(v) = w$. Hence f is not invertible if it is singular. The following theorem extends this fact to give precise conditions for invertibility.

Theorem 1. Let f be a linear map $U \to V$.
   (a) f is invertible if and only if im $(f) = V$ and nullity $(f) = 0$.
   (b) If U is finite-dimensional, then f is invertible if and only if dim $(V) = $ dim (U) and nullity $(f) = 0$.

Proof. (a) Suppose f is invertible. Then im $(f) = V$, for otherwise there would be vectors in v with no corresponding u. Since the equation $f(u) = v$ has a unique solution, nullity $(f) = 0$ by 2(b) of the linear operator equation theorem (section 5D).
   Now suppose that im $(f) = V$ and nullity $(f) = 0$. Since im $(f) = V$, for each v in V there is a u in U with $f(u) = v$. Since nullity $(f) = 0$, there is only one such u by 2(a) of the linear operator equation theorem.
   (b) Suppose f is invertible. Then nullity $(f) = 0$ and im $(f) = V$ by (a); hence rank $(f) = $ dim (V). But rank $(f) + $ nullity $(f) = $ dim (U) (Theorem 5C-2), hence dim $(V) = $ dim (U).
   Now suppose that dim $(U) = $ dim (V) and nullity $(f) = 0$. By Theorem 5C-2 again, rank $(f) = $ dim $(U) = $ dim (V). Hence dim (im $(f)) = $ dim (V). But V is finite-dimensional because it has the same dimension as U, hence a subspace of V cannot have the same dimension as V unless it equals V. Therefore im $(f) = V$. We can now deduce from (a) that f is invertible.   □

Corollary 1. The map $\mathbb{K}^p \to \mathbb{K}^p$ corresponding to a p by p matrix (where $\mathbb{K} = \mathbb{R}$ or $\mathbb{C}$) is invertible if and only if the matrix is.

Proof. This follows from Theorem 1(b), since a square matrix is invertible if and only if its nullity is 0.   □

Warning 1. The finite-dimensional restriction in Theorem 1(b) is essential. The following example shows that if U and V are infinite-dimensional, then $f: U \to V$ can be non-invertible even when nullity $(f) = 0$. This is one respect in which infinite-dimensional and finite-dimensional spaces behave really differently.

Example 2. Let S be the space of all polynomials. Define $F: S \to S$ by $F(p) = q$ where

$q(t) = tp(t)$ for all polynomials p. If $F(p) = 0$ then $tp(t) = 0$ for all t; therefore p is the zero polynomial, hence $\ker(F) = \{0\}$. Thus nullity$(F) = 0$. But F is not invertible: let q be the constant polynomial defined by $q(t) = 1$ for all t; then there is no p in S such that $F(p) = q$. Thus dim (S) and dim (T) are equal (they are both infinite), and F has nullity zero, but it is not invertible.

Warning 2. Theorem 1(a) says that f is invertible if and only if it is nonsingular and its image is its codomain. Now, the term 'nonsingular' is often applied to matrices with the extra restriction that the matrix is square. In this usage, a matrix is invertible if and only if it is nonsingular (which implies that it is square). But for mappings, nonsingular does not imply invertible.

Mappings for which the image equals the codomain are sometimes called **surjective**. Then f is invertible if and only if it is surjective and nonsingular.   □

This completes our discussion of rank, nullity and related matters. The next section introduces a new idea, which in one form or another will occupy a good deal of the rest of this book.

5F   EIGENVECTORS

Much of linear algebra is based, directly or indirectly, on a single idea: solving sets of linear algebraic equations. There is a network of related ideas: elementary operations, linear dependence, basis and dimension, kernel and image, rank and nullity – they all develop naturally out of solving linear equations. This section introduces a new and equally important group of ideas.

Example 1. Define $f: \mathbb{R}^2 \to \mathbb{R}^2$ by $f(x, y) = (3x - y, 3y - x)$. This linear map corresponds in the usual way to the matrix

$$\begin{pmatrix} 3 & -1 \\ -1 & 3 \end{pmatrix}$$

Look at it geometrically as an operator which moves each point (x, y) to a new point $f(x, y)$. All linear operators map subspaces to subspaces, as shown in section 5B, so f maps each line through O into a subspace of $\mathbb{R}^2$. We shall use these subspaces to build up a picture of the mapping.

Every line through O has the form $\{(at, bt): t \in \mathbb{R}\}$ for some numbers a and b. If $a = 0$ the line is vertical, otherwise its slope is b/a. The image of the point (at, bt) under f is

$$f(at, bt) = (3at - bt, 3bt - at)$$

As t varies, this traces out a line through O with slope $(3b - a)/(3a - b)$. Writing $b/a = m$ for the slope of the original line, and m' for the slope of its image under f, we have

$$m' = \frac{3m - 1}{3 - m}$$

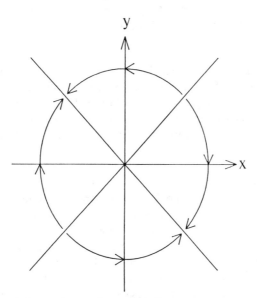

Fig. 5.5 – How lines through O transform under a certain linear map.

Plugging numbers into this formula shows that if $m > 1$ or $m < -1$ then $m' > m$, while if $-1 < m < 1$ then $m' < m$. If $m = \pm 1$ then $m' = m$. Thus the lines of slope ± 1 map into themselves, while all other lines are rotated as shown in Fig. 5.5.

Although the lines of slope ± 1 map into themselves, individual points do not. Since $f(x, x) = (2x, 2x)$, points on the line of slope 1 are mapped to points twice as far from the origin on the same line. Since $f(x, -x) = (4x, -4x)$, points on the line of slope -1 are mapped to points four times as far from the origin. The transformation can be described as stretching by a factor of 2 along the line $y = x$, combined with stretching by a factor of 4 along the line $y = -x$, and rotations of points lying on other lines.

These two stretchings are the essential features of this map, as we shall now show. Let p and q be vectors along the lines of slope 1 and -1 respectively; then $\{p, q\}$ is a basis for $\mathbb{R}^2$, and $f(p) = 2p$, $f(q) = 4q$. Let c and d be the components of a vector u with respect to this basis. Then $f(u) = f(cp + dp) = cf(p) + df(q) = 2cp + 4dq$. Thus the image of every vector in the plane can be deduced from the information that f stretches by 2 and by 4 in the p and q directions respectively.   □

The useful idea of directions which remain unchanged by a mapping can be applied to any linear transformation of a vector space into itself. But if f maps V to a different space W, then v and $f(v)$ are in different spaces and it makes no sense to ask if $f(v)$ is in the same direction as v. The idea applies only to mappings of the type $V \to V$.

Definition 1. Let $f : V \to V$ be a linear transformation. If

$$f(v) = \lambda v \quad \text{and} \quad v \neq 0$$

where λ is a scalar, then v is called an **eigenvector** of f, and λ is the corresponding

eigenvalue. The eigenvalue λ and the eigenvector v are said to **belong** or **correspond** to each other.   ☐

Eigenvectors of f are vectors on which f acts in a particularly simple way: it multiplies them by the number λ. If v is an eigenvector then $f(v)$ is parallel to v. An eigenvalue is the factor by which an eigenvector is multiplied under the action of f.

We exclude the zero vector because of this interpretation in terms of direction – the zero vector has no direction. Another view of the condition $v \neq 0$ will emerge from Theorem 1 below.

Examples 2. (a) In Example 1, $(1, 1)$ is an eigenvector with eigenvalue 2. In fact (a, a) is an eigenvector with eigenvalue 2 for any nonzero number a. Also $(a, -a)$ is an eigenvector with eigenvalue 4 for any $a \neq 0$. The analysis in Example 1 shows that these are the only eigenvectors and eigenvalues of f.

(b) Define $g: \mathbb{R}^3 \to \mathbb{R}^3$ by $g(x, y, z) = (x - y, x + y, 0)$. We cannot analyse this three-dimensional map by the method of Example 1. But it is obvious that $g(0, 0, z) = (0, 0, 0)$ for all z, hence $g(0, 0, z) = 0(0, 0, z)$. This is the eigenvalue equation with $\lambda = 0$; we see that 0 is an eigenvalue with eigenvector $(0, 0, z)$ for any $z \neq 0$ (there is no reason why eigen*values* should not be zero; it is eigen*vectors* which must be nonzero).

(c) Define $h: \mathbb{R}^3 \to \mathbb{R}^3$ by $h(x, y, z) = (4x, 4y, 0)$. We have $h(0, 0, z) = (0, 0, 0) = 0(0, 0, z)$, so 0 is an eigenvalue with eigenvectors $(0, 0, z)$ for any $z \neq 0$. Also $h(x, y, 0) = (4x, 4y, 0) = 4(x, y, 0)$, so 4 is another eigenvalue, with eigenvectors $(x, y, 0)$ for any x, y.

Exercise 1. If zero is an eigenvalue of f, then f is not invertible.   ☐

Definition 1 can be interpreted algebraically as well as geometrically. Regard $f(v) = \lambda v$ as an equation for v in which λ can take various values. The equation is always satisfied by $v = 0$. For some values of λ this is the only solution. But there may be certain values of λ for which there are nonzero solutions; these values are the eigenvalues of the operator f, and the corresponding nonzero solutions are the eigenvectors. This point of view can be formulated as follows.

Notation. For any space V, write i for the identity map $V \to V$ defined by $i(v) = v$ for all $v \in V$.

Theorem 1. A scalar λ is an eigenvalue of a linear map $f: V \to V$ if and only if the map $f - \lambda i$ is singular. The nonzero members of $\ker(f - \lambda i)$ are the eigenvectors belonging to eigenvalue λ.

Proof. By definition, λ is an eigenvalue of f if and only if there is a $v \neq 0$ such that $f(v) = \lambda v$, which is equivalent to $f(v) - \lambda i(v) = 0$, or

$$(f - \lambda i)v = 0$$

This is precisely the condition for $f - \lambda i$ to be singular. The eigenvectors v are the nonzero vectors mapped to 0 by $f - \lambda i$.   ☐

It follows that the set of eigenvectors corresponding to any one eigenvalue is almost a subspace (not quite, because it does not include 0).

Corollary (eigenspace). If λ is an eigenvalue of a linear operator $f: V \to V$, then the set of all eigenvectors belonging to λ, together with the zero vector, forms a subspace of V. It is called the **eigenspace** belonging to λ.

Proof. Theorem 1 shows that the eigenvectors are the nonzero elements of the kernel of $f - \lambda i$. Hence the subspace $\ker(f - \lambda i)$ consists of the eigenvectors together with the zero vector. $\qquad\qquad\square$

In Example 1, there are two eigenspaces, the lines with slope ± 1. In Example 2(c), $\{(0,0,z)\}$ is the eigenspace belonging to eigenvalue 0, and $\{(x,y,0)\}$ is the eigenspace for eigenvalue 4. The eigenspace $\{(x,y,0)\}$ is two-dimensional; there are two linearly independent eigenvectors belonging to eigenvalue 4. We say that 4 is a 'double eigenvalue'.

Definition 2. The **geometric multiplicity** of an eigenvalue is the dimension of its eigenspace. A **simple** or **nondegenerate** eigenvalue is one with multiplicity 1; a **double** eigenvalue is one with multiplicity 2, and so on. $\qquad\qquad\square$

The term 'geometric multiplicity' is used to distinguish this from a different kind of multiplicity which we shall meet later, in Chapter 7.

An eigenspace for an operator f has the property that for every vector s in S we have $f(s) \in S$. Thus f maps points in S into points in S; we say that S is invariant under f.

Definition 3. An **invariant subspace** for a linear mapping $f: V \to V$ is a subspace S of V such that $f(s) \in S$ for all $s \in S$. We say that S is invariant **under** f. $\qquad\square$

Every eigenspace of an operator is an invariant subspace; this follows at once from the definitions. But not every invariant subspace is an eigenspace. For example, the whole space V is always an invariant subspace; the definition is obviously satisfied in this case. But it is not generally an eigenspace.

Example 3. In Example 2(b), the eigenspace $\{(0,0,z)\}$ is invariant under g. The subspace $\{(x,y,0)\}$ is also invariant, since $g(x,y,0) = (x',y',0)$ where $x' = x - y$, $y' = x + y$. But $\{(x,y,0)\}$ is not an eigenspace; the vector $(1,1,0)$, for example, is not an eigenvector since $g(1,1,0) = (0,2,0)$ which is not proportional to $(1,1,0)$. $\qquad\square$

We have introduced eigenvectors in purely mathematical terms, as vectors on which a linear mapping acts particularly simply. But they play a central role in physics, as the next section shows.

*5G   EIGENVALUES AND VIBRATING SYSTEMS

This section outlines how linear algebra is used in mechanics. Even if you are not interested in mechanics, please read the next two paragraphs, because they give a new perspective on eigenvalues.

Most mechanical and electrical systems vibrate. The rate of vibration is described by a number called the frequency. Most systems can vibrate at several different frequencies; they are the eigenvalues of a certain linear operator. Musical instruments are based on these natural vibrations; when you listen to music, you experience eigenvalues directly, through your ears.

There is a physical interpretation for eigenvectors too. A complicated mechanical system may move in a complicated way. But there are special types of motion which are quite simple to describe and analyse. Physically speaking they are vibrations, and mathematically speaking they are eigenvectors of a certain linear operator.

The rest of the section sets out these ideas in detail. If you are not interested in the workings of the physical world in which you live, then skip to the next section.

Vibrations in mechanical systems

A mechanical system can be described by a set of coordinates which specify the state of the system. They may or may not be the Cartesian coordinates of points of the system, and there may be a finite or an infinite number of them.

Consider first the case with finitely many coordinates, for example, a machine with a finite number of parts, each described by a few numbers giving its location and orientation. Let $x_1, \ldots, x_n$ be the coordinates of the system, and write x for the n-vector with entries $x_1, \ldots, x_n$. The state of the system is thus described by an element of $\mathbb{R}^n$. As the system moves, its coordinates change; they are functions of time and can be written $x_r(t)$. The vector x itself thus depends on time. Any system, however complicated, is represented mathematically by a single point moving in $\mathbb{R}^n$.

The laws of mechanics describe how it moves. Newton's laws say that the accelerations (second time-derivatives of the coordinates) depend on the forces, and the forces depend on the state of the system. Thus for each r, x_r'' depends on all the coordinates (where $''$ denotes the second derivative with respect to time). Writing x'' for the n-vector whose rth entry is x_r'', we have

$$x'' = F(x)$$

where $F(x)$ is an n-vector depending on x; its entries are related to the forces in the system. Thus F is a vector function of a vector argument, that is, a mapping $\mathbb{R}^n \to \mathbb{R}^n$. It will not usually be linear. But a linear mapping can be derived as follows.

Most systems have a state of equilibrium, in which all the coordinates are constant (think of a bridge as an example of a mechanical system). When the system is slightly disturbed from that state, its movements will often remain close to that state (think of a bridge vibrating after a gust of wind). Then the values of x are always near an equilibrium value. As explained at the end of section 5A, a nonlinear mapping can often be approximated by a linear mapping. The equation above can then be replaced by

$$x'' = f(x) \tag{1}$$

where f is a linear mapping.

The general solution of this equation is quite complicated, but there are special

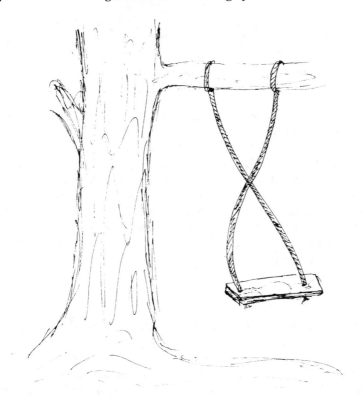

Fig. 5.6 – A system which is oscillating, but not in a normal mode.

solutions which are simple. They correspond to the whole system oscillating in unison. Each coordinate x_r is proportional to $\sin(pt + \delta)$, where t is time, p is a constant giving the frequency of the oscillation (the same for all the coordinates), and δ is a constant called the phase of the oscillation. A system behaving like this is said to be in a **normal mode of oscillation**.

Consider, for example, a child's swing (see Fig. 5.6). Given a random push, it will twist and swing in a complicated fashion. But in normal use it swings to and fro in a regular way, which is a normal mode of oscillation.

When the system is in a normal mode, we have

$$x_r = a_r \sin(pt + \delta_r) \quad \text{for } r = 1, \dots n \tag{2}$$

where p is a constant giving the frequency, and for each r the constants a_r and δ_r give the magnitude and phase respectively of the oscillation in the rth coordinate. Differentiating (2) gives

$$x_r'' = -p^2 a_r \sin(pt + \delta_r) = -p^2 x_r$$

for each r. Hence $x'' = -p^2 x$. But $x'' = f(x)$ by (1), so

$$f(x) = -p^2 x$$

This is just the eigenvalue equation; we have justified the claim at the beginning of this section that vibration frequencies are essentially eigenvalues. The eigenvectors of the linear operator f correspond to the normal modes of oscillation, and the eigenvalues are minus the squares of the vibration frequencies.

Vibrating strings

The theory above applies to systems with a finite number of coordinates. Similar ideas apply more generally.

Consider a guitar string, for example. It contains infinitely many points; it is, in a sense, the '$n = \infty$' case of the theory above. The frequency of its vibrations is related to eigenvalues of an operator on an infinite-dimensional space, as follows.

Take an x-axis along the string, so that the ends of the string are at $x = 0$ and $x = a$. Suppose it vibrates in the xy plane, each point oscillating through a small distance in the y-direction; let $Y(x, t)$ be the y-displacement at the point x at time t. The acceleration of a point of the string is then $\partial^2 Y/\partial t^2$. Newton's laws, under various reasonable assumptions, lead to

$$c^2 \frac{\partial^2 Y}{\partial x^2} = \frac{\partial^2 Y}{\partial t^2} \tag{3}$$

where c is a number depending on the tension and the weight of the string. Equation (3) is called the wave equation.

In the same way as above, we can look for normal-mode solutions in which all points of the string vibrate sinusoidally with the same frequency, so that $Y(x, t)$ is proportional to $\sin(pt)$ for each x. Thus $Y(x, t) = u(x)\sin(pt)$ for some function u. Substituting into (3) gives

$$D^2 u = (-p^2/c^2)u \tag{4}$$

where $D^2 u$ denotes the second derivative of $u(x)$. This is an eigenvalue equation again; vibration frequencies are related to eigenvalues in much the same way as in the finite-dimensional theory above.

We shall now look more closely at the details. The string occupies the interval $[0, a]$ of the x-axis. Let $S[0, a]$ be the vector space of all functions which are smooth on $[0, a]$ (see Appendix D). The second derivative of a smooth function is smooth, so D^2 maps $S[0, a]$ into itself. Now, for any real k, $D^2[\sin(kx)] = (-k^2)\sin(kx)$. Hence the function $\sin(kx)$ is an eigenvector for eigenvalue $-k^2$. Similarly, for any k, $\exp(kx)$ is an eigenvector for eigenvalue k^2. Thus every real number is an eigenvalue. But this means that all frequencies are possible, and when you pluck a guitar string you will hear a terrible noise consisting of a mixture of all possible frequencies. This is not so; something is wrong with our theory.

The trouble is that we have not used the fact that the ends of our string are fixed. This means that $Y(0, t) = Y(a, t) = 0$ for all t, hence

$$u(0) = u(a) = 0 \tag{5}$$

These conditions must be applied to equation (4).

How does (5) fit into the algebraic structure? It can be incorporated into a vector space by defining V to be the set of all smooth functions satisfying (5); it is easy to show that V is a vector space. Then D^2 is a linear operator on V, and its eigenvectors are functions $u \in V$ satisfying $D^2 u = \lambda u$. Exponential functions satisfy this equation but do not belong to V because they do not satisfy (5). The function $\sin(kx)$ satisfies $D^2 u = \lambda u$ if $\lambda = -k^2$, and will satisfy (5) if $\sin(ka) = 0$, that is, $ka = n\pi$ for some integer n. Hence we have an infinite set of eigenvalues $\lambda = -(\pi/a)^2, -(2\pi/a)^2, \ldots$, leading to a set of vibration frequencies $c\pi/a, 2c\pi/a, \ldots$. Instead of all frequencies being possible, we now have a definite lowest frequency $c\pi/a$, and then integer multiples of it. This agrees precisely with the observed behaviour of vibrating strings; we have a very satisfactory interpretation of vibration frequencies in terms of eigenvalues.

There is, however, a difficulty. Eigenvalues belong to operators of the type $V \to V$. Our operator D^2 does not map V to itself, because u'' may not satisfy (5). The arguments above are therefore unsound.

The arguments are sensible, but they do not fit into our mathematical framework; it is not powerful enough to deal with differential operators on infinite-dimensional spaces. The framework can be extended and strengthened. The details are difficult, but the end result is a satisfactory rigorous version of the theory of eigenvalues of differential operators. See, for example, Kreider *et al.* (1966) and Griffel (1981).

5H   EIGENVALUE PROBLEMS

The problem of finding eigenvectors and eigenvalues of a given operator is called the **eigenvalue problem**. Section 5G shows that eigenvalue problems are of some practical value, but they are interesting from the purely mathematical point of view too, as you will see in later chapters.

The example at the beginning of section 5F was solved by a rather laborious method. This section uses a more efficient technique, which applies to matrix operators. Many of the operators arising in practice are of this type; and we shall see in Chapter 9 that in a finite-dimensional space every operator can, in a certain sense, be regarded as a matrix operator. So dealing with matrices is not a great restriction.

Given a real n by n matrix A, we define the corresponding operator $a: \mathbb{R}^n \to \mathbb{R}^n$ by $a(v) = Av$ for all v in $\mathbb{R}^n$, where v is written as a column vector. We shall refer to the eigenvectors and eigenvalues of the operator a as the eigenvectors and eigenvalues of the matrix. They are the real solutions of the eigenvalue equation $Av = \lambda v$. In the same way, given a complex matrix A, its eigenvalues and eigenvectors are the complex solutions of the equation $Av = \lambda v$.

Example 1. Consider the matrix

$$A = \begin{pmatrix} 0 & -1 \\ 1 & 0 \end{pmatrix}$$

Its eigenvalues are numbers λ such that $Av = \lambda v$ for some nonzero 2-vector v. We have

$$\begin{pmatrix} 0 & -1 \\ 1 & 0 \end{pmatrix} \begin{pmatrix} x \\ y \end{pmatrix} = \lambda \begin{pmatrix} x \\ y \end{pmatrix}$$

or

$$-y = \lambda x \quad \text{and} \quad x = \lambda y \tag{1}$$

Hence $(-y)y = \lambda xy = \lambda(\lambda y)y$, and $x^2 = \lambda yx = \lambda(-\lambda x)x$, so

$$\text{either} \quad x = y = 0 \quad \text{or} \quad \lambda^2 = -1 \tag{2}$$

No real number can satisfy $\lambda^2 = -1$, so $x = y = 0$. But eigenvectors must be nonzero; therefore this matrix has no eigenvalues.

There is a subtle point here. The entries of A can be regarded as complex numbers; they happen to have imaginary part zero, but they still belong to the field $\mathbb{C}$. So we can regard A as a complex matrix, corresponding to an operator on the space $\mathbb{C}^2$; its eigenvalues are then complex numbers λ such that the equation $Av = \lambda v$ is satisfied by some nonzero v in $\mathbb{C}^2$. Again, a 'complex number' here may in special cases be real.

If we use complex numbers, then (1) and (2) can be satisfied for nonzero (x, y). We have $\lambda^2 = -1$, so $\lambda = \pm i$. For $\lambda = i$, (1) gives $-y = ix$ and $x = iy$, which are satisfied for any x if $y = -ix$. Hence the eigenvectors corresponding to eigenvalue i are the vectors $x(1, -i)$ for all nonzero numbers x. The eigenspace is $\text{Sp}\{(1, -i)\}$. A similar easy calculation shows that the eigenspace for eigenvalue $-i$ is $\text{Sp}\{(1, i)\}$. $\quad\square$

This example shows that a matrix with real entries may be regarded either as a real matrix or as a complex matrix whose entries happen to have imaginary part zero, and the solution of the eigenvalue problem is different in these two cases. In order to avoid confusion, we use the following terminology.

Definition 1. A **matrix over** $\mathbb{R}$ means a matrix whose entries are real numbers. A **matrix over** $\mathbb{C}$ means a matrix whose entries belong to the field of complex numbers. In general, a matrix over any field $\mathbb{K}$ is a matrix whose elements belong to $\mathbb{K}$. $\quad\square$

Thus a matrix with real entries may be regarded as a matrix over $\mathbb{R}$ or as a matrix over $\mathbb{C}$.

Definition 2(a). The eigenvalues and eigenvectors of an n by n matrix A over $\mathbb{K}$ are defined to be the eigenvalues and eigenvectors of the mapping $a: \mathbb{K}^n \to \mathbb{K}^n$ defined by $a(v) = Av$ for all v in $\mathbb{K}^n$. $\quad\square$

This definition is obviously equivalent to the following.

Definition 2(b). A real number λ is said to be an eigenvalue of an n by n matrix A over $\mathbb{R}$ if there is a real n-vector $v \neq 0$ such that $Av = \lambda v$. A complex number λ is said to be an eigenvalue of an n by n matrix B over $\mathbb{C}$ if there is a complex n-vector $v \neq 0$ such that $Bv = \lambda v$. The n-vectors v in each case are called eigenvectors of A. $\quad\square$

Example 1 above shows that a real matrix can have eigenvalues over $\mathbb{C}$ while having no eigenvalues over $\mathbb{R}$. This cannot happen the other way round.

Proposition 1. If A is a real matrix, then every eigenvalue of A over $\mathbb{R}$ is also an eigenvalue of A over $\mathbb{C}$.

Proof. If the equation $Av = \lambda v$ has real solutions, they can be regarded as complex solutions which happen to have zero imaginary part. Hence there is a nonzero v in $\mathbb{C}^n$ satisfying $Av = \lambda v$, so λ is an eigenvalue of A over $\mathbb{C}$. $\quad\square$

The question now is how to solve eigenvalue problems, that is, how to find eigenvalues and eigenvectors. Example 1 was very easy. For larger matrices a more systematic method is needed.

We are looking for nonzero solutions of $Av = \lambda v$, that is, $Av - \lambda v = 0$, or

$$(A - \lambda I)v = 0 \tag{3}$$

where I is the unit matrix, satisfying $Iv = v$. The solution set of (3) is the null space of $A - \lambda I$; eigenvalues are values of λ for which $A - \lambda I$ is singular, that is, nullity$(A - \lambda I) > 0$. We can find nullity$(A - \lambda I)$ by Gaussian elimination, as follows.

Examples 2. Consider the matrix

$$A = \begin{pmatrix} 1 & 0 & 0 \\ 0 & 1 & 1 \\ 1 & -1 & 1 \end{pmatrix}$$

(a) The eigenvalues of A over $\mathbb{R}$ are real numbers λ such that $A - \lambda I$ is a singular real matrix. Now,

$$\begin{pmatrix} 1 & 0 & 0 \\ 0 & 1 & 1 \\ 1 & -1 & 1 \end{pmatrix} - \lambda \begin{pmatrix} 1 & 0 & 0 \\ 0 & 1 & 0 \\ 0 & 0 & 1 \end{pmatrix} = \begin{pmatrix} 1-\lambda & 0 & 0 \\ 0 & 1-\lambda & 1 \\ 1 & -1 & 1-\lambda \end{pmatrix}$$

When $\lambda = 1$, the first row is zero; hence the rows are linearly dependent, so the rank is less than 3. Rank + nullity $= 3$, so the nullity is positive, and $A - 1I$ is singular. Thus 1 is an eigenvalue of A. The corresponding eigenvectors are the solutions of $(A - 1I)v = 0$, that is,

$$\begin{pmatrix} 0 & 0 & 0 \\ 0 & 0 & 1 \\ 1 & -1 & 0 \end{pmatrix} v = 0$$

This system is easily solved: $v_3 = 0$ and $v_1 - v_2 = 0$, so $v = (t, t, 0)$ is an eigenvector for any $t \neq 0$. Thus the eigenspace corresponding to $\lambda = 1$ is $\mathrm{Sp}\{(1, 1, 0)\}$.

What about other eigenvalues? We look for them by reducing $A - \lambda I$ to echelon form, so that its rank is obvious. We have already dealt with the case $\lambda = 1$. If $\lambda \neq 1$, elementary row operations give

$$\begin{pmatrix} 1-\lambda & 0 & 0 \\ 0 & 1-\lambda & 1 \\ 1 & -1 & 1-\lambda \end{pmatrix} \rightarrow \begin{pmatrix} 1 & 0 & 0 \\ 0 & 1-\lambda & 1 \\ 1 & -1 & 1-\lambda \end{pmatrix} \rightarrow \begin{pmatrix} 1 & 0 & 0 \\ 0 & 1-\lambda & 1 \\ 0 & -1 & 1-\lambda \end{pmatrix}$$

$$\rightarrow \begin{pmatrix} 1 & 0 & 0 \\ 0 & 1-\lambda & 1 \\ 0 & 0 & 1-\lambda+(1-\lambda)^{-1} \end{pmatrix}$$

If $\lambda \neq 1$, then the (3, 3) entry is $[(1 - \lambda)^2 + 1]/(1 - \lambda)$, which is nonzero. Hence this matrix is in echelon form with all rows nonzero; hence the rows are linearly independent, the rank is 3, and the nullity is zero. Thus 1 is the only eigenvalue, and we have solved the eigenvalue problem for A over the real numbers.

(b) Now consider the eigenvalue problem for A over $\mathbb{C}$. The argument above still shows that 1 is an eigenvalue, with eigenspace $\mathrm{Sp}\{(1,1,0)\}$, and the calculations for the case $\lambda \neq 1$ are still valid. But the interpretation is different in the complex case, because the $(3,3)$ entry $[(1-\lambda)^2 + 1]/(1-\lambda)$ in the echelon form of $A - \lambda I$ can now vanish. In fact, when $\lambda = 1 \pm i$, the third row of $A - \lambda I$ is zero, so $A - \lambda I$ is singular. Hence A has three eigenvalues over $\mathbb{C}$: 1, $1 + i$, and $1 - i$.

For $\lambda = 1 + i$, we have

$$A - \lambda I = \begin{pmatrix} 1 & 0 & 0 \\ 0 & -i & 1 \\ 0 & 0 & 0 \end{pmatrix}$$

so the eigenvector equation $(A - \lambda I)v = 0$ gives $v_1 = 0$ and $-iv_2 + v_3 = 0$. Hence the eigenvectors for $\lambda = 1 + i$ are $(0, t, it)$ for any $t \neq 0$, and this eigenspace is $\mathrm{Sp}\{(0,1,i)\}$. A similar calculation shows that the eigenspace for eigenvalue $1 - i$ is $\mathrm{Sp}\{(0,1,-i)\}$. This completes the solution of the eigenvalue problem for this matrix. □

This matrix has either 1 or 3 eigenvalues, depending on whether we work over the real or the complex field. It is natural to wonder just how many eigenvalues a matrix can have; is it a coincidence that this 3 by 3 matrix has no more than 3 eigenvalues? The following theorem gives the answer. It says that eigenvectors belonging to different eigenvalues are linearly independent; it follows that an operator on an n-dimensional space cannot have more than n eigenvalues.

Theorem 1 (independence of eigenvectors). Let $e_1, \ldots, e_r$ be eigenvectors of a linear operator $f: V \to V$, belonging to different eigenvalues $\lambda_1, \ldots, \lambda_r$. Then $\{e_1, \ldots, e_r\}$ is linearly independent. □

The proof is not difficult; it will be found in section 5L.

Corollary. An operator on an n-dimensional space cannot have more than n eigenvalues.

Proof. If there were more than n eigenvalues, then Theorem 1 would give a set of more than n linearly independent vectors in an n-dimensional space, which is impossible. □

Theorem 1 leads to new ideas about eigenvectors. If an operator f on an n-dimensional space has n different eigenvalues, then there is a linearly independent set of n eigenvectors. But any n linearly independent vectors in an n-dimensional space form a basis. Therefore we can use the eigenvectors of f as basis vectors; this seems a natural basis to use when dealing with f. We shall see in later chapters (and also in section 5J) that there are indeed great advantages in using an eigenvector basis.

*5I   ROTATIONS AND ORTHOGONAL MATRICES

This section works entirely in the space $\mathbb{R}^n$. It uses ideas which do not strictly belong to the theory set out in Chapters 3 and 5. If you insist on ideological purity, skip this section. But if you want to reach useful results quickly, then read on.

We shall consider mappings of a special type, which can be interpreted as rotations in the space $\mathbb{R}^n$. They are interesting for their own sake; and they play a special role in linear algebra, as we shall see in later chapters.

The mathematical description of rotations

Suppose three perpendicular axes are given in three-dimensional space. For each point P, there is a corresponding vector v in $\mathbb{R}^3$, whose entries are the coordinates of P with respect to these axes. They can also be regarded as components of the position vector of P with respect to unit vectors along the coordinate axes.

A rotation of the space is a transformation which moves points to new positions; we imagine points in the space moving, while the axes stay fixed. In terms of the vectors v, a rotation is a map $\mathbb{R}^3 \to \mathbb{R}^3$. It may or may not be a linear map. Consider, for example, the map T which rotates all points through an angle of $90°$, say, about the line $x = y = 1$. The origin is moved to the point $(2, 0, 0)$ by this rotation, so $T(0, 0, 0) = (2, 0, 0)$. Hence T is not linear (since every linear transformation maps the zero vector to itself). If we are to use the theory of linear transformations, we must consider only rotations about lines through the origin, which leave the origin fixed.

Rotations of this type are given by matrices acting on column vectors. For example, consider the matrix

$$M = \begin{pmatrix} \cos\theta & \sin\theta & 0 \\ -\sin\theta & \cos\theta & 0 \\ 0 & 0 & 1 \end{pmatrix} \tag{1}$$

Let P be the point with coordinates x, y, z with respect to perpendicular axes. If v is the column vector with entries x, y, z, then the entries of Mv are $x\cos\theta + y\sin\theta$, $y\cos\theta - x\sin\theta, z$. It is easy to verify, by drawing a diagram, that these are the coordinates of the point obtained by rotating P through an angle θ about the z-axis. Thus M gives a rotation about the z-axis. Rotations about the other two axes are given by matrices similar to M.

But what about the general case? It is not clear how to write down a matrix for rotation about a general axis, or how to deal with higher dimensions. To generalise, we must look more carefully at the general character of rotations.

A rotation which leaves the origin fixed has the property of swinging position vectors through some angle, without changing their lengths. Lengths and angles are crucial to the concept of rotation; but they have not appeared in our general treatment of vector spaces, and will not do so until Chapter 11. But in the special case of real 3-vectors, we can use the results of section 1E, giving simple formulas for lengths and angles, in terms of the components of a 3-vector with respect to an 'orthonormal' basis consisting of unit vectors at right angles to each other.

Orthonormal sets of vectors are thus closely linked with rotations: orthonormal bases lead to simple expressions for lengths and angles, the concepts underlying the idea of a rotation. In fact the link is even closer than this. Consider an orthonormal set S of vectors in $\mathbb{R}^3$. If a rotation is applied, the lengths of the vectors in S do not change, so they remain normalised. And all vectors are rotated through the same angle, so perpendicular vectors stay perpendicular. Hence S is still orthonormal after the rotation. Thus:

> a rotation transforms an orthonormal set into another orthonormal set   (2)

This simple fact is the key to the mathematical description of rotations, as we shall soon see.

Orthogonal matrices

We now generalise to n dimensions. Section 1F set up the geometry of $\mathbb{R}^n$: the length of an n-vector is the number $\|x\| = \sqrt{(x \cdot x)}$, and the angle between x and y is $\cos^{-1}[x \cdot y/\|x\|\|y\|]$.

We saw in Exercise 2D-3 that the dot product can be expressed neatly in matrix notation: if x and y are column vectors, then their dot product is $x^T y$. Hence lengths and angles are given by

$$\|x\| = \sqrt{(x^T x)} \quad \text{and} \quad \theta = \cos^{-1}[x^T y/\|x\|\|y\|] \tag{3}$$

We now have the tools needed to study rotations in n-space. We shall investigate the linear maps which satisfy condition (2), which we take to be the essential property of a rotation.

A natural way of specifying a linear map $\mathbb{R}^n \to \mathbb{R}^n$ is by an n by n matrix A; a column vector x is mapped to Ax. Let $\{e_1, \ldots, e_n\}$ be the usual basis for $\mathbb{R}^n$; it is orthonormal by Exercise 1F-1. If A represents a rotation, then (2) says that the set $\{Ae_1, Ae_2, \ldots, Ae_n\}$ is orthonormal. But Ae_i is just the ith column of A. Hence if A represents a rotation, its columns are orthonormal n-vectors.

Example 1. The matrices

$$\begin{pmatrix} 3/5 & 4/5 \\ -4/5 & 3/5 \end{pmatrix} \quad \text{and} \quad \begin{pmatrix} 7/9 & 4/9 & 4/9 \\ 4/9 & 1/9 & -8/9 \\ 4/9 & -8/9 & 1/9 \end{pmatrix}$$

have orthonormal columns. So has the 3 by 3 matrix M written down near the beginning of this section.

Exercise 1. An amazing coincidence: the matrices in Example 1 have orthonormal rows as well as columns. Try to write down other examples of matrices with orthonormal columns, and see whether they too have orthonormal rows. Can you explain the results?   □

Consider, then, a square matrix A whose columns $a_1, \ldots, a_n$ are orthonormal vectors satisfying

$$(a_j)^T a_k = \delta_{jk} \tag{4}$$

for all j, k. The left-hand side of (4) is just the (j, k) entry of the matrix product $A^T A$, and the right-hand side is the (j, k) entry of the unit matrix I. Hence (4) is equivalent to the elegant equation

$$A^T A = I \qquad (5)$$

Matrices satisfying (5) are called 'orthogonal'. The name reflects the fact that the columns are orthogonal vectors. Since the columns are in fact orthonormal, it might be more logical to call the matrix 'orthonormal', but we shall follow the established terminology.

Definition 1. A real n by n matrix A is called **orthogonal** if $A^T A = I$.

Exercise 2. Show that if A is orthogonal, then so is A^T. Now explain the results of Exercise 1.

Exercise 3. For readers who know about groups: the set of all orthogonal n by n matrices is a group under matrix multiplication. This group is important in geometry, physics, chemistry, and geology – mineral crystals are classified according to an elaborate scheme based on group theory. See Coxeter (1981). ☐

Definition 1 was based on condition (2) above, expressing the fact that lengths and right angles remain unchanged. Does it follow that all other angles remain unchanged?

The question is easily answered. Equation (3) shows that angles remain unchanged provided that lengths and products of the form $x^T y$ remain unchanged. This is assured by the following theorem.

Theorem 1 (orthogonal matrices). If A is an n by n orthogonal matrix, then
  (i) for any n-vector x, $\|Ax\| = \|x\|$;
  (ii) for any n-vectors x and y, $(Ax)^T(Ay) = x^T y$.

Proof. (i) follows at once from (ii), on taking $y = x$.
  (ii) $(Ax)^T Ay = x^T A^T Ay = x^T Iy = x^T y$. ☐

This theorem says that applying an orthogonal matrix to a vector does not change its length, it just rotates it about the origin, through an angle which is the same for all the vectors, so that the angle between two vectors says the same. Thus orthogonal matrices behave like rotations in three-space.

Theorem 1 says that every orthogonal matrix gives a rotation. What about the converse? Does every possible rotation correspond to an orthogonal matrix? The answer is yes; we shall show how to construct an orthogonal matrix which rotates the standard basis into any other possible orientation, that is, into any other orthonormal basis for $\mathbb{R}^n$.

Theorem 2. Given any orthonormal set S of n vectors in $\mathbb{R}^n$, there is an orthogonal matrix which transforms the standard basis into the set S. Its columns are the n-vectors in S.

Proof. Given any orthonormal set $\{v_1,\ldots,v_n\}$, let A be the matrix with columns $v_1,\ldots,v_n$; it is orthogonal because its columns are orthonormal. By the column lemma, $Ae_i = v_i$, where e_i is the ith standard basis vector in $\mathbb{R}^n$.   □

Remark. If you think geometrically, you will see that there is something missing here. A linear map taking one orthonormal set into another might be a reflection rather than a rotation. In section 11G we will return to this point, and divide the class of orthogonal matrices into rotation and reflection matrices. But that needs tools which are not yet available.

*5J   THE POWER METHOD

This optional section depends on no other optional section. It deals with an elegant method for finding eigenvalues numerically.

Suppose A is an n by n matrix over $\mathbb{K}$ with n eigenvalues $\lambda_1,\ldots,\lambda_n$ and corresponding eigenvectors $e_1,\ldots,e_n$. The argument at the end of section 5H shows that $\{e_1,\ldots,e_n\}$ is a basis for the space $\mathbb{K}^n$. Now, suppose that there is one eigenvalue whose magnitude is larger than all the others. We can number the eigenvalues so that this one is the first:

$$|\lambda_1| > |\lambda_r| \quad \text{for all } r > 1$$

λ_1 is called the **leading eigenvalue**.

There might not be a leading eigenvalue. For example, the eigenvalues might be $1, -1, 0.8, -0.5$, so that there are two eigenvalues of equal magnitude larger than the others. But this is an exceptional case; most matrices do have a leading eigenvalue, and the rest of this section is based on that assumption.

For any vector x, writing it in terms of its components $c_1,\ldots$ with respect to the basis $\{e_1,\ldots,e_n\}$ gives

$$x = c_1 e_1 + \ldots + c_n e_n$$

Hence

$$Ax = c_1 A e_1 + \ldots + c_n A e_n$$

$$= c_1 \lambda_1 e_1 + \ldots + c_n \lambda_n e_n$$

because the e_r are eigenvectors of A. Again, by linearity we have

$$A^2 x = \sum_r c_r \lambda_r A e_r = \sum_r c_r (\lambda_r)^2 e_r$$

Multiplying by A repeatedly in this way, we have for any k,

$$A^k x = \sum_r c_r (\lambda_r)^k e_r \tag{1}$$

There are two ways of looking at this formula. If the eigenvalues and eigenvectors of A are known, (1) can be used to evaluate $A^k x$ for any vector x without calculating the powers of A. If, on the other hand, the eigenvalues of A are unknown, (1) leads to a simple method of finding them.

We have assumed that $|\lambda_1| > |\lambda_r|$ for all $r > 1$. It follows that all the powers of $|\lambda_1|$

are greater than the powers of $|\lambda_r|$ for $r > 1$, and if k is large, then $|\lambda_1|^k$ will be much larger than the powers of the other eigenvalues. Thus the first term in equation (1) will be much larger than all the others, which can therefore be neglected, giving

$$A^k x \simeq c_1 (\lambda_1)^k e_1 \quad \text{for large } k \tag{2}$$

where '$\simeq$' means 'is approximately equal to'.

The right-hand side of (2) is an eigenvector belonging to the leading eigenvalue, unless $c_1 = 0$. But x is a vector chosen at random, and it is unlikely that its e_1 component will be exactly zero. So starting with almost any vector, we can find a good approximation to the leading eigenvector simply by multiplying x by a power of A. Better and better approximations can be found by using higher and higher powers. The eigenvalue also emerges from this calculation: it is the factor by which the vector is scaled when it is multiplied by A.

There is one practical complication. If $|\lambda_1| > 1$, then the right-hand side of (2) grows rapidly as k increases, and the numbers soon get too large to handle. On the other hand, if $|\lambda_1| < 1$, then the numbers become inconveniently small. This problem is easily solved by rescaling the vector at each stage so as to keep its entries of reasonable size; for example, one could make its largest entry equal to 1 at each stage.

We now sum up the above discussion in the form of an algorithm. Note that we use x_k to denote the kth vector in a sequence of approximations; it does not mean the kth entry of an n-vector.

The power method for finding the leading eigenvalue and eigenvector of an n by n matrix with n eigenvalues, to a specified accuracy, is as follows.

Take any vector in the space; call it x_0.

Repeat the following steps, first taking $k = 1$, then $2, \ldots$. Stop when two successive steps give the same answer to within the desired accuracy.

(a) Calculate $y_k = Ax_{k-1}$.

(b) Let b_k be the entry of y_k with the largest modulus. If all entries of y_k are zero, start again with a different x_0.

(c) Let $x_k = y_k / b_k$. Then x_k and b_k are the kth approximations to the eigenvector and eigenvalue respectively. □

This, you may feel, is all very well, but it only applies to a matrix with n eigenvalues and a unique leading eigenvalue. How can one tell whether a given a matrix has these qualifications? It isn't easy; but it doesn't matter. The algorithm can be applied to any square matrix. If the matrix does not satisfy the conditions, then the method may or may not work, but if it does, then it gives an eigenvalue and an eigenvector. Consider the following example.

Example 1. Take

$$A = \begin{pmatrix} 0 & 5 & -6 \\ -4 & 12 & -12 \\ -2 & -2 & 10 \end{pmatrix}$$

The vector

$$\begin{pmatrix} 1 \\ 1 \\ 1 \end{pmatrix}$$

is as good a choice as any other for the starting vector x_0. Table 5.1 shows the results of seven steps of the power method, and shows that 16 is an eigenvalue, with eigenvector $(0.5, 1.0, -0.5)$. The conclusion that 16 is an eigenvalue follows directly from the calculations: the last row of the table shows that $(0.5, 1.0, -0.5)$ satisfies the eigenvector equation to three-figure accuracy, with eigenvalue 16; more steps would give higher accuracy. When the method works, the results justify themselves, and are independent of the theoretical reasoning at the beginning of this section.

Table 5.1 – Results of the power method; b_i is the ith approximation to the eigenvalue.

i	b_i	Entries of the vector x_i		
1	6.0	−0.1667	−0.6667	1.0
2	−19.33	0.4828	1.0	−0.6034
3	17.31	0.4981	1.0	−0.5191
4	16.25	0.4998	1.0	−0.5044
5	16.05	0.4999	1.0	−0.5010
6	16.01	0.5000	1.0	−0.5003
7	16.00	0.5000	1.0	−0.5000

Example 2. The matrix

$$\begin{pmatrix} 0 & -1 \\ 1 & 0 \end{pmatrix}$$

has no real eigenvectors, as we saw in section 5H. Therefore the power method, applied with real numbers, must fail. In fact, starting with $(1, 1)$, the vectors $x_1, x_2, \ldots$ are $(-1, 1), (-1, -1), (1, -1), (1, 1), \ldots$. The cycle repeats indefinitely and never settles down to an eigenvector in the way that the vectors in Table 5.1 do. It is easy to verify that the method fails in the same way for any initial vector.   □

Thus the method works when there are eigenvalues, and breaks down when there are no eigenvalues. This is very satisfactory. However, there are drawbacks. The method only works for a matrix with a unique leading eigenvalue; it can be rather slow to converge; and it gives the leading eigenvalue only. But it can be improved.

The inverse power method

The power method described above gives the largest eigenvalue of a matrix. The smallest eigenvalue can be found by turning the matrix upside down so that the smallest eigenvalue becomes the largest. 'Largest' here means 'of largest modulus', and similarly for 'smallest'.

More precisely, if A is an invertible matrix, then its eigenvalues λ_i are all nonzero, and the eigenvalues of A^{-1} are $(\lambda_i)^{-1}$. (Proof: the eigenvalues are nonzero by Exercise 5F-1; if $Ax = \lambda x$ then $x = A^{-1}\lambda x$ so $\lambda^{-1}x = A^{-1}x$.) Hence the largest eigenvalue of A^{-1} is the reciprocal of the smallest eigenvalue of A.

It follows that the smallest eigenvalue of A can be found by applying the power method to A^{-1}. This technique is called the 'inverse power method'. The procedure is the same as the algorithm above, except that (i) b_k is an approximation to the reciprocal of the smallest eigenvalue of A, and (ii) step (a) is replaced by $y_k = A^{-1}x_{k-1}$. In practice, A^{-1} is not computed explicitly; one solves the equation $Ay_k = x_{k-1}$ for y_k by Gaussian elimination, which boils down to the same thing but is quicker.

The inverse power method has the same limitations as the original power method. But there is a clever technique which overcomes these defects.

The shifted inverse power method

The power method depends on the existence of a leading eigenvalue λ_1, with $|\lambda_1/\lambda_r| > 1$ for $r > 1$. The discussion at the beginning of this section shows that when $|\lambda_1/\lambda_r|$ is large, the method gives high accuracy after relatively few steps. The inverse power method computes the eigenvalue λ_n of smallest modulus; inverting the above argument shows that it converges rapidly when $|\lambda_n/\lambda_r|$ is much less than 1 for all $r < n$. We shall now see how to transform the given matrix A so as to reduce $|\lambda_n/\lambda_r|$, and therefore improve the performance of the method.

It is easy to change the eigenvalues of a matrix: subtracting a number s from each diagonal entry reduces all the eigenvalues by s. In other words, the eigenvalues of $A - sI$ are $\lambda_i - s$. (Proof: $Ax = \lambda x$ is equivalent to $(A - sI)x = \lambda x - sx = (\lambda - s)x$.) This procedure is called, for obvious reasons, **shifting** the eigenvalues. If s is close to the smallest eigenvalue λ_n of A, then the smallest eigenvalue of $A - sI$ will be nearly zero, and much smaller than the others. The inverse power method applied to the matrix $A - sI$ will therefore converge very quickly to $\lambda_n - s$.

We thus have the shifted inverse power method: find an approximation s to the smallest eigenvalue, and then apply the inverse power method to the matrix $A - sI$ obtained by subtracting s from the diagonal entries of A.

How does one find s? One method is to begin with the ordinary inverse power method, to get a rough estimate of the eigenvalue, and then use shifting to improve the accuracy. There are other ways of estimating eigenvalues: see sections 7C and 12C.

Other virtues of the shifted inverse power method

The power method works for matrices with a unique leading eigenvalue; the inverse method needs a unique eigenvalue of smallest magnitude, and fails if there are two smallest eigenvalues with equal magnitudes but different signs. The shifted method deals effortlessly with this case. If λ and $-\lambda$ are the smallest eigenvalues, apply a shift s, and the new matrix has eigenvalues $\lambda - s$ and $-(\lambda + s)$; if s is close to λ, then $|\lambda - s|$ will be much less then $|\lambda + s|$, and the inverse method will converge quickly to λ.

Furthermore, the method is not restricted to the largest or smallest eigenvalues. It can be used to find any eigenvalue of a matrix, provided that a rough estimate of the eigenvalue can be found (by the method of section 7C, for example). If λ_{ap} is an approximation to an eigenvalue λ, then shifting by λ_{ap} transforms λ to $\lambda - \lambda_{ap}$, which will be smaller than all the other shifted eigenvalues if λ_{ap} is close to λ. The inverse power method will then determine $\lambda - \lambda_{ap}$ accurately, giving λ accurately.

It should now be clear how flexible and useful the shifted inverse power method is. But we have merely scratched the surface of the problem of computing eigenvalues; for a full account, see Golub and Van Loan (1983).

*5K   THE ALGEBRA OF LINEAR TRANSFORMATIONS

This optional section depends on no other optional section. It deals with mathematical properties of the set of all linear transformations $V \to W$, where V and W are vector spaces over the same field $\mathbb{K}$.

Many of the examples in this chapter have been matrix transformations on $\mathbb{R}^n$ or $\mathbb{C}^n$. Matrices have a rich algebraic structure; they can be added together, multiplied by each other, and multiplied by numbers. This structure can be extended to linear operators in a natural way.

Digression for physicists. Why are abstract operators worth so much trouble? Well, consider quantum mechanics. It is the best available theory of the physical world; and it is based firmly on linear algebra (see section 12K). Linear operators are used in quantum mechanics to represent measurable quantities, such as positions, velocities, etc., in the same way that real numbers are used in elementary mechanics. We therefore need to be able to manipulate operators algebraically.   □

An obvious thing to do with two mathematical objects is to add them.

Definition 1 (addition of operators). If f and g are operators $V \to W$, their sum $f + g$ is defined by $(f + g)(x) = f(x) + g(x)$ for all x in V.   □

In other words, $f + g$ maps any vector into the sum of its images under f and g. In the same way we can define a scalar multiple of an operator: it maps any vector into a multiple of its image under the operator.

Definition 2 (scaling an operator). Let f be an operator $V \to W$, where V and W are vector spaces over $\mathbb{K}$. For any scalar k the operator kf is defined by $(kf)(x) = kf(x)$ for all x in V.

Example 1. Let f and g by the linear operators $\mathbb{K}^q \to \mathbb{K}^p$ defined $f(x) = Ax$ and $g(x) = Bx$, where A and B are p by q matrices over $\mathbb{K}$. Then $f + g$ maps x to $Ax + Bx = (A + B)x$. Hence $f + g$ is the operator corresponding to the matrix $A + B$. Similarly, for any scalar k, kf is the operator corresponding to the matrix kA.   □

We have shown how linear operators can be added together and multiplied by scalars. This is the characteristic property of a vector space.

Theorem 1 (space of linear transformations). Let V and W be vector spaces over $\mathbb{K}$. The set of all linear transformations from V to W is itself a vector space over $\mathbb{K}$.

Proof. We must show that the sum of two linear transformations is another linear transformation, and similarly for scaling. This is easy; see section 5L.   ☐

We will use this fact in Chapter 14; for the present, regard it as an attractive piece of mathematical scenery, to be admired in passing but not lingered over.

Linear mappings can be multiplied together as well as added. The idea is simple: given a vector x, apply a mapping g, giving a new vector $g(x)$, and then apply another mapping f. The product of f and g corresponds to applying the two maps in succession. Care is needed with the domains; $g(x)$ must belong to the domain of f, otherwise f cannot be applied to $g(x)$. Thus, if g maps $U \to V$, then the domain of f must include V if the product fg is to be defined. The situation is illustrated in Fig. 5.7.

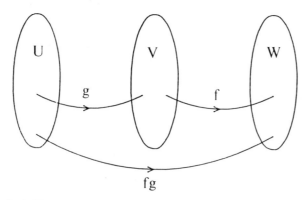

Fig. 5.7 – The product of two maps.

Definition 3 (product of operators). If $f: V \to W$ and $g: U \to V$, then the product $fg: U \to W$ is defined by $(fg)(x) = f(g(x))$ for all x in U.   ☐

Note that the product fg corresponds to applying the mapping g first, then f. The order is inverted because of the convention of writing function symbols to the left of their arguments; $fg(x) = f(g(x))$ means 'apply f to the result of applying g to x'.

Under the conditions of Definition 3, fg is defined but gf may not make sense. Even when fg and gf are both defined, they will not in general be equal. This is just like matrix multiplication; in fact, matrix multiplication is, as we shall see, a special case of Definition 3.

Definition 3 applies to any operators. But our main interest is in linear operators; it is natural to ask whether the product of two linear operators is linear.

Theorem 2. The product of two linear operators is a linear operator.

Proof. Let $f: V \to W$ and $g: U \to V$ be linear operators. Then for any x in U and any scalar k we have

$$(fg)(kx) = f(g(kx)) = f(kg(x)) \quad \text{because } g \text{ is linear}$$
$$= kf(g(x)) \quad \text{because } f \text{ is linear}$$
$$= k(fg)(x)$$

Similarly, for any x, y in U,

$$(fg)(x + y) = f(g(x + y)) = f(g(x) + g(y)) \quad \text{because } g \text{ is linear}$$
$$= f(g(x)) + f(g(y)) \quad \text{because } f \text{ is linear}$$
$$= (fg)(x) + (fg)(y)$$

Thus fg is linear. ☐

Exercise 1. If $f: \mathbb{R}^m \to \mathbb{R}^p$ corresponds in the usual way to the p by m matrix A, and $g: \mathbb{R}^n \to \mathbb{R}^m$ corresponds to the m by n matrix B, then fg corresponds to the matrix AB. ☐

Everything in this section so far has been straightforward, not to say obvious. However, there is a nontrivial theorem about operator multiplication: it concerns the rank of a product.

Given two operators of known rank, what can be said about the rank of their product? One might perhaps guess that the rank of the product is the product of the ranks.

Exercise 2. This guess is wrong. (Hint: see Example 2D-1.) ☐

To understand what is going on, look at a picture. Suppose $f: V \to W$ has rank r, and $g: U \to V$ has rank s. Then g maps U into the s-dimensional subspace $\text{im}(g)$ of V (see Fig. 5.8). Now, f maps V into the r-dimensional subspace $\text{im}(f)$ of W. The image of the product operator fg is the set of all vectors of the form $f(g(u))$ for u in U; that is, it is the set

$$\{f(v'): v' \in \text{im}(g)\}$$

This is a subset of $\text{im}(f)$, as is clearly shown by Fig. 5.8, and therefore has dimension $\leq \dim[\text{im}(f)] = r$. Thus $\dim[\text{im}(fg)] \leq r$; that is, $\text{rank}(fg) \leq r$.

The rank of fg is also related to rank (g). The image of the operator fg is the image under f of the subspace $\text{im}(g)$. Now, $\text{im}(g)$ has dimension $s = \text{rank}(g)$. Linear

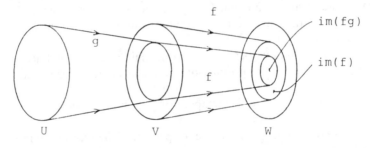

Fig. 5.8 – The rank of a product.

mappings cannot increase the dimension of a space (Theorem 5B-2). Hence $\dim(\text{im}(fg)) \leq s$, i.e., $\text{rank}(fg) \leq s$.

This discussion can be summed up very simply: the rank of a product cannot exceed the rank of either factor.

Theorem 3 (rank of product). The rank of the product of two operators is less than or equal to the rank of each factor.   □

The proof is contained in the paragraphs above; a more concise version will be found in section 5L.

This theorem is the most that can be said in general about ranks of products. The product of operators of ranks r and s can in general have any rank $\leq r$ and $\leq s$.

Now that we have discussed multiplying operators, we can look at inverses from a different point of view. In section 2E we defined the inverse of a matrix A to be a matrix B such that $AB = BA = I$. The first part of the following result says that the inverse of a mapping f satisfies similar equations. The second part says, in effect, that f is invertible if it has a left inverse and a right inverse, and the rest extends other familiar properties of matrix inverses to operators in general.

Proposition 1 (properties of the inverse). Let f be a linear map $U \to V$.

(a) If f is invertible, then $ff^{-1} = i_V$ and $f^{-1}f = i_U$, where i_S denotes the identity map on a space S.

(b) If there are maps $g: V \to U$ and $h: V \to U$ such that $fg = i_V$ and $hf = i_U$, then f is invertible, and $g = h = f^{-1}$.

(c) If f is invertible and linear, then so is f^{-1}.

(d) If f is invertible, and $g: T \to U$ is another invertible map, then fg is invertible and $(fg)^{-1} = g^{-1}f^{-1}$.   □

The proof is straightforward; see section 5L.

5L   PROOFS OF THEOREMS

Theorem 5C-2 (rank plus nullity theorem). If $f: V \to W$ is a linear map, and V is finite-dimensional, then

$$\text{rank}(f) + \text{nullity}(f) = \dim(\text{domain of } f)$$

Proof. We must show that $\dim[\text{im}(f)] = n - \dim[\text{ker}(f)]$, where $n = \dim(V)$. Let $\dim[\text{ker}(f)] = k$. Since $\text{ker}(f)$ is a subspace of V, we have $k \leq n$. Let $\{b_1, \ldots, b_k\}$ be a basis for $\text{ker}(f)$. By Theorem 3H-2 there are elements $b_{k+1}, \ldots, b_n$ of V such that $\{b_1, \ldots, b_n\}$ is a basis for V. We shall prove that the image of $\{b_{k+1}, \ldots, b_n\}$ under f is a basis for $\text{im}(f)$.

First we show that $\{f(b_{k+1}), \ldots, f(b_n)\}$ spans $\text{im}(f)$. Let x be any member of $\text{im}(f)$, then $x = f(v)$ for some v in V. Let c_i be the components of v with respect to the basis $\{b_1, \ldots, b_n\}$. Then

$$x = f(c_1 b_1 + \ldots + c_n b_n)$$

Since f is linear,

$$x = c_1 f(b_1) + \ldots + c_n f(b_n) \tag{1}$$

Since $b_1, \ldots, b_k$ belong to the null space of f, the first k terms in (1) are zero, giving

$$x = c_{k+1} f(b_{k+1}) + \ldots + c_n f(b_n)$$

Thus every element of $\text{im}(f)$ is a linear combination of $\{f(b_{k+1}), \ldots, f(b_n)\}$; hence this set spans $\text{im}(f)$.

To show that it is a basis, we must prove that it is linearly independent. Let $a_{k+1}, \ldots, a_n$ be scalars such that

$$a_{k+1} f(b_{k+1}) + \ldots + a_n f(b_n) = 0 \tag{2}$$

Then $f(a_{k+1} b_{k+1} + \ldots + a_n b_n) = 0$ because f is linear. Hence

$$a_{k+1} b_{k+1} + \ldots + a_n b_n \in \ker(f)$$

Let $c_1, \ldots, c_k$ be the components of this vector with respect to the basis $\{b_1, \ldots, b_k\}$ for $\ker(f)$. Then

$$c_1 b_1 + \ldots + c_k b_k - a_{k+1} b_{k+1} - \ldots - a_n b_n = 0 \tag{3}$$

But $\{b_1, \ldots, b_n\}$ is a basis for V, and is therefore linearly independent, so the coefficients in (3) must be zero. Hence all the a's are zero, so the coefficients in (2) are zero, proving that $\{f(b_{k+1}), \ldots, f(b_n)\}$ is linearly independent. Therefore it is a basis for $\text{im}(f)$, which is thus $(n-k)$-dimensional.   □

Theorem 5H-1 (independence of eigenvectors). Let $e_1, \ldots, e_r$ be eigenvectors of a linear operator $f: V \to V$, belonging to different eigenvalues $\lambda_1, \ldots, \lambda_r$. Then $\{e_1, \ldots, e_r\}$ is linearly independent.

Proof. We use induction. Assuming that $\{e_1, \ldots, e_k\}$ is linearly independent, we shall prove that $\{e_1, \ldots, e_{k+1}\}$ is linearly independent.

Suppose that

$$c_1 e_1 + \ldots + c_{k+1} e_{k+1} = \sum_1^{k+1} c_i e_i = 0 \tag{4}$$

Applying f gives

$$\sum_1^{k+1} c_i \lambda_i e_i = 0 \tag{5}$$

Subtracting λ_{k+1} times (4) from (5) gives

$$\sum_1^{k+1} (\lambda_i - \lambda_{k+1}) c_i e_i = 0 \tag{6}$$

The last term on the left-hand side of (6) is zero. Hence (6) says that a linear combination of the linearly independent set $\{e_1, \ldots, e_k\}$ equals zero; hence the coefficients are zero. Since $\lambda_{k+1} \neq \lambda_i$, we must have $c_i = 0$ for $i = 1, \ldots, k$. It now follows from (5) that $c_{k+1} = 0$ too. This proves that if the first k eigenvectors are linearly independent, then so are the first $k+1$.

The case $k = 1$ is trivial: $\{e_1\}$ is linearly independent because e_1 is an eigenvector and therefore nonzero. This completes the proof.   □

Theorem 5K-1 (space of linear transformations). Let V and W be vector spaces over $\mathbb{K}$. The set of all linear transformations from V to W is itself a vector space over $\mathbb{K}$.

Proof. Let L be the set of linear transformations $V \to W$. We must show that for any f and g in L, $f + g \in L$ and $kf \in L$ for all scalars k; and that the laws of vector algebra are satisfied. The proof is simple, but it is easy to get lost because it is rather long.

(i) *Proof that $f + g \in L$.* The addition rule $(f + g)(v) = f(v) + g(v)$ gives a member of W for each v in V, so $f + g$ is a mapping $V \to W$. We must show that it is linear. For any v in V and k in $\mathbb{K}$, we have

$$(f + g)(kv) = f(kv) + g(kv) \qquad \text{by definition of } f + g$$

$$= kf(v) + kg(v) \qquad \text{because } f \text{ and } g \text{ are linear}$$

$$= k[f(v) + g(v)]$$

$$= k(f + g)(v) \qquad \text{by definition of } f + g$$

Hence $f + g$ satisfies the scaling part of the linearity definition.

Now let x and y be any elements of V. Then

$$(f + g)(x + y) = f(x + y) + g(x + y) \qquad \text{by definition of } f + g$$

$$= f(x) + f(y) + g(x) + g(y) \qquad \text{by linearity}$$

$$= (f + g)(x) + (f + g)(y) \qquad \text{by definition of } f + g$$

This completes the proof that $f + g$ is a linear map, and therefore belongs to L.

(ii) *Proof that $kf \in L$ for all scalars k.* For any v in V and any scalar m we have

$$(kf)(mv) = kf(mv) \qquad \text{by definition of } kf$$

$$= kmf(v) \qquad \text{because } f \text{ is linear}$$

$$= mkf(v) = m(kf)(v) \qquad \text{by definition of } kf$$

Hence kf satisfies the scaling rule for linear operators. Again, for any x and y in V we have, using similar reasoning,

$$(kf)(x + y) = kf(x + y) = kf(x) + kf(y) = (kf)(x) + (kf)(y)$$

so kf satisfies the addition rule for linear operators. This completes the proof that $kf \in L$ for all f in L and scalars k.

(iii) *Proof that the vector space axioms are satisfied.* Commutativity of addition is proved as follows: for any v in V we have

$$(f + g)(v) = f(v) + g(v) = g(v) + f(v) = (g + f)(v)$$

using the fact that addition of the vectors $f(v), g(v)$ in the space W is commutative. Thus $(f + g)(v) = (g + f)(v)$ for all v, which means that $f + g$ and $g + f$ are the same mapping, that is, $f + g = g + f$.

The zero element of L is the mapping O which maps every vector in V into the zero element of W. Using arguments similar to those above, it is easy to verify that

O has the properties of the zero element, and that all the other vector space axioms are satisfied. ☐

Theorem 5K-3 (rank of product). The rank of the product of two operators is less than or equal to the rank of each factor.

Proof. Suppose f maps $V \to W$ and g maps $U \to V$. Then

$$\text{im}(fg) = \{f(g(u)): u \in U\} = \{f(v): v \in \text{im}(g)\} \subset \{f(v): v \in V\}$$

Hence $\dim[\text{im}(fg)] \leq \dim[\text{im}(f)]$ so $\text{rank}(fg) \leq \text{rank}(f)$. Also,

$$\text{rank}(fg) = \dim[\text{im}(fg)] = \dim\{f(v): v \in \text{im}(g)\}$$
$$= \dim[f(\text{im}(g))]$$
$$\leq \dim[\text{im}(g)] \quad \text{by Theorem 5B-2}$$

Hence $\text{rank}(fg) \leq \text{rank}(g)$. ☐

Proposition 5K-1 (properties of the inverse). Let $f: U \to V$ be a linear map.
   (a) If f is invertible, then $ff^{-1} = i_V$ and $f^{-1}f = i_U$, where i_S denotes the identity map on a space S.
   (b) If there are maps $g: V \to U$, and $h: V \to U$ such that $fg = i_V$ and $hf = i_U$, then f is invertible, and $g = h = f^{-1}$.
   (c) If f is invertible and linear, then so is f^{-1}.
   (d) If f is invertible, and $g: T \to U$ is another invertible map, then fg is invertible and $(fg)^{-1} = g^{-1}f^{-1}$.

Proof. (a) follows immediately from the definition; for any u in U, $f^{-1}(f(u))$ is the vector which maps to $f(u)$ under f, namely, u. Hence $f^{-1}f$ is the identity map. Similarly, $f(f^{-1}(v)) = f(\text{the vector which maps to } v) = v$ so ff^{-1} is the identity.
   (b) Given any v in V, let $u = g(v)$; then $f(u) = f(g(v)) = i_V(v) = v$. There is only one u such that $f(u) = v$, for if there is a u' satisfying $f(u') = v$, then $u = i_U(u) = h(f(u)) = h(v) = h(f(u')) = i_U(u') = u'$. This proves that f is invertible.
   Multiplying the equations $fg = i_V$ and $hf = i_U$ on the left and the right respectively by f^{-1}, we have $g = h = f^{-1}$.
   (c) If f is invertible, then $ff^{-1} = i_V$ and $f^{-1}f = i_U$ by (a). Hence f^{-1} is invertible by (b).
   To show that f^{-1} is linear, take any v, v' in V, and write $u = f^{-1}(v)$, $u' = f^{-1}(v')$. Then for all scalars a, a' we have

$$f^{-1}(av + a'v') = f^{-1}(af(u) + a'f(u'))$$
$$= f^{-1}(f(au + a'u')) \quad \text{because } f \text{ is linear}$$
$$= au + a'u' \quad \text{by (a)}$$
$$= af^{-1}(v) + a'f^{-1}(u')$$

Hence f^{-1} is linear.
   (d) $(g^{-1}f^{-1})fg = g^{-1}g = i_T$, and similarly $fg(g^{-1}f^{-1}) = i_V$. The result follows from (b). ☐

PROBLEMS FOR CHAPER 5

Hints and answers for problems marked [a] will be found on page ???.

Section 5A

[a]1. Which of the following mappings are linear?
   (a) $A: \mathbb{R}^3 \to \mathbb{R}^2: A(x, y, z) = (x, z)$;
   (b) $A: \mathbb{R}^4 \to \mathbb{R}^3: A(x, y, z, w) = (x, 1, w)$;
   (c) $A: \mathbb{R}^4 \to \mathbb{R}^3: A(x, y, z, w) = (x, 0, w)$;
   (d) $A: \mathbb{C}^2 \to \mathbb{C}^2: A(z, w) = (w, 2z - w)$;
   (e) $A: V \to W: Af = 3f' - f$, where W is the space of all $\mathbb{R} \to \mathbb{R}$ functions and V is the subspace of W consisting of differentiable functions;
   (f) $A: V \to R: Am = \det(m)$, where V is the space of all 2 by 2 real matrices and $\det(m) = m_{11}m_{22} - m_{12}m_{21}$.

[a]2. A, B, C are linear maps $\mathbb{C}^2 \to \mathbb{C}^2$. Find the images of $(1, 0)$ and $(0, 1)$ under A, B, C, given the following information:

$$A(i, 2) = (3, 1 + i) \qquad A(i, 0) = (1, 0) \qquad B(i, 0) = (-i, 0)$$
$$B(0, i) = (0, -i) \qquad C(1, 1) = (2, 1) \qquad C(-1, 1) = (6, 3)$$

[a]3. Define $A: \mathbb{R}^2 \to \mathbb{R}^2$ by $A(x, y) = (ax + by, cx + dy)$ where a, b, c, d are real constants. Show that A is linear. Can you find values of a, b, c, d which give the following transformations?
   (a) reflection in the x-axis;
   (b) reflection in the y-axis;
   (c) reflection in the line $y = (\tan \theta)x$;
   (d) magnification by a factor of 10;
   (e) rotation about the origin through an angle θ;
   (f) rotation about the point $(0, 1)$ through an angle θ.

4. You are given a black box with two knobs and three dials. A 'knob' is a control which can be set to any desired value. A 'dial' gives a reading controlled by a mechanism inside the box. The only thing you know about the mechanism is that the output (dial readings) is a linear transformation of the input (knob settings).
   What experiments would you do in order to enable you to predict the output from any given settings of the knobs? Embody your answer in a formula.

[a]5. Problem: given that a linear transformation maps $(0, 1)$ to $(1, 2)$ and maps $(1, 0)$ to $(2, 1)$, find the image of $(-2, -3)$. Criticise the following solution: 'From the given data it is clear that the transformation adds 1 to each component. Hence the answer is $(-1, -2)$.'

[a]6. A map $a: \mathbb{R}^3 \to \mathbb{R}^3$ is defined by $a(x, y, z) = (cx + dz, ey + fz, gz)^T$ where $c, \ldots, g$ are given numbers. Express a in the form $a(v) = Av$ for all v, where $v = (x, y, z)^T$ and A is a matrix which you should determine.

Sections 5B, 5C

ᵃ7. Find bases for the image spaces and null spaces (= kernels) of the following maps. What is their rank and nullity?
  (a) $A: \mathbb{R}^2 \to \mathbb{R}^2$, $A(x, y) = (x + y, x + y)$;
  (b) $B: \mathbb{R}^3 \to \mathbb{R}^4$, $B(x, y, z) = (x + 2y - z, y + z, x + y - 2z, x - 3z)$;
  (c) $D: \mathbb{C}^2 \to \mathbb{C}^2$, $D(z, w) = (-iw, iz)$.

ᵃ8. $f: U \to V$ is a linear map. Are the following true or false?
  (a) If nullity $(f) = \dim(U)$ then $f = 0$.
  (b) If rank $(f) = \dim(V)$ then $f = 0$.
  (c) If f is singular then nullity $(f) = 0$.
  (d) If rank $(f) = \dim(V)$ then f is nonsingular.

ᵃ9. Return to problem 4. Describe the kernel and image of this transformation in simple nonmathematical language.

10. P is the space of all real polynomials. Define a map $X: P \to P$ by $Xp = q$ where $q(x) = xp(x)$. Find the image and kernel of X.
  Similarly for the operator D which maps each polynomial to its derivative.

11. Find the rank and nullity of the mapping f from the space of 2 by 2 real matrices to itself, defined by $f(A) = MA$ where

$$M = \begin{pmatrix} 1 & -3 \\ -3 & 9 \end{pmatrix}$$

12. Find the kernel of the skew-symmetric matrix of Chapter 4, Problem 26.

13. The space of all infinite sequences is defined in Chapter 3, Problem 3; call it S. Define $L: S \to S$ by $L(x_1, x_2, \ldots) = (x_2, x_3, \ldots)$; it shifts a sequence one place to the left. Similarly define a right shift operator $R: S \to S$ by $R(x_1, x_2, \ldots) = (0, x_1, x_2, \ldots)$. Show that L and R are linear. What can you say about their ranks and nullities?
  (See also Problem 41.)

14. True or false? 'If N and I are any two subspaces of a k-dimensional space V such that $\dim(N) + \dim(I) = k$, then there is a linear operator with null space N and image space I.'

ᵃ15. The sum of two operators $A, B: V \to W$ is defined in the obvious way by $(A + B)x = Ax + Bx$ for all x. Show that im $(A + B) \subset$ im $(A) +$ im (B), where the sum of subspaces is as defined in section 3J. Hence find inequalities relating the rank and nullity of $A + B$ to those of A and B.

ᵃ16. V is an n-dimensional space, and $a: V \to V$ is a linear map. Under what conditions does im $(a) = \ker(a)$?

Section 5D

ᵃ17. True or false?

(a) If f is a singular mapping then the equation $f(x) = b$ has infinitely many solutions.

(b) If the equation $g(x) = b$ has a unique solution then g is nonsingular.

(c) The equation $h(x) = c$ has no solutions unless $c \in \text{im}(h)$.

(d) If $d \in \ker(a)$ then the equation $a(x) = d$ has infinitely many solutions.

(e) Five linear equations for real numbers $x_1, \ldots, x_5$ can always be solved to determine them uniquely.

(f) Five linear equations for complex numbers $x_1, \ldots, x_5$ can always be solved to determine them uniquely.

(g) Six linear equations for real numbers $x_1, \ldots, x_5$ cannot be solved; there is no solution.

ᵃ18. Let W be the space of all $\mathbb{R} \to \mathbb{R}$ functions, and V the subspace consisting of all twice-differentiable functions f satisfying $f(0) = f(1) = 0$. Define an operator $A: V \to W$ by $Af = f'' + kf$, where k is a real constant. Find the nullity of A. (The answer will depend on the value of k.)

Given a function u, what can you say about the number of solutions of the equation $f'' + kf = u$ which satisfy the boundary conditions $f(0) = f(1) = 0$?

19. The equation $AX - XB = C$ occurs in control engineering; A, B, C, X are n by n matrices. Working in the vector space of n by n matrices, prove that

(a) there is a unique solution X for every C provided that there is no nonzero matrix Y satisfying $AY = YB$; and

(b) if there is a nonzero Y satisfying $AY = YB$, then there are matrices C for which the equation $AX - XB = C$ has no solution. (See also Chapter 10, Problem 17)

20. $C[0, 1]$ is the complex space of continuous functions on the interval $[0, 1]$, and λ is a complex number. Define $T: C[0, 1] \to C[0, 1]$ as follows. For any f in $C[0, 1]$, Tf is the function whose value at x is

$$(Tf)(x) = f(x) - \lambda \int_0^1 (3xy - 5x^2 y^2) f(y) \, dy$$

(a) Show that T is linear, and that $\ker(T)$ contains functions of the form $ax + bx^2$ where a and b are constants.

(b) Consider the case $\lambda = 4$. Show that $\ker(T)$ consists of the functions $a(x - x^2)$ for all complex a. Deduce that the problem of finding a function f such that $Tf = \sin(x)$ cannot have a unique solution.

(c) Show that for all values of λ except ± 4, the map T is nonsingular.

Section 5E

ᵃ21. Are the following maps invertible?

(a) $A: \mathbb{C}^2 \to \mathbb{C}^2$ defined by $A(x, y, z) = (x + y + 17z, x - z, 3x - y + 15z)$;

(b) $D: P \to P$ defined by $Dp = p'$, the derivative of p, where P is the space of all real polynomials;

(c) $J: P \to P$ defined by $Jp = q$ where $q(x) = \int_0^x p(t)\, dt$;

(d) $X: P \to P$ defined by $Xp = r$ where $r(x) = xp(x)$;

(e) DX where D, X are as in (b), (d) above;

(f) XD in the notation of (e).

[a]22. (a) Give an example of a map between finite-dimensional spaces which is nonsingular yet not invertible.

(b) A nonsingular map can always be made invertible by redefining the codomain, as follows. For any linear map $f: U \to V$, write $W = \text{im}(f)$; then f maps U to W. If f is nonsingular, prove that for each $w \in W$ there is exactly one $u \in U$ such that $f(u) = w$; hence the map $f: U \to W$ is invertible.

23. (a) Let S be the space of all smooth functions, as defined in Appendix D. Let D^2 be the operator which maps a function into its second derivative. Show that the map $D^2: S \to S$ is singular.

(b) Let $T = \{f \in S: f(0) = f'(0) = 0\}$. Show that the map $D^2: T \to S$ is nonsingular.

(c) (For readers familiar with differentiating integrals.) Define an operator E on smooth functions by $Ef = g$ where $g(x) = \int_0^x (x - y) f(y)\, dy$. Show that $g(0) = g'(0) = 0$ for any f, and $g'' = f$. Deduce that E maps S to T, and $E^{-1} = D^2$.

(d) It is reasonable for the inverse of differentiation to be integration; one might expect the inverse of double differentiation to be double integration, but (c) gives the inverse as a single integral, in the case where the conditions $f(0) = f'(0) = 0$ are applied. Can you generalise, and find a single integral form for the inverse of D^n (with appropriate conditions at $x = 0$)?

24. (For readers familiar with double integrals.) Define an operator $J: C[0, 1] \to C[0, 1]$ by $Jf = g$ where $g(x) = \int_0^1 (3xy - 5x^2 y^2) f(y)\, dy$. Thus the operator of Problem 20 is $T = I - \lambda J$.

Show that J has rank 2, and is not invertible. Find a formula giving $J^2 f$ for any function f.

Sections 5F, 5G, 5H

[a]25. (a) Use Gaussian elimination on $A - \lambda I$ to find the eigenvalues and eigenvectors of the $\mathbb{R}^2 \to \mathbb{R}^2$ operator corresponding to the matrix

$$A = \begin{pmatrix} 2 & -1 \\ 1 & 2 \end{pmatrix}$$

(b) Similarly, find the eigenvalues and eigenvectors of the $\mathbb{C}^2 \to \mathbb{C}^2$ operator given by the same matrix A.

[a]26. Show that every square matrix has the same eigenvalues as its transpose, with the same geometric multiplicities.

27. (a) Show that every magic square (see Chapter 3, Problem 5) has an eigenvector $(1 \quad 1 \quad \ldots \quad 1)^{\mathrm{T}}$.

(b) In renaissance Europe, n by n magic squares with integer entries $1, 2, \ldots, n^2$ were used in astrology and medicine (see Rouse Ball & Coxeter (1974)). Prove that a matrix of this type has an eigenvalue $(\frac{1}{2})n(n^2 + 1)$.

^a28. If λ is an eigenvalue of a matrix A with eigenvector v, show that
 (a) λ^2 is an eigenvalue of A^2 with eigenvector v;
 (b) λ^k is an eigenvalue of A^k with eigenvector v, for every positive integer k;
 (c) for any polynomial p, $p(\lambda)$ is an eigenvalue of $p(A)$ with eigenvector v;
 (d) if A is invertible, then λ^{-1} is an eigenvalue of A^{-1} with eigenvector v.

^a29. $a: V \to V$ is a linear map. Is im (a) an invariant subspace for a? What about ker (a)?

30. Let A and B be p by q and q by p matrices respectively. Prove that if v is an eigenvector of AB with a nonzero eigenvalue, then Bv is nonzero and is an eigenvector of BA with the same eigenvalue.

Deduce that AB and BA have the same eigenvalues (except that 0 may be an eigenvalue of one but not the other).

^a31. Let f_n, r_n be the number of foxes and rabbits respectively in a certain region in year number n. It is reasonable to suppose that f_{n+1} depends on f_n and r_n (more rabbits means healthier foxes, hence more foxes next year). Similarly r_{n+1} will depend on r_n and f_n (more foxes this year means fewer rabbits surviving to next year). A very simple model embodying these ideas is

$$u_{n+1} = Au_n \quad \text{where} \quad u_n = \begin{pmatrix} f_n \\ r_n \end{pmatrix}$$

and A is a 2 by 2 matrix with $A_{12} > 0$ and $A_{21} < 0$.
 (a) What would be the effect on the foxes and rabbits if the following were true?
   (i) A has an eigenvalue equal to 1.
   (ii) A has two positive eigenvalues, 1 and p, with $p < 1$.
   (iii) A has two eigenvalues greater than 1.
   (iv) A has two eigenvalues p and q with $0 < p < 1 < q$.
 (b) It is reasonable to take $0 < A_{11} < 1$ and $A_{22} > 1$. Give biological interpretations of these assumptions.

32. Consider a stochastic matrix A (defined in Example 2C-4). This problem uses eigenvalues to prove the facts (d) and (e) below, which seem at first sight to have nothing to do with eigenvalues.
 (a) Show that 1 is an eigenvalue of A, and that $(1, 1, \ldots, 1)$ is an eigenvector of A^{T} with eigenvalue 1.
 (b) Use Example 2C-4 to argue (without writing down any equations) that a stochastic matrix cannot have an eigenvector with non-negative entries belonging to an eigenvalue different from 1.

(c) Show that if $(1, 1, \ldots, 1)$ is an eigenvector of B^T with eigenvalue 1, and the entries of B are non-negative, then B is stochastic.

(d) Show that if p is a polynomial with non-negative coefficients, and $p(1) = 1$, then $p(A)$ is stochastic if A is. (Hint: Problem 28.)

(e) Show that if A is an invertible stochastic matrix, then the columns of A^{-1} add to 1. Is A^{-1} necessarily stochastic?

Section 5I

a33. True or false?

(a) If an n by n matrix M transforms every orthonormal basis for $\mathbb{R}^n$ into another orthonormal basis, then M is orthogonal.

(b) If an n by n matrix M transforms an orthonormal basis B for $\mathbb{R}^n$ into another orthonormal basis, then M is orthogonal.

a34. Every rotation in $\mathbb{R}^3$ is a rotation about an axis: there is a line of points which remain fixed, and other points rotate in planes perpendicular to the axis. For example, the matrix (1) in section 5I gives a rotation about the z-axis.

Algebraically, a line of fixed points satisfying $Ax = x$ is an eigenspace belonging to eigenvalue 1. Hence find the axis of the rotation given by the matrix

$$(1/7) \begin{pmatrix} -6 & 3 & -2 \\ -3 & -2 & 6 \\ 2 & 6 & 3 \end{pmatrix}$$

35. Orthogonal transformations in 3-space can be pictured as rotations about an axis (see Problem 34). Things are different in 4-space.

Verify that the matrix below is orthogonal. Try to describe in geometrical language the corresponding mapping $\mathbb{R}^4 \to \mathbb{R}^4$.

$$\begin{pmatrix} \cos(u) & \sin(u) & 0 & 0 \\ -\sin(u) & \cos(u) & 0 & 0 \\ 0 & 0 & \cos(v) & \sin(v) \\ 0 & 0 & -\sin(v) & \cos(v) \end{pmatrix}$$

(There is an essential difference between even and odd dimensions – see Chapter 11, Problem 37)

Section 5J

a36. Apply the power method to the matrix

$$\begin{pmatrix} 2 & -1 & 0 & 0 \\ -1 & 2 & -1 & 0 \\ 0 & -1 & 2 & -1 \\ 0 & 0 & -1 & 2 \end{pmatrix}$$

with starting vector $(1 \quad 1 \quad 1 \quad 1)^T$. Do the calculations by hand. Do not normalise the vector at each stage, because it is easier to work with whole numbers.

After a few iterations, you may be able to spot a pattern and deduce the exact eigenvalue.

37. What happens if the power method is applied to a matrix which does not have a dominant eigenvalue? Find out by experimenting with the matrix

$$\begin{pmatrix} -1 & 4 \\ 12 & 1 \end{pmatrix}$$

38. Suppose the n by n matrix A has n different eigenvalues with $|\lambda_1| > |\lambda_2| > \ldots > |\lambda_n|$. The leading eigenvalue λ_1 can be found by the power method. Show that λ_2 can be found by applying the power method with starting vector $(A - \lambda_1 I)x$ for any n-vector x. What about the other eigenvalues?

(Note: this method is very nice in theory, but in practice it breaks down because of numerical roundoff errors. The QR method of section 11K is better for computing higher eigenvalues.)

39. The shifted inverse power method is too complicated for hand calculation. If you like computer programming, design programs for the three versions of the power method (ordinary, inverse, and shifted inverse).

(See Burden and Faires (1985) for algorithms; Mason (1984) gives BASIC programs.)

Section 5K

ª40. True or false? For any linear maps $f, g: V \to W$,
   (a) rank $(f + g) \le$ rank $(f) +$ rank (g);
   (b) rank $(f + g) \ge$ rank $(f) -$ rank (g);
   (c) rank $(f + g) \ge$ rank $(f) -$ rank $(g)|$;
   (d) $(f + g)(S) = f(S) + g(S)$ for all subspaces S of V, in the notation of Definitions 5B-1 and 3K-1.

ª41. Return to Problem 13. Show that one and only one of LR and RL is the identity, and hence one of L and R is the left inverse of the other. Are they invertible?

42. Give an example of a linear map $f: V \to V$ (for some vector space V) such that $f^2 = 0$ but $f \ne 0$. Prove that for any such map, $i - f$ is invertible, where i is the identity map on V.

43. A linear map $f: V \to V$ is called **nilpotent** if some power of f is zero. Prove that every nilpotent map is singular.

44. Prove that rank $(AB) \ge$ rank $(A) +$ rank $(B) - q$ for all p by q matrices A and q by s matrices B. Hence show that for every n by n matrix A, rank $(A^k) \ge k$ rank $(A) - n(k - 1)$.

45. An **idempotent** on a vector space V is a linear map $e: V \to V$ such that $e^2 = e$.

(a) The zero and identity maps on any V are obviously idempotents; they are called the trivial idempotents. Write down a nontrivial idempotent on $\mathbb{R}^2$.

(b) Write i for the identity map. Prove that if e is an idempotent, then so is $i - e$, and $e(i - e) = (i - e)e = 0$.

(c) Prove that for any idempotent e, $\ker(e) = \text{im}(i - e)$, and $\ker(e) \cap \text{im}(e) = \{0\}$.

(d) If you have read section 3K, show that $V = \ker(e) \oplus \text{im}(e)$ for every idempotent $e: V \to V$.

6

Determinants

This chapter introduces a useful piece of machinery for identifying singular matrices and solving eigenvalue problems. The first three sections introduce determinants and various ways of evaluating them. The rest of the chapter is optional; it deals with other aspects of the theory of determinants.

6A   TRANSFORMATIONS OF THE PLANE

The properties of linear systems hinge on the question of when a matrix is singular. In Chapter 5 we answered it algebraically, by reducing the matrix to echelon form. We shall now look at it geometrically, beginning in $\mathbb{R}^2$, where the geometry is simple.

Consider a straight line in $\mathbb{R}^2$, parallel to a vector v and passing through a point with position vector u. This line is the set $\{u + tv : t \in \mathbb{R}\}$ (see Chapter 1 Problem 4). Now apply the mapping given by the matrix

$$A = \begin{pmatrix} a & b \\ c & d \end{pmatrix}$$

The image of the line is the set $\{Au + tAv : t \in \mathbb{R}\}$, which is a line through Au parallel to Av. All lines parallel to v map into lines parallel to Av, so every parallel pair of lines is mapped to another parallel pair. Hence a square is mapped into a parallelogram, as shown in Fig. 6.1.

This picture reveals the rank of A. If rank$(A) = 0$, then A maps $\mathbb{R}^2$ to the zero-dimensional subspace $\{0\}$, so the parallelogram degenerates to the single point 0. If rank$(A) = 1$, then A maps $\mathbb{R}^2$ to a one-dimensional subspace, which is a line through the origin; the parallelogram degenerates into a line rather than a point. If A is invertible, then its rank is 2, neither of the above special cases arises, and we have a genuine parallelogram with nonzero area. The area of the parallelogram is the

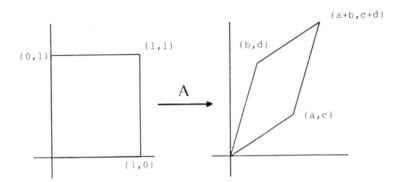

Fig. 6.1 – Image of the unit square.

geometrical idea corresponding to the algebraic concepts of rank and invertibility; the area is zero if and only if rank $(A) < 2$.

    The area of the parallelogram is twice the area of the shaded triangle in Fig. 6.2. We calculate it by subtracting a rectangle and two small right-angled triangles from the large right-angled triangle, giving

$$2[(a+b)(c+d)/2 - bc - bd/2 - ac/2] = ad - bc \tag{1}$$

The calculation assumes that a, b, c, d are all positive and $d/b > c/a$ so that the point (b, d) is above and to the left of (a, c), as in the diagram. If $d/b = c/a$ then all the vertices of the parallelogram lie on the same line, and the area is zero, agreeing with (1). It is not hard to show that in all other cases the area of the parallelogram is still given by $ad - bc$, give or take a minus sign.

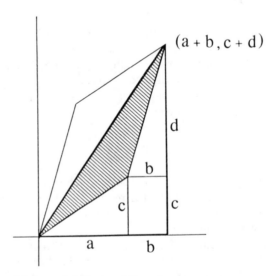

Fig. 6.2 – Calculation of the area of the image of the unit square.

    Thus the number $ad - bc$ is zero if and only if the matrix A is singular. Since it determines the character of A, it is called the 'determinant' of A.

Definition 1. The **determinant** of the matrix

$$A = \begin{pmatrix} a & b \\ c & d \end{pmatrix}$$

is the number $ad - bc$. It is denoted by $\det(A)$ or $|A|$ or

$$\begin{vmatrix} a & b \\ c & d \end{vmatrix}$$

□

Beware: the notation $|A|$ looks like the modulus of A, but its properties are quite different. For example, the determinant need not be positive. Areas, on the other hand, are positive numbers, so the area of the parallelogram is given by the magnitude of the determinant, $|\det(A)|$ where $|\ \ |$ denotes the modulus. (One could write it as $\|A\|$, where the outer bars denote the modulus and the inner bars denote the determinant; but that would be ridiculous.)

The determinant gives a simple way of deciding when a 2 by 2 matrix is singular: calculate $\det(A)$, then A is singular if and only if $\det(A) = 0$.

We introduced $\det(A)$ as the area of the image of the unit square. We shall now generalise step by step to relate $\det(A)$ to areas of arbitrary regions.

Larger squares obviously map into larger parallelograms. Multiplying all dimensions in Fig. 6.2 by k gives a square of side k and area k^2, mapping into a parallelogram with area $|kakd - kbkc| = k^2|ad - bc|$. Thus $|ad - bc|$ is the ratio of the areas of the parallelogram and the square.

Because the transformation is linear, the location of the square in the plane makes no difference, as we shall now prove. Set

$$S = \{(x, y): 0 \le x, y \le k\} \quad \text{and} \quad P = \{As: s \in S\}$$

Then S is a square with one corner at the origin, and P is its image under A. for any fixed vector d, the set $S' = \{d + s: s \in S\}$ is the square S shifted through d. Its image is given by

$$P' = \{A(d + s): s \in S\} = \{Ad + As: s \in S\} = \{Ad + u: u \in P\}$$

This is simply P shifted by the vector Ad. Hence P' has the same area as P, just as S' has the same area as S, and it follows that the area ratio does not depend on the location of the square.

Finally, observe that for any figure built up from squares, its image under A is built up from the corresponding parallelograms, each of which has area $|ad - bc|$ times the area of its corresponding square. Any region can be approximated by a large number of small squares (imagine it drawn on graph paper). So we reach the following conclusion.

Determinants as magnification factors. If the transformation corresponding to the matrix A is applied to any region of the plane, its image has area $|\det(A)|$ times the area of the original region. In short, the transformation multiplies areas by a factor $|\det(A)|$.

□

The discussion above gives an intuitive argument, not a strict proof. We cannot claim to have proved anything about areas, because we have not defined precisely what is meant by the area of a region in the plane. It is possible, though not easy, to give such a definition and put things on a rigorous basis. But there is no need for that here. The informal treatment above is enough to give some geometrical insight into determinants, as a background for the algebraic theory which follows.

The aim of the next section is to extend the idea of determinants to higher dimensions. We have derived the formula $ad - bc$ for a 2 by 2 determinant by working with Fig. 6.2. To draw the corresponding diagram in three dimensions is not easy; to draw it in higher dimensions...?

Clearly, blind imitation of the pattern of this section is little help in the case of n by n matrices. In the next section we shall generalise from the 2 by 2 case in a less crude way. First we shall study the algebraic properties of the 2 by 2 determinant defined above. Then we shall use the algebraic approach to define n by n determinants.

6B n BY n DETERMINANTS

We begin by setting out the algebraic properties of the 2 by 2 determinant

$$\begin{vmatrix} a & b \\ c & d \end{vmatrix} = ad - bc$$

Proposition 1 (properties of 2 by 2 determinants)
(a) Det $(I) = 1$, where I denotes the unit matrix.
(b) Det $(A) = 0$ if A has two identical rows.
(c) Det is a linear function of each row, that is,

$$\det \begin{pmatrix} ar_1 + bs_1 \\ r_2 \end{pmatrix} = a \det \begin{pmatrix} r_1 \\ r_2 \end{pmatrix} + b \det \begin{pmatrix} s_1 \\ r_2 \end{pmatrix}$$

and similarly for a matrix whose second row is a linear combination of two other rows. Here r_1, s_1, etc. denote row vectors with two entries.

Exercise 1. Prove Proposition 1.   □

This theorem expresses simple geometrical facts. Property (a) states the obvious fact that the identity transformation multiplies areas by a factor of 1. Property (b) says that a matrix with two identical rows maps a square into a parallelogram of zero area; this also is obvious because such a matrix is singular, and maps the plane into a line or a point. The geometrical aspect of property (c) will be discussed later.

When we generalise to the n-dimensional case, elementary operations will play a key role. The following theorem sets out their effect on determinants.

Proposition 2 (further properties of 2 by 2 determinants)
(d) Interchanging two rows of a matrix changes the sign of its determinant.
(e) Multiplying one row by k has the effect of multiplying the determinant by k.
(f) Adding a multiple of one row to another row does not change the determinant.
(g) If one row consists entirely of zeros, the determinant vanishes.

Proof. These facts follow immediately from the definition $\det(A) = ad - bc$. But we shall not use this formula. We shall deduce (d) to (g) from Proposition 1, in a way that generalises easily to the n-dimensional case where the formula $ad - bc$ is not available. We prove part (f) first, because (d) can be deduced from it.

(f) Using (b) and (c) of Proposition 1, we have

$$\det\begin{pmatrix} r_1 + kr_2 \\ r_2 \end{pmatrix} = \det\begin{pmatrix} r_1 \\ r_2 \end{pmatrix} + k\det\begin{pmatrix} r_2 \\ r_2 \end{pmatrix} = \det\begin{pmatrix} r_1 \\ r_2 \end{pmatrix}$$

The same method applies when the elementary operation is applied to the second row rather than the first.

(e) Put $b = 0$ in Proposition 1(c).

(d) Using (f), we have

$$\det\begin{pmatrix} r_1 \\ r_2 \end{pmatrix} = \det\begin{pmatrix} r_1 - r_2 \\ r_2 \end{pmatrix} = \det\begin{pmatrix} r_1 - r_2 \\ r_1 \end{pmatrix} = \det\begin{pmatrix} -r_2 \\ r_1 \end{pmatrix}$$

In the second step here, we added the first row to the second. The result now follows from (e) with $k = -1$.

(g) If one row of A is $(0,0)$, let A' be the matrix obtained by replacing the zero row by $(1,1)$. Since $(0,0) = 0(1,1)$, (e) gives $\det(A) = 0\det(A') = 0$. $\square$

Property (f), which is a consequence of (c), can be interpreted geometrically as follows. Consider a Type III elementary operation applied to the identity matrix I. The result is the matrix

$$\begin{pmatrix} 1 & k \\ 0 & 1 \end{pmatrix}$$

Exercise 2. This matrix maps the unit square into the parallelogram with vertices $(0,0),(0,1),(k,1),(1+k,1)$ shown in Fig. 6.3. $\square$

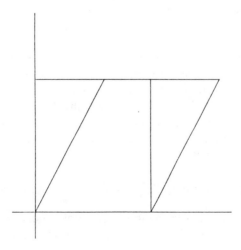

Fig. 6.3 – Shearing a square.

Property (f) expresses the fact that the area of a square is not changed by shearing it as in Fig. 6.3. This is one of the basic properties of area discussed in elementary geometry.

Propositions 1 and 2 give the algebraic expression of the geometrical ideas of section 6A. We now use the algebraic approach to define the determinant of a general square matrix.

Definition 1. The **determinant** of a square matrix A is a number $\det(A)$ or $|A|$ with the following properties:

(a) $\det(I) = 1$;

(b) $\det(A) = 0$ if A has two identical rows;

(c) det is a linear function of each row.   □

Property (c) is to be interpreted as in Proposition 1. Spelling it out in laborious detail: if the ith row of A equals $cu + dv$, where c and d are scalars and u and v are n-vectors, then $\det(A) = c\det(A') + d\det(A'')$, where A' and A'' are the matrices obtained from A by replacing its ith row by u and v respectively.

Determinants are calculated by means of elementary operations, as we shall see below. The following theorem gives the effect of the operations on a determinant.

Theorem 1 (properties of n by n determinants)

(d) Interchanging two rows of a matrix changes the sign of its determinant.

(e) Multiplying one row of a matrix by k has the effect of multiplying its determinant by k.

(f) Adding a multiple of one row of a matrix to another row does not change its determinant.

(g) If one row consists entirely of zeros, then the determinant is zero.

Proof. This theorem is the same as Proposition 2 except that it is not restricted to the 2 by 2 case. Proposition 2 was proved by deducing (d) to (g) from the properties (a), (b), (c) established in Proposition 1. Exactly the same arguments show that Theorem 1 follows from the properties (a), (b), (c) set out in Definition 1; the only difference is that the matrices must be written out with n rows instead of 2.   □

Theorem 1 gives the effect of elementary row operations on determinants. Now, every matrix can be transformed to reduced echelon form by means of elementary row operations. For a square matrix A, the reduced echelon form has a row of zeros if A is singular, and is the unit matrix if A is nonsingular. Hence the reduced echelon form has determinant 0 or 1 (by (g) and (a) respectively). It follows that $\det(A)$ is zero when A is singular. This is the main point of determinants for our purposes: they give a test for singularity.

Theorem 2 (singular matrices). $\mathrm{Det}(A) = 0$ if and only if A is singular.

Proof. Theorem 1 shows that elementary operations multiply a determinant by a nonzero factor. Hence if A is row-equivalent to B, then $\det(A) = 0$ if and only if $\det(B) = 0$.

In section 4I we showed that if A is nonsingular, it is row-equivalent to I with determinant 1; otherwise it is row-equivalent to a matrix with a row of zeros and hence determinant 0. It follows that $\det(A)$ is zero in the second but not the first case. □

The following examples illustrate the calculation of determinants, using elementary operations and Theorem 1.

Example 1. Applying Type III elementary operations, and using Theorem 1(f) at each step, we have

$$
\begin{vmatrix} 1 & 2 & 3 \\ 2 & 3 & 5 \\ 3 & 4 & 2 \end{vmatrix} =
\begin{vmatrix} 1 & 2 & 3 \\ 0 & -1 & -1 \\ 0 & -2 & -7 \end{vmatrix} =
\begin{vmatrix} 1 & 2 & 3 \\ 0 & -1 & -1 \\ 0 & 0 & -5 \end{vmatrix}
$$

$$
=
\begin{vmatrix} 1 & 0 & 1 \\ 0 & -1 & -1 \\ 0 & 0 & -5 \end{vmatrix} =
\begin{vmatrix} 1 & 0 & 0 \\ 0 & -1 & 0 \\ 0 & 0 & -5 \end{vmatrix}
$$

Applying Theorem 1(e) to the second and third rows gives

$$
(-1)(-5) \begin{vmatrix} 1 & 0 & 0 \\ 0 & 1 & 0 \\ 0 & 0 & 1 \end{vmatrix} = 5 \quad \text{by Definition 1(a)}
$$

Example 2.

$$
\begin{vmatrix} 1 & 2 & 3 \\ 2 & 3 & 4 \\ 3 & 4 & 5 \end{vmatrix} =
\begin{vmatrix} 1 & 2 & 3 \\ 0 & -1 & -2 \\ 0 & -2 & -4 \end{vmatrix} =
\begin{vmatrix} 1 & 2 & 3 \\ 0 & -1 & -2 \\ 0 & 0 & 0 \end{vmatrix} = 0
$$

using Theorem 1(f) three times and then (g). □

Remark. If you are not interested in logical subtleties, ignore this paragraph. If you are, you may have noticed a weakness in the arguments above. A determinant was defined as a number with certain properties (a),(b),(c). We then deduced other properties (d) to (g) which allowed us to evaluate the number. In Example 1, for instance, *if* the determinant has the properties (a), (b), (c), *then* it must have the value 5. But there might conceivably be *no* number having properties (a),(b),(c); the definition might be inconsistent. In this case all the arguments collapse. The logic will be made watertight in section 6D, which gives an explicit formula for the general n by n determinant and proves that the definition is consistent.

Exercise 3. (For readers of the above remark.) Refute the argument: 'Definition 1 cannot be inconsistent, at least for 3 by 3 determinants, because the method of Examples 1 and 2 gives perfectly sensible and consistent values for 3 by 3 determinants.'

6C OTHER WAYS OF EVALUATING DETERMINANTS

In the last section we evaluated $\det(A)$ by reducing A to the unit matrix. Some of this work is unnecessary. One only needs to reduce to the echelon form (which for square matrices is upper triangular, as defined in Example 3D-6).

Theorem 1 (triangular determinants). The determinant of an upper triangular matrix is the product of its diagonal entries.

Proof. (a) Suppose first that A is an n by n upper triangular matrix with all diagonal entries nonzero. Then A can be reduced to the unit matrix as follows. First, each diagonal entry a_{ii} is transformed to 1 by dividing the ith row by a_{ii}. This multiplies the determinant by $1/a_{ii}$, for $i = 1, \ldots, n$. Then all the entries above the main diagonal are reduced to zero by Type III operations, which do not change the determinant (Theorem 6B-1). The result is the unit matrix, with determinant 1. Hence $\det(A)(1/a_{11})\ldots(1/a_{nn}) = 1$, so $\det(A)$ is the product of the diagonal entries a_{ii}.

(b) Now suppose that A is upper triangular with a diagonal entry zero, say $A_{jj} = 0$. We shall show that $\det(A) = 0$.

The jth row has at least j leading zeros. Apply Gauss reduction to all the rows below the jth, giving a new matrix B, say. The $(j+1)$th row of B has at least $j+1$ leading zeros, the $(j+2)$th row has at least $j+2$ leading zeros, and so on; hence the nth row has at least n zeros. Thus the last row of B consists entirely of zeros, so $\det(B) = 0$ by Theorem 6B-1. But Gauss reduction multiplies determinants by a nonzero factor, hence $\det(A) = 0$. ☐

Example 1. In Example 6B-1, we can write down the answer after the second step. There is no need to reduce to the unit matrix. ☐

The eigenvalue problems in Chapter 5 led to matrices whose entries are functions of λ. Gauss reduction is a clumsy way of evaluating such a determinant; but there is a better way. It involves new ideas, which we shall introduce in terms of the 3 by 3 case.
Consider the determinant

$$P = \begin{vmatrix} a & b & c \\ d & e & f \\ g & h & i \end{vmatrix}$$

The first row is $(a, b, c) = a(1, 0, 0) + b(0, 1, 0) + c(0, 0, 1)$. Since the determinant is linear in the first row, we have

$$\begin{vmatrix} a & b & c \\ d & e & f \\ g & h & i \end{vmatrix} = a\begin{vmatrix} 1 & 0 & 0 \\ d & e & f \\ g & h & i \end{vmatrix} + b\begin{vmatrix} 0 & 1 & 0 \\ d & e & f \\ g & h & i \end{vmatrix} + c\begin{vmatrix} 0 & 0 & 1 \\ d & e & f \\ g & h & i \end{vmatrix} \quad (1)$$

Consider the first determinant on the right of (1). Subtracting multiples of the first row from the others gives

$$\begin{vmatrix} 1 & 0 & 0 \\ 0 & e & f \\ 0 & h & i \end{vmatrix} = \begin{vmatrix} 1 & 0 & 0 \\ 0 & e & f \\ 0 & 0 & i - fh/e \end{vmatrix} = ei - fh \quad \text{by Theorem 1, if } e \neq 0$$

(2)

If $e = 0$ we interchange the last two rows of (2):

$$\begin{vmatrix} 1 & 0 & 0 \\ 0 & 0 & f \\ 0 & h & i \end{vmatrix} = - \begin{vmatrix} 1 & 0 & 0 \\ 0 & h & i \\ 0 & 0 & f \end{vmatrix} = -hf$$

by Theorem 1 again. In both cases we have

$$\begin{vmatrix} 1 & 0 & 0 \\ 0 & e & f \\ 0 & h & i \end{vmatrix} = ei - fh = \begin{vmatrix} e & f \\ h & i \end{vmatrix}$$

For the second term on the right of (1), we have

$$\begin{vmatrix} 0 & 1 & 0 \\ d & e & f \\ g & h & i \end{vmatrix} = - \begin{vmatrix} d & e & f \\ 0 & 1 & 0 \\ g & h & i \end{vmatrix} = - \begin{vmatrix} d & e & f \\ 0 & 1 & 0 \\ 0 & 0 & i - fg/d \end{vmatrix} = -(di - fg)$$

provided that $d \neq 0$. If $d = 0$, start again and interchange the first and last rows, getting

$$- \begin{vmatrix} g & h & i \\ 0 & e & f \\ 0 & 1 & 0 \end{vmatrix} = \begin{vmatrix} g & h & i \\ 0 & 1 & 0 \\ 0 & e & f \end{vmatrix} = \begin{vmatrix} g & h & i \\ 0 & 1 & 0 \\ 0 & 0 & f \end{vmatrix} = fg$$

In both cases the answer equals

$$-(di - fg) = - \begin{vmatrix} d & f \\ g & i \end{vmatrix}$$

A similar argument shows that

$$\begin{vmatrix} 0 & 0 & 1 \\ d & e & f \\ g & h & i \end{vmatrix} = \begin{vmatrix} d & e \\ g & h \end{vmatrix}$$

Putting all this together, we have

$$P = \begin{vmatrix} a & b & c \\ d & e & f \\ g & h & i \end{vmatrix} = a \begin{vmatrix} e & f \\ h & i \end{vmatrix} - b \begin{vmatrix} d & f \\ g & i \end{vmatrix} + c \begin{vmatrix} d & e \\ g & h \end{vmatrix}$$

(3)

This is a very useful formula. It expresses a 3 by 3 determinant in terms of 2 by 2 determinants which can be written down at once. It can be generalized as follows. Each term on the right of (3) is the product of an entry of the first row with a determinant obtained from P by deleting the row and column containing that entry. There is also a plus or a minus sign. The 2 by 2 determinants are called 'minors', and when the sign is included they are called 'cofactors'. Thus the cofactor corresponding to b is

$$-\begin{vmatrix} d & f \\ g & i \end{vmatrix}$$

Things work in the same way for n by n determinants.

Definition 1. The **(i,j)-minor** of an n by n matrix is the $(n-1)$ by $(n-1)$ determinant obtained by deleting the ith row and the jth column (the row and column containing the (i,j) entry). The **cofactor** of the (i,j) entry is $(-1)^{i+j}$ times the (i,j)-minor.

Example 2. The cofactor of a in the determinant P of (3) equals its minor

$$\begin{vmatrix} e & f \\ h & i \end{vmatrix}.$$

The cofactor of d is

$$-\begin{vmatrix} b & c \\ h & i \end{vmatrix}$$

because $(-1)^{2+1} = -1$.   □

The signs $(-1)^{i+j}$ form a simple pattern:

$$
\begin{array}{ccccccc}
+ & - & + & - & + & . & . & . \\
- & + & - & + & - & . & . & . \\
+ & - & + & - & + & . & . & . \\
- & + & - & + & - & . & . & . \\
+ & - & + & - & + & . & . & . \\
. & . & . & . & . & . & .
\end{array}
$$

The pattern can be used to find the signs; for example, the second row shows that the cofactors of the entries in the second row of a 3 by 3 determinant are respectively, $-$, $+$ and $-$ the corresponding minors. This agrees with the signs in (3).

    The manipulations leading to (3) generalise to n by n matrices. One can generalise in another direction too. There is nothing special about the first row. Indeed, any two rows of a determinant can be interchanged, at the cost of a minus sign. Hence there should be a formula like (3) involving the second or third row instead of the first. Both these generalisations are contained in the following useful result.

Theorem 2 (expansion by rows). Every determinant equals the sum of the entries in any one row multiplied by the corresponding cofactors.   □

This theorem can be proved by the method used above to deduce (3), but a neater proof will be given in section 6D.

Example 3. To evaluate

$$\begin{vmatrix} 1 & 3 & -2 \\ 0 & -1 & 2 \\ 4 & 1 & 1 \end{vmatrix}$$

it is easiest to expand by the second row: the first term is then zero, giving $(-1)(1+8) - 2(1-12) = 13$.   □

   This is an efficient way of evaluating 3 by 3 determinants. For larger determinants, Gauss reduction is often better, but it is largely a matter of taste.
   Finally, we shall discuss a property of determinants which is simple but far from obvious.

Theorem 3 (row–column symmetry). The determinant of any matrix equals the determinant of its transpose.   □

You can easily verify this by working out examples, but at this stage it is not at all clear why it holds in general. A proof will be given in section 6D.
   Transposing a matrix interchanges rows and columns. Thus Theorem 3 shows that the general properties of the rows of a determinant also apply to the columns.

Corollary 1 (column properties). A determinant is a linear function of its columns, it vanishes if two columns are identical, and elementary operations on columns have the same effect as elementary operations on rows.

Proof. Suppose B is obtained by interchanging two columns of the square matrix A. Then B^T is obtained from A^T by interchanging two rows, and we have $\det(B) = \det(B^T) = -\det(A^T) = -\det(A)$. The other properties are proved in the same way.
                                                                              □

Theorem 3 can be used to extend Theorem 2, as follows.

Theorem 4 (expansion by rows or columns). Every determinant equals the sum of the entries in any row or column multiplied by the corresponding cofactors.   □

This is easily proved in the same way as Corollary 1. It gives a very flexible way of evaluating determinants; one can often choose a row or column with many zeros, so as to simplify the calculation.
   One more property of determinants should be mentioned here: their relation to matrix multiplication. If A and B are n by n matrices, then $\det(AB) = \det(A)\det(B)$. This is reasonable and simple to state, but not quite so easy to prove: see section 6E.
   We have now dealt with all the properties of determinants which are needed in later chapters. The rest of the chapter is optional; it gives a different approach to the whole subject.

*6D   THE PERMUTATION DEFINITION OF DETERMINANTS

This optional section develops the theory of determinants from a different point of view, which has two advantages. It makes some of the properties of determinants easier to prove. And, unlike the definition in section 6B, it gives an explicit general formula for det (A). This lays to rest the doubts expressed at the end of section 6B.

We begin by looking closely at a general 3 by 3 matrix A. Expanding by the first row gives

$$\det (A) = a_{11}a_{22}a_{33} - a_{11}a_{23}a_{32} - a_{12}a_{21}a_{33} + \text{three other terms} \qquad (1)$$

This is the sum of products of the form $\pm a_{1i}a_{2j}a_{3k}$, where i, j, k are the numbers 1, 2, 3 arranged in some order. In order to understand the signs in (1), we must study the properties of such rearrangements.

Definition 1. A permutation of the numbers $1, 2, \dots, n$ is a rearrangement of them. More precisely, a permutation p of the set $S = \{1, \dots, n\}$ is a mapping from S to itself such that $p(i) \neq p(j)$ when $i \neq j$. This implies that every number in S appears exactly once in the list $p(1), p(2), \dots, p(n)$. The number $p(i)$ is called the **image** of i under the permutation p.

We include the 'rearrangement' which consists of leaving them in their original order; this **identity** permutation is defined by $p(i) = i$ for all i in S.   □

The number of different permutations of the set $\{1, \dots, n\}$ is $n(n-1)(n-2)\dots 2.1$, usually written $n!$. This is easy to see: $p(1)$ can be chosen in n ways, leaving $n-1$ choices for $p(2)$ (because $p(2)$ cannot equal $p(1)$), $n-2$ choices for $p(3)$, and so on, giving $n(n-1)\dots = n!$ possibilities altogether.

A permutation can be specified by writing the numbers $1, \dots, n$ in a row with the images $p(1), \dots, p(n)$ underneath.

Example 1

$$\begin{pmatrix} 1 & 2 & 3 & 4 \\ 3 & 1 & 2 & 4 \end{pmatrix}$$

is a permutation of 1, 2, 3, 4. Call this permutation f; then $f(1) = 3, f(2) = 1, f(3) = 2$ and $f(4) = 4$.   □

Definition 2. An inverted pair in a permutation p is a pair of numbers whose order is changed by p. In other words, it is a pair of numbers i, j with $i < j$ and $p(i) > p(j)$.

Example 2. In Example 1, $\{1, 2\}$ and $\{1, 3\}$ are inverted pairs. They can be spotted by drawing lines joining each number in the top row of the permutation symbol to the same number in the bottom row, thus:

1   2   3   4

3   1   2   4

Each intersection corresponds to an inverted pair: if the lines ending at two numbers r and s, say, on the bottom row cross each other, then the numbers vertically above r and s form an inverted pair (a moment's thought should make this obvious). Thus the crossed lines ending 1 and 3 show that $\{1, 2\}$ is an inverted pair, and the crossed lines ending at 2 and 3 show that $\{1, 3\}$ is an inverted pair.　　□

Definitions 3. A permutation is said to be **odd** if it has an odd number of inverted pairs, and **even** otherwise. We define the **sign** of a permutation p by $\operatorname{sgn}(p) = +1$ if p is even, and $\operatorname{sgn}(p) = -1$ if p is odd.

Example 3. The permutation of Example 1 is even. The identity permutation is even; it clearly has no inverted pairs. The permutation

$$\begin{pmatrix} 1 & 2 & 3 & 4 \\ 4 & 1 & 2 & 3 \end{pmatrix}$$

is odd.　　□

The set of all permutations of $\{1, \dots, n\}$ splits into two classes, the even and the odd permutations. Given a permutation in one class, a permutation in the other class is obtained by an interchange, as defined below.

Definition 4. Two permutations p and q are said to be **related by an interchange** if there are two numbers i and j such that $p(i) = q(j)$, $p(j) = q(i)$, and $p(k) = q(k)$ for all numbers k other than i and j.　　□

Expressed in plain language, this says that p and q are the same except that the images of i and j are interchanged. In terms of the permutation symbol, q is obtained by interchanging the ith and jth entries in the bottom row of p, and vice versa.

Proposition 1 (interchanges). If p and q are related by an interchange, then $\operatorname{sgn}(p) = -\operatorname{sgn}(q)$.

Proof. If two adjacent elements are interchanged, then the sign changes. This is clear from the diagram in Example 2: if the lines leading to two adjacent entries on the bottom row are not crossed, then interchanging the entries crosses them; and if the lines were originally crossed, then interchanging the entries uncrosses them.

Now suppose two elements i, j on the bottom row of a permutation symbol are separated by a single element m. Performing the three adjacent interchanges (im), (ij), (jm) in succession has the effect of interchanging i and j, leaving m in the middle. This operation multiplies the sign by $(-1)^3 = -1$. It is easy to extend this argument by induction, showing that any interchange can be achieved by an odd number of adjacent interchanges. This completes the proof.　　□

Exercise 1. Write out the induction proof referred to above.　　□

We can now return to determinants, armed with a knowledge of permutations. The expression (1) for a 3 by 3 determinant has 6 terms, corresponding to the $3! = 6$

possible permutations of $\{1, 2, 3\}$. Each term is of the form $\pm a_{1i}a_{2j}a_{3k}$, where $\{i, j, k\}$ is a permutation of $\{1, 2, 3\}$; and the sign of the term is the sign of that permutation. The generalisation to n by n determinants is fairly obvious; it is given in the following theorem.

Theorem 1 (determinant in terms of permutations). Let A be an n by n matrix. Then

$$\det(A) = \sum_p \text{sgn}(p) a_{1p(1)} a_{2p(2)} \cdots a_{np(n)} \tag{2}$$

summed over the $n!$ permutations of $\{1, \ldots, n\}$.

Proof. We must prove that the value of $\det(A)$ given by (2) has the properties listed in Definition 6B-1:
  (a) $\det(I) = 1$;
  (b) $\det(A) = 0$ if A has two identical rows;
  (c) $\det$ is a linear function of each row.
  Proof of (a). The term in (2) corresponding to the identity permutation is $a_{11}a_{22} \ldots a_{nn} = 1$ when $A = I$. All the other terms are zero because they have at least one factor a_{ij} with $j \neq i$, which is zero. Hence $\det(I) = 1$.
  Proof of (b). The terms in (2) cancel out in pairs if two rows of A are equal. It is easy to see this by writing out the six terms for a 3 by 3 determinant. The argument below is provided for readers who are not satisfied with a typical example and prefer laborious general proofs.
  Suppose the ith and the jth rows are the same, so that

$$a_{ir} = a_{jr} \quad \text{for all } r \tag{3}$$

For any permutation p, define a new permutation $q = C(p)$ to be the same as p except that the images of i and j are interchanged, that is, $q(i) = p(j)$ and $q(j) = p(i)$. The permutations p and $C(p)$ have opposite signs by Proposition 1. We shall show that each term in (2) cancels out against the term related to it by the interchange C.
  Let e be an even permutation, and set $f = C(e)$. Then $e(i) = f(j)$, $e(j) = f(i)$, and $e(k) = f(k)$ if $k \neq i, j$. Hence all the factors in the e and the f terms in (2) are the same except the ith and the jth, which are $a_{ie(i)}$ and $a_{je(j)}$ in the e term, and $a_{if(i)}$ and $a_{jf(j)}$ in the f term. But $a_{ie(i)} = a_{if(j)} = a_{jf(j)}$ by (3), and similarly $a_{je(j)} = a_{if(i)}$. Thus all factors in the e term are the same as those in the f term; since they have opposite signs, they add to 0. Every even term in (2) cancels out against an odd term in this way, giving $\det(A) = 0$.
  Proof of (c). Suppose the ith row is a linear combination of two n-vectors b and c:

$$a_{ij} = Bb_j + Cc_j \quad \text{for } j = 1, \ldots, n$$

for some numbers B and C. Each term in (2) contains just one entry from the ith row, and is therefore the sum of two terms, proportional to B and C respectively. Collecting these terms together gives

$$\det(A) = B \det(A_1) + C \det(A_2)$$

where A_1 and A_2 are the matrices obtained from A by replacing the ith row by b and c respectively. This is the required linearity property. $\square$

Equation (2) is often taken as the definition of determinants, and everything else deduced from it. I feel that there is no need to grapple with permutations before understanding and using determinants, and the approach of sections 6A, B, C is simpler and clearer; but it is a matter of taste. There is no doubt, however, that (2) is almost useless for calculating determinants in practice. For a 20 by 20 determinant (very small by the standards of modern computing practice), evaluating the formula (2) requires about 10^{20} arithmetical operations, while the Gauss reduction method of section 6C requires only about 3000. For a fast computer, the Gauss reduction method takes around 0.003 seconds, while the permutation method would take over a million years.

In sections 6B and 6C we deduced many useful properties from Definition 6B-1, but left one or two unproved, because they are easier to prove using permutations. In particular, the fact that $\det(A) = \det(A^T)$ follows directly from equation (2); in the 3 by 3 case it is easily verified by writing out the terms. To prove it in general requires a little more theory.

We define the **inverse** of a permutation p to be the permutation p^{-1} which reverses its effect, so that p followed by p^{-1} gives the identity. In other words, if $p(i) = j$ then $p^{-1}(j) = i$.

Proposition 2. A permutation and its inverse have the same sign.

Proof. If r, s is an inverted pair for a permutation p, then

$$(r - s)[p(r) - p(s)] < 0$$

Set $t = p(r)$ and $u = p(s)$; then the above inequality becomes

$$[p^{-1}(t) - p^{-1}(u)](t - u) < 0$$

Hence u, t is an inverted pair for p^{-1}. For each inverted pair of p there is an inverted pair for p^{-1}, and vice versa by a similar argument, so p and p^{-1} have the same number of inverted pairs and hence the same sign.   □

Theorem 2 (determinant of transpose). $\det(A^T) = \det(A)$ for all square matrices A.

Proof. Since $(A^T)_{ij} = (A)_{ji} = a_{ji}$ in the notation of this section, we have

$$\det(A^T) = \sum_p \operatorname{sgn}(p) a_{p(1)1} a_{p(2)2} \cdots a_{p(n)n}$$

In each term the ith factor is $a_{p(i)i} = a_{jq(j)}$ where $j = p(i)$ and q is the inverse of p. The factors can now be rearranged to give

$$\det(A^T) = \sum_p \operatorname{sgn}(p) a_{1q(1)} a_{2q(2)} \cdots a_{nq(n)} \quad \text{where } q = p^{-1}$$

$$= \sum_q \operatorname{sgn}(q) a_{1q(1)} a_{2q(2)} \cdots a_{nq(n)} = \det(A)$$

since $\operatorname{sgn}(p) = \operatorname{sgn}(p^{-1})$ as shown above, and when p runs through all $n!$ permutations so does p^{-1}.   □

Finally we shall prove the expansion of an n by n determinant in terms of $(n-1)$ by $(n-1)$ determinants. It is very boring, and should be skipped by all but the keenest readers.

Theorem 3 (expansion in cofactors). Every determinant equals the sum of the entries in any one row or column multiplied by the corresponding cofactors.

Proof. First consider expanding by the nth row. In the sum (2), some terms will have a_{n1} as a factor, some will have a_{n2} as a factor, and so on. Each term contains just one of these entries, so the $n!$ terms are divided into n groups, the first group having a_{n1} as a common factor, the second group having a_{n2}, and so on. The number of terms in each group is $(n!)/n = (n-1)!$.

Consider the nth group. It can be written as $a_{nn}A_{nn}$ where A_{nn}, the coefficient of a_{nn} in $\det(A)$, is given by

$$A_{nn} = \sum \text{sgn}(p)a_{1p(1)}\cdots a_{(n-1)p(n-1)} \tag{4}$$

summed over all permutations p such that $p(n) = n$. These are just the permutations of $\{1,\ldots,n-1\}$, and the terms in the sum are products of entries of the $(n-1)$ by $(n-1)$ matrix obtained by deleting the last row and column of A. Hence A_{nn} is the determinant of that matrix, which equals the (n, n) cofactor (see Definition 6C-1). Thus we have proved the following:

the coefficient of a_{nn} in $\det(A)$ is the (n, n) cofactor $\tag{5}$

We deal with the jth group of terms by reduction to the case above. Consider the matrix A' obtained from A by exchanging the jth column with the column to its right $n-j$ times until it reaches the end. Writing C_r for the rth column of A, we have

$$A' = (C_1 \quad C_2 \ldots C_{j-1} \quad C_{j+1} \ldots C_n \quad C_j)$$

Let M_{rs} and M'_{rs} denote the (r, s)-minors of A and A' respectively. Then

$$M'_{nn} = M_{nj} \tag{6}$$

since they are both obtained from A by deleting the nth row and the jth column.

Since A' is obtained from A by $n-j$ interchanges,

$$\det(A') = (-1)^{n-j}\det(A) = (-1)^{n+j}(-1)^{-2j}\det(A)$$

so

$$\det(A') = (-1)^{n+j}\det(A) \tag{7}$$

Let A_{nj} denote the coefficient of a_{nj} in $\det(A)$. Since a_{nj} is the (n, n) entry of A', it follows from (5) that the coefficient of a_{nj} in $\det(A')$ is M'_{nn}. It now follows from (7) that

$$A_{nj} = (-1)^{n+j}M'_{nn}$$
$$= (-1)^{n+j}M_{nj}$$

by (6), and this is the (n, j) cofactor of A. Thus the coefficient in (2) of each entry in the last row is the cofactor of that entry, which completes the proof that a matrix can be expanded in terms of the last row.

To deal with expansion about the ith row, move the ith row to the end by $n - i$ interchanges and use an argument similar to the above. Finally, expansions about columns can be reduced to expansions about rows by Theorem 2.   □

This completes our account of the theory of determinants, except for two things. The next section proves that the determinant of a product is the product of the determinants, and the following section discusses the relation of determinants to matrix inversion.

*6E   THE MULTIPLICATION THEOREM

This optional section uses the theory of elementary matrices developed in section 4L.
    The fact that $\det(AB) = \det(A)\det(B)$ follows in the 2 by 2 case from the geometrical ideas of section 6A: two transformations in succession give a net magnification equal to the product of the individual magnifications. For the general n by n case we shall give an algebraic proof, showing that the multiplication rule applies to elementary matrices, and then using the fact that any matrix can be built up from elementary matrices.

Lemma 1. If E is an n by n elementary matrix, then $\det(EA) = \det(E)\det(A)$ for any n by n matrix A.

Proof. (i) If E corresponds to a row interchange, then $\det(EA) = -\det(A)$ for any A. Taking $A = I$ here gives $\det(E) = -1$. Hence $\det(EA) = -\det(A) = \det(E)\det(A)$, which proves the lemma for this type of elementary operation.
    (ii) If E corresponds to multiplying a row by k, then $\det(EA) = k\det(A)$ by Theorem 6B-1. Taking $A = I$ again gives $\det(E) = k$, hence $\det(EA) = \det(E)\det(A)$.
    (iii) If E corresponds to a Type III elementary operation, then $\det(EA) = \det(A)$ by Theorem 6B-1. Again, taking $A = I$ gives $\det(E) = 1$, so we have $\det(EA) = \det(E)\det(A)$ in this case too.   □

Theorem 1 (multiplication of determinants). For any two n by n matrices A and B, $\det(AB) = \det(A)\det(B)$.

Proof. If A is singular, then so is AB by the rank-of-product theorem, and $\det(AB) = 0 = \det(A)\det(B)$ in this case. If A is nonsingular, then it is a product of elementary matrices $E_1 \ldots E_r$ by Theorem 4L-3, so

$$\det(AB) = \det(E_1 E_2 \ldots E_r B)$$
$$= \det(E_1)\det(E_2 \ldots E_r B) \quad \text{by Lemma 1}$$
$$= \det(E_1)\det(E_2)\det(E_2 \ldots E_r B)$$
$$= \ldots = \det(E_1)\det(E_2)\ldots\det(E_r)\det(B)$$
$$= \det(E_1)\det(E_2)\ldots\det(E_{r-2})\det(E_{r-1}E_r)\det(B) \quad \text{by Lemma 1}$$
$$= \det(E_1)\det(E_2)\ldots\det(E_{r-3})\det(E_{r-2}E_{r-1}E_r)\det(B)$$
$$= \ldots = \det(E_1 E_2 \ldots E_r)\det(B) = \det(A)\det(B)$$ □

To prove this simple fact directly from the permutation definition is quite difficult; the easy proof above illustrates the power of the elementary-matrix decomposition of nonsingular matrices.

*6F   THE ADJUGATE MATRIX

This optional section is not part of our main line of development. It deals with an extension of the theory of cofactors which is often mentioned, and occasionally useful.

We have seen that a determinant equals the sum of the entries in a row or column multiplied by the cofactors. Using the cofactors of a different row or column gives zero, as the following theorem shows.

Theorem 1. If A_{ij} denotes the (i,j) cofactor of the matrix $A = (a_{ij})$, then

$$\sum_k a_{ik} A_{jk} = \delta_{ij} \det(A) \tag{1}$$

and

$$\sum_k a_{ki} A_{kj} = \delta_{ij} \det(A) \tag{2}$$

Proof. We have dealt with the case $i = j$ in Theorem 6D-3. So suppose $i \neq j$, and let A' be the matrix obtained from A by replacing the jth row by the ith row. In other words

$$a'_{jk} = a'_{ik} = a_{ik} \quad \text{for all } k$$

and all other entries of A and A' are the same. Then A' has two identical rows, so $\det(A') = 0$. Since A and A' are the same except for the jth row, the (j,k) cofactors of A and A' are the same. Therefore expanding A' by the jth row gives

$$\sum_k a'_{jk} A_{jk} = 0$$

But $a'_{jk} = a_{ik}$, so we have proved (1). The same method applied to columns proves (2).   □

Definition 1. The **adjugate** adj(A) of a square matrix A is the transpose of the matrix of cofactors of A. That is, $[\text{adj}(A)]_{ij} = A_{ji}$ for all i, j, where A_{ji} denotes the (j, i) cofactor of A.

Note. Some authors use the term 'adjoint', which is confusing because 'adjoint matrix' also means something quite different (see Chapter 11).   □

Equation (1) now gives $\sum_k a_{ik} [\text{adj}(A)]_{kj} = \delta_{ij} \det(A)$. Since δ_{ij} is the (i,j) entry of the unit matrix I, we have

$$A \cdot \text{adj}(A) = \det(A) \cdot I \tag{3}$$

If $\det(A) \neq 0$, this gives an expression for the inverse of A.

Theorem 2. If A is invertible, then $A^{-1} = \text{adj}\,(A)/\det\,(A)$.

Proof. This follows from (3) on dividing by $\det(A)$.   □

This is a nice explicit formula for the inverse of a matrix. But it is a clumsy way of calculating inverses in practice; the method of section 4I is much more efficient.

Exercise 1. Take a 3 by 3 matrix and calculate its inverse by the elementary-operation method of section 4I and by the adjugate-matrix method. Estimate how many arithmetical operations would be needed to do the same for a 5 by 5 matrix.

PROBLEMS FOR CHAPTER 6

Hints and answers for problems marked [a] will be found on page 282.

Sections 6A, 6B, 6C

[a]1. Use row or column operations, and/or expansion in terms of cofactors, to evaluate the following:

(a)
$$\begin{vmatrix} 1 & 2 & 5 \\ 2 & 3 & -1 \\ 1 & 3 & 16 \end{vmatrix}$$
(b)
$$\begin{vmatrix} 2 & 1 & 3 & 2 \\ 3 & 0 & 1 & -2 \\ 1 & -1 & 4 & 3 \\ 2 & 2 & -1 & 1 \end{vmatrix}$$

2. Invent more examples like Problem 1.

3. Show that
$$\begin{vmatrix} a & -b & -a & b \\ b & a & -b & -a \\ c & -d & c & -d \\ d & c & d & c \end{vmatrix} = 4(a^2 + b^2)(c^2 + d^2)$$

[a]4. The numbers 5198, 1035, 8257 and 3956 are divisible by 23. Show that
$$\begin{vmatrix} 5 & 1 & 9 & 8 \\ 1 & 0 & 3 & 5 \\ 8 & 2 & 5 & 7 \\ 3 & 9 & 5 & 6 \end{vmatrix}$$
is divisible by 23.

[a]5. (a) For a fixed $n > 1$, let M be the space of all n by n real matrices. Define a map $f: M \to \mathbb{R}$ by $f(A) = \det(A)$ for all A in M. Is f linear? If so, what is its kernel?
  (b) Given vectors $u_1, \ldots, u_{n-1}$ in $\mathbb{R}^n$, define $f: \mathbb{R}^n \to \mathbb{R}$ by $f(v) = \det(A(v))$, where $A(v)$ is the matrix with columns $v, u_1, \ldots, u_{n-1}$. Is f linear? If so, what is its kernel?

6. Consider the statement: 'For any n by n matrix A, $\det(A)$ is a sum of terms, each of which is $\pm$ a product of n entries of A, no two of which are from the same row or column.'

Prove it for $n = 3$ by means of the cofactor expansion. Then prove it for $n = 4$, expanding by the last column and using the result for $n = 3$. Use induction to extend the result to all n.

7. (a) Using the results of section 6A, show that the area of the triangle with vertices $(0,0), (x_1, y_1), (x_2, y_2)$ is

$$\pm (1/2) \begin{vmatrix} x_1 & y_1 \\ x_2 & y_2 \end{vmatrix}$$

(b) Show that the area of the triangle with vertices (x_i, y_i), $i = 1, 2, 3$, is

$$\pm (1/2) \begin{vmatrix} x_1 & y_1 & 1 \\ x_2 & y_2 & 1 \\ x_3 & y_3 & 1 \end{vmatrix}$$

(c) Can you extend this result to give the volume of a tetrahedron?

a 8. Prove that every skew-symmetric matrix (see Chapter 2, Problem 4) is singular.

9. M is the n by n matrix with every entry equal to a except for the diagonal entries, which all equal $a + b$. Show that $\det(M) = b^{n-1}(b + na)$.

a 10. An n by n matrix A is defined by $A = xe + I$, where x is a given column vector and $e = (1, 1, 1, \ldots, 1)$. Show that $\det(A) = 1 + x_1 + x_2 + \ldots + x_n$.

Hence evaluate the n by n determinant with (i, j) entry a_i if $i \neq j$ and all diagonal entries equal to y.

a 11. A is the n by n matrix with (r, s) entry $p^{(r-1)(s-1)}$, where p is a complex number satisfying $p^n = 1$. By considering the product $\bar{A}^T A$, find the modulus of $\det(A)$.

a 12. For any n by n matrix A, construct a **bordered determinant** D by adding a border round two sides as follows:

$$D = \begin{vmatrix} a_{11} & a_{12} & a_{13} & \cdot & \cdot & \cdot & a_{1n} & x_1 \\ a_{21} & a_{22} & a_{23} & \cdot & \cdot & \cdot & a_{2n} & x_2 \\ \cdot & \cdot & \cdot & \cdot & \cdot & \cdot & & \cdot \\ a_{n1} & a_{n2} & a_{n3} & \cdot & \cdot & \cdot & a_{nn} & x_n \\ x_1 & x_2 & x_3 & \cdot & \cdot & \cdot & x_n & z \end{vmatrix}$$

Show that $D = z \det(A) - \sum_{ij} A_{ij} x_i x_j$, where A_{ij} is the (i, j) cofactor of A.

13. The concept of rank can be expressed in terms of determinants as follows. A **kth order minor** of a p by q matrix A is a determinant formed by choosing k rows and

k columns of A and discarding the rest. Note that a minor as defined in section 6C is a minor of order $n-1$ according to this definition.

Define the **determinantal rank** of A as the largest number k such that A has a nonzero minor of order k.

(a) Take a simple example, and verify that determinantal rank equals rank as defined in Chapter 4.

(b) Give a general argument showing that the two notions of rank agree.

(Historically, the determinant definition came first, and the theory of Chapter 4 was developed later.)

14. Given n numbers $x_1,\ldots,x_n$, define the **Vandermonde matrix** $V(x_1,x_2,\ldots,x_n)$ to be the n by n matrix with (i,j) entry $(x_j)^{i-1}$. Write $DV(x_1,\ldots,x_n)$ for its determinant.

(a) For $n=2$ and 3, verify that $DV(x_1,\ldots,x_n)=\Pi_{i>j}(x_i-x_j)$, where $\Pi_{i>j}$ means the product of all the terms obtained by giving i and j all values from 1 to n subject to $i>j$.

(b) Use induction to prove that the formula in (a) holds for all n. Deduce that $V(x_1,x_2,\ldots,x_n)$ is nonsingular provided that the numbers $x_1,\ldots,x_n$ are all different.

(c) Given n numbers $x_1<x_2<\ldots x_n$, set $D=DV(x_1,\ldots,x_n)$. For any number x let $D_i(x)$ be the polynomial obtained by replacing x_i by x in D. Set $L_i(x)=D_i(x)/D$ for $i=1,\ldots,n$; it is just a constant times the polynomial D_i.

Prove that $L_i(x_j)=\delta_{ij}$. Deduce that $\{L_1,\ldots,L_n\}$ is linearly independent, and therefore a basis for the space of polynomials of degree less than n.

15. (a) Use the functions L_i of Problem 14(c) to solve the following problem. The values of a function f at n points are given: $f(x_i)=y_i$ for $i=1,\ldots,n$. Find a polynomial which equals f at the points $x_1,\ldots,x_n$. (It is called an 'interpolating polynomial', and may be thought of as a convenient approximation to the function f. (For a different approach, see Chapter 11, Problem 26. For the dangers of polynomial interpolation see Burden and Faires (1985).))

(b) Take $f(x)=x^k$ in part (a), for $k=0,1,\ldots,n-1$, and deduce that $X=V(x_1,\ldots,x_n)L$, where X and L are the column vectors with kth entries x^{k-1} and $L_k(x)$ respectively.

(c) Hence justify the following rule for finding the inverse of $V(x_1,\ldots,x_n)$: its (i,j) entry is the coefficient of x^{j-1} in the polynomial L_i. (For another approach, see Problem 22.)

16. If you like the results of Problems 7–12, go to your library and look for older books on determinants and matrices; you are likely to find many beautiful formulas.

Sections 6D, 6E, 6F

17. Use the multiplication theorem to prove that the determinant of a real orthogonal matrix is either 1 or -1.

Verify that the matrix M in section 5I, equation (1), has determinant 1, while the matrix R obtained by changing the sign of M_{33} has determinant -1.

(The matrix R gives a rotation combined with reflection in the xy plane, changing the sign of the z component. In general, pure rotation matrices have determinant 1, while reflections correspond to orthogonal matrices with determinant -1. For more information, see Mirsky (1955).)

18. This problem generalises the cofactor expansion for evaluating determinants, using the general minors defined in Problem 13.

For a square matrix A, the **cofactor** of a minor M is defined as $(-1)^S$ times the minor obtained by deleting from A those rows and columns which appear in M; S is the sum of all the row numbers and column numbers appearing in M. Check that the cofactor of an $(n-1)$th-order minor as defined here agrees with Definition 6C-1.

Then $\det(A)$ can be evaluated as follows. Take any integer k between 1 and n; chose any k rows, take all the different kth-order minors $M_1, M_2, \ldots$ which include those rows, then $\det(A) = M_1 C_1 + M_2 C_2 + \ldots$, where C_i is the cofactor of the minor M_i. This result is called Laplace's expansion theorem.

Solve Problems 1(b) and 3 using Laplace's expansion, taking the first two rows. Can you extend the theory of section 6D to give a proof of Laplace's expansion?

[a]19. True or false? If a square matrix A is partitioned as

$$A = \begin{pmatrix} P & Q \\ 0 & R \end{pmatrix}$$

where P and R are square (but Q need not be), then $\det(A) = \det(P)\det(R)$.
(The answer comes easily from the Laplace expansion of Problem 18.)

20. Use the adjugate-matrix method to find the inverse of the matrix of Chapter 4, Problem 26.

21. Prove that for every nonsingular n by n matrix A, $\det(\mathrm{adj}(A)) = (\det(A))^{n-1}$, and $\mathrm{adj}(\mathrm{adj}(A)) = kA$ where k is a number (which you should find).

22. Use the adjugate to derive the expression in Problem 15(c) for the inverse of a Vandermonde matrix.

23. The **permanent** of an n by n matrix A is defined by

$$\mathrm{per}(A) = \sum_p a_{1p(1)} a_{2p(2)} \cdots a_{np(n)}$$

summed over all permutations p of $\{1, \ldots, n\}$. Prove that
   (i) $\mathrm{per}(A)$ is a linear function of each row;
   (ii) $\mathrm{per}(A)$ is unchanged if two rows are interchanged;
   (iii) $\mathrm{per}(A^T) = \mathrm{per}(A)$.

Call a permutation p 'totally moving' if for each i, $p(i) \neq i$. Prove that the number of totally moving permutations of $\{1, \ldots, n\}$ is $\mathrm{per}(J)$, where J is the n by n matrix with 0 in the main diagonal and 1 everywhere else. (The theory of permanents is in many ways parallel to that of determinants; they are useful in combinatorial mathematics. See Minc (1978).)

15

7

Eigenvalue problems and the characteristic equation

Chapter 5 defined an eigenvalue of a square matrix A as a number λ such that $A - \lambda I$ is singular. We now return to eigenvalue problems, armed with the determinant as a test for singularity of matrices. Sections 7A and 7B show how to solve eigenvalue problems using determinants. The next two sections are optional; section 7C discusses an elegant and simple way of obtaining rough estimates of eigenvalues, and section 7D shows how eigenvalues control the behaviour of the iteration methods for linear systems introduced in section 4K. The last section contains proofs of some results from earlier sections.

7A   THE CHARACTERISTIC EQUATION

An eigenvalue of an n by n matrix A is a scalar λ such that $A - \lambda I$ is singular. A square matrix is singular if and only if its determinant vanishes, therefore λ is an eigenvalue of A if and only if

$$\det(A - \lambda I) = 0 \tag{1}$$

that is,

$$\begin{vmatrix} a_{11} - \lambda & a_{12} & a_{13} & \cdots & a_{1n} \\ a_{21} & a_{22} - \lambda & a_{23} & \cdots & a_{2n} \\ a_{31} & a_{32} & a_{33} - \lambda & \cdots & a_{3n} \\ \cdot\cdot & \cdot\cdot & \cdot\cdot & \cdots & \cdot\cdot \\ a_{n1} & a_{n2} & a_{n3} & \cdots & a_{nn} - \lambda \end{vmatrix} = 0$$

This is called the 'characteristic equation' of the matrix A.

In the 2 by 2 case we have

$$\begin{vmatrix} a_{11}-\lambda & a_{12} \\ a_{21} & a_{22}-\lambda \end{vmatrix} = (a_{11}-\lambda)(a_{22}-\lambda) - a_{12}a_{21} = 0$$

a quadratic equation for λ. It is easily solved, giving at most two solutions; hence a 2 by 2 matrix can have at most two eigenvalues.

For a 3 by 3 matrix, expanding $\det(A-\lambda I)$ gives a cubic equation in λ; an n by n matrix gives a polynomial of degree n.

Definition 1. The **characteristic polynomial** of an n by n matrix is the polynomial $p(x) = \det(A-xI)$. The **characteristic equation** of A is $p(x)=0$, where p is its characteristic polynomial.   □

Finding eigenvalues thus means solving a polynomial equation. Appendix F shows that every polynomial of degree n can be expressed as a product of n linear factors:

$$p(x) = c(x-\lambda_1)(x-\lambda_2)..(x-\lambda_n) \tag{2}$$

The numbers λ_i are the roots of the equation $p(x)=0$; they are the eigenvalues of the matrix for which p is the characteristic polynomial.

It is essential to allow the λ's to be complex. Consider the polynomial x^2+1, for example. The equation $x^2+1=0$ has no real solutions, and x^2+1 cannot be expressed in the form (2) using real numbers. The elegant linear factorisation (2) holds for complex numbers, but fails if restricted to reals. This fact has implications for eigenvalue problems.

In section 5H we considered a real matrix with no real eigenvalues. One wonders whether there are complex matrices with no complex eigenvalues. The answer is no.

Theorem 1. Every square matrix has at least one eigenvalue over $\mathbb{C}$.

Proof. Its characteristic equation, like every polynomial equation (see Appendix F), has at least one solution in $\mathbb{C}$.   □

Example 1. Consider the matrix

$$A = \begin{pmatrix} 1 & -1 & 4 \\ 3 & 2 & -1 \\ 2 & 1 & -1 \end{pmatrix}$$

Its characteristic polynomial is

$$p(x) = \det(A-xI) = \begin{vmatrix} 1-x & -1 & 4 \\ 3 & 2-x & -1 \\ 2 & 1 & -1-x \end{vmatrix}$$

We use the methods of section 6B. Adding the last row to the first gives

handwritten note: multiplying any row by k, multiplies row by; 3−x

$$p(x) = \begin{vmatrix} 3-x & 0 & 3-x \\ 3 & 2-x & -1 \\ 2 & 1 & -1-x \end{vmatrix} = (3-x) \begin{vmatrix} 1 & 0 & 1 \\ 3 & 2-x & -1 \\ 2 & 1 & -1-x \end{vmatrix}$$

Expanding by the first row gives

$$p(x) = (3-x)[(2-x)(-1-x) + 1 + 3 - 2(2-x)]$$
$$= (3-x)[x^2 + x - 2] = (3-x)(x+2)(x-1)$$

The characteristic equation $p(x) = 0$ has three real roots, and the eigenvalues are 3, −2, and 1.

The eigenvectors v are the solutions of the equation $(A - \lambda I)v = 0$, that is, the vectors in the null space of the matrix $A - \lambda I$. We find the null space by Gauss–Jordan reduction. For $\lambda = 3$ we have

$$\begin{pmatrix} -2 & -1 & 4 \\ 3 & -1 & -1 \\ 2 & 1 & -4 \end{pmatrix} \rightarrow \begin{pmatrix} -2 & -1 & 4 \\ 0 & -\frac{5}{2} & 5 \\ 0 & 0 & 0 \end{pmatrix} \rightarrow \begin{pmatrix} 1 & \frac{1}{2} & -2 \\ 0 & 1 & -2 \\ 0 & 0 & 0 \end{pmatrix} \rightarrow \begin{pmatrix} 1 & 0 & -1 \\ 0 & 1 & -2 \\ 0 & 0 & 0 \end{pmatrix}$$

Since row operations do not change the null space, the eigenvectors (x, y, z) belong to the null space of the last matrix above, and therefore satisfy $x - z = 0$ and $y - 2z = 0$. Here z can be given any value, and then the equations are satisfied if $y = 2z$ and $x = z$. The eigenvectors are therefore $(a, 2a, a)$ for any number $a \neq 0$; the eigenspace for $\lambda = 3$ is one-dimensional.

The same method gives the eigenvectors for the other eigenvalues. For $\lambda = -2$ they are $(-a, a, a)$ for any a, and for $\lambda = 1$ they are $(-a, 4a, a)$. This completes the solution of the eigenvalue problem for A over the real numbers. There are three real eigenvalues (the most that a 3 by 3 matrix can have), each with a one-dimensional eigenspace.

Consider now the eigenvalue problem for this matrix over $\mathbb{C}$. Eigenvalues over $\mathbb{R}$ are also eigenvalues over $\mathbb{C}$ (see section 5H), so 3, −2, 1 are eigenvalues over $\mathbb{C}$, with the eigenvectors given above; the only difference is that a can now be complex. For this matrix, the eigenvalue problems over $\mathbb{R}$ and $\mathbb{C}$ have the same solution.

Example 2. Now consider the matrix

$$B = \begin{pmatrix} 1 & 1 & -1 & -1 \\ 1 & 1 & 1 & 1 \\ 1 & -1 & 1 & -1 \\ 1 & -1 & -1 & 1 \end{pmatrix}$$

Its characteristic polynomial is

$$p(x) = \begin{vmatrix} 1-x & 1 & -1 & -1 \\ 1 & 1-x & 1 & 1 \\ 1 & -1 & 1-x & -1 \\ 1 & -1 & -1 & 1-x \end{vmatrix} = \begin{vmatrix} 2-x & 2-x & 0 & 0 \\ 1 & 1-x & 1 & 1 \\ 1 & -1 & 1-x & -1 \\ 0 & 0 & x-2 & 2-x \end{vmatrix}$$

by adding the second row to the first and subtracting the third from the fourth. Factoring $(2-x)$ out of the first and the last rows and then subtracting the first column from the second gives

$$p(x) = (2-x)^2 \begin{vmatrix} 1 & 0 & 0 & 0 \\ 1 & -x & 1 & 1 \\ 1 & -2 & 1-x & -1 \\ 0 & 0 & -1 & 1 \end{vmatrix}$$

Expanding by the first row gives

$$p(x) = (2-x)^2 \begin{vmatrix} -x & 1 & 1 \\ -2 & 1-x & -1 \\ 0 & -1 & 1 \end{vmatrix} = (2-x)^2 \begin{vmatrix} -x & 2 & 1 \\ -2 & -x & -1 \\ 0 & 0 & 1 \end{vmatrix}$$

so the characteristic equation $p(\lambda) = 0$ reads

$$(\lambda - 2)^2 (\lambda^2 + 4) = 0 \qquad (3)$$

The only real solution is $\lambda = 2$, and the matrix has one real eigenvalue. The eigenspace is $\ker(B - 2I)$. The method of Example 1 gives

$$\begin{pmatrix} -1 & 1 & -1 & -1 \\ 1 & -1 & 1 & 1 \\ 1 & -1 & -1 & -1 \\ 1 & -1 & -1 & -1 \end{pmatrix} \rightarrow \begin{pmatrix} 1 & -1 & 0 & 0 \\ 0 & 0 & 1 & 1 \\ 0 & 0 & 0 & 0 \\ 0 & 0 & 0 & 0 \end{pmatrix}$$

Thus $x - y = 0$ and $z + w = 0$, where x, y, z, w are the components of an eigenvector. Hence $(y, y, -w, w)$ is an eigenvector of A, belonging to eigenvalue 2, for any numbers y, w not both zero (if $y = w = 0$, then v is the zero vector, which is a member of the eigenspace but not an eigenvector). The eigenspace is $\{y(1, 1, 0, 0) + w(0, 0, -1, 1): y, w \in \mathbb{R}\} = \mathrm{Sp}\{(1, 1, 0, 0), (0, 0, -1, 1)\}$; it is two-dimensional.

Now consider the eigenvalue problem over the complex numbers. We must include the complex roots of (3): $\lambda = \pm 2i$. Their eigenspaces are found in the same way as above. The eigenspace belonging to $\lambda = 2i$ is the null space of

$$\begin{pmatrix} 1 - 2i & 1 & -1 & -1 \\ 1 & 1 - 2i & 1 & 1 \\ 1 & -1 & 1 - 2i & -1 \\ 1 & -1 & -1 & 1 - 2i \end{pmatrix}$$

Using row operations as above gives the eigenspace $\mathrm{Sp}\{(i, -i, 1, 1)\}$; it is one-dimensional. The eigenspace for eigenvalue $-2i$ is $\mathrm{Sp}\{(-i, i, 1, 1)\}$.

We have two one-dimensional eigenspaces and one two-dimensional eigenspace. There is therefore a linearly independent set of four eigenvectors, forming a basis for the space $\mathbb{C}^4$. Working over $\mathbb{R}$ we found only two linearly independent eigenvectors. Hence there is an eigenvector basis if we work over the complex numbers but not if we work over the reals. Eigenvector bases are very useful, as we saw in section 5J and will see again in Chapter 8; so working over $\mathbb{C}$ is better than working over $\mathbb{R}$.

□

We have just considered a real matrix with complex eigenvalues; now for a complex matrix with real eigenvalues.

Example 3. Let

$$A = \begin{pmatrix} 3 & 1-i \\ 1+i & 4 \end{pmatrix}$$

Its characteristic equation is $(3-x)(4-x)-(1-i)(1+i)=0$, or $x^2-7x+10=0$, so the eigenvalues are 5 and 2.

The eigenvectors belonging to eigenvalue 5 satisfy

$$\begin{pmatrix} -2 & 1-i \\ 1+i & -1 \end{pmatrix}\begin{pmatrix} x \\ y \end{pmatrix} = \begin{pmatrix} 0 \\ 0 \end{pmatrix}$$

These equations are easily solved, giving an eigenvector $(1, 1+i)$ (any nonzero multiple is also an eigenvector, of course). Similarly, $(i-1, 1)$ is an eigenvector belonging to the eigenvalue 2.

In general, a matrix with complex entries is likely to have complex eigenvalues. This matrix is an exception. In Chapter 12 we shall see that it belongs to a large and important class of exceptions.

7B MULTIPLE EIGENVALUES

In the last section we solved the eigenvalue problem for

$$B = \begin{pmatrix} 1 & 1 & -1 & -1 \\ 1 & 1 & 1 & 1 \\ 1 & -1 & 1 & -1 \\ 1 & -1 & -1 & 1 \end{pmatrix}$$

We found that its characteristic equation is

$$(x-2)^2(x^2+4)=0 \qquad (1)$$

2 is an eigenvalue, with a two-dimensional eigenspace $\mathrm{Sp}\{(1,1,0,0),(0,0,-1,1)\}$. There are two other eigenvalues: 2i, with eigenspace $\mathrm{Sp}\{(i,-i,1,1)\}$; and $-2i$, with eigenspace $\mathrm{Sp}\{(-i,i,1,1)\}$.

In the language of section 5F, this matrix has a double eigenvalue, with geometric multiplicity 2. Now, there is another sense in which it is double: the factor $(x-2)$ occurs squared in the characteristic equation (1), so 2 is a double root of the equation.

Definition 1. The **algebraic multiplicity** of an eigenvalue λ is the power of $(x-\lambda)$ occurring in the characteristic polynomial $p(x)$. An eigenvalue of algebraic multiplicity 1 is called **algebraically simple.**

□

We can now say that in the example above, the algebraic multiplicity of each eigenvalue equals its geometric multiplicity: we have one double and two simple eigenvalues, in both senses. It is natural to wonder whether the two multiplicities are always equal.

Exercise 1. Solve the eigenvalue problem for the matrix

$$\begin{pmatrix} 1 & 1 \\ 0 & 1 \end{pmatrix}$$

□

This example shows that the above suggestion is definitely wrong: the matrix has an algebraically double but geometrically simple eigenvalue.

Yet it is hard to avoid feeling that there should be some relation between the two kinds of multiplicity. One relation follows from Theorem 5H-1 as follows. Eigenvectors belonging to different eigenvalues are linearly independent. There cannot be more than n linearly independent vectors in an n-dimensional space, hence the sum of the dimensions of all the eigenspaces for an n by n matrix cannot exceed n. Thus the sum of the geometric multiplicities of all the eigenvalues cannot exceed n. But if we work over $\mathbb{C}$, the sum of the algebraic multiplicities equals n, because the characteristic polynomial has n linear factors. Hence the sum of the geometric multiplicities is less than or equal to the sum of the algebraic multiplicities.

There is a stronger result: the geometric multiplicity of each eigenvalue is less than or equal to the algebraic multiplicity. But this is not so easy to prove; it needs the machinery introduced in Volume 2.

*7C GERSHGORIN'S THEOREM

This optional section depends on no other optional section. It gives an elegant method for estimating eigenvalues.

Solving the characteristic equation is a good way of finding the eigenvalues of a matrix which is not too large. But for a 50 by 50 matrix, say (quite small by the standards of current computing practice), it means solving a polynomial equation of degree 50. This is not at all easy.

One alternative is the iterative method described in section 5J. This gives approximations to the eigenvalues. The approximation can always be improved by more calculation. But there is no fully reliable way of knowing how accurate the approximation is at any given stage.

The method of this section leads to results of a different kind. The power method gives information like 'λ is somewhere around 2.36', while Gershgorin's theorem gives information like 'λ is between 2.1 and 2.9'. The Gershgorin result may be less accurate, but it has the merit of being precise and rigorous. For example, if some other method gives a value of 2.07, the Gershgorin result implies that there must be a mistake somewhere, whereas no firm conclusion could be drawn from the power-method statement.

Gershgorin's theorem has the further advantage of simplicity, involving very little calculation. It can therefore be used whenever a quick estimate of eigenvalues is needed. Iterative methods, for example, often need to start from a rough approximation to the eigenvalues; Gershgorin's theorem can be useful here.

We start from the following simple idea. For a diagonal matrix, the diagonal entries are the eigenvalues (this follows at once from the characteristic equation). If a matrix is 'almost diagonal' in some sense, then the eigenvalues should be almost equal to the diagonal entries.

But what does 'almost diagonal' mean? One might say that a matrix is almost diagonal if the diagonal entries are much larger than the others. But it is not quite as simple as that. Consider the product of an n by n matrix A with a column vector x. The first entry of Ax is $A_{11}x_1 + \ldots + A_{1n}x_n$. If n is large, then the $n-1$ terms involving off-diagonal entries may swamp the first term even if each term individually is much smaller than the first. In order for the diagonal term to dominate, it must exceed the sum of the magnitudes of the other elements in its row.

Definition 1. An n by n matrix A is **strictly diagonally dominant** if

$$|A_{ii}| > \sum_{j \neq i} |A_{ij}| \quad \text{for } i = 1, \ldots, n \tag{1}$$

where $\sum_{j \neq i}$ denotes the sum over all values of j from 1 to n excluding $j = i$.

A is **diagonally dominant** if it satisfies condition (1) with $>$ replaced by $\geq$.   □

Matrices satisfying (1) are sometimes called 'strictly row diagonally dominant' because each diagonal entry dominates its row. One can similarly define 'column diagonal dominance' where each diagonal entry dominates the sum of the other entries in its column; but we shall not need this idea.

Exercise 1. Every diagonal entry of a strictly diagonally dominant matrix is nonzero.
    □

From now on we abbreviate 'strictly diagonally dominant' to **SDD**.

An extreme example of an SDD matrix is a diagonal matrix with nonzero entries down the diagonal and zeros everywhere else. Such matrices are nonsingular. We shall show that the same applies to all SDD matrices. The method uses Gaussian elimination to reduce the matrix to diagonal form without changing the diagonal dominance.

Lemma 1. If A is SDD, so is the matrix obtained by one step of the Gauss reduction procedure, taking the $(1, 1)$ entry as pivot.   □

The best way to convince yourself of this is to work out a few examples. The proof will be found in section 7E.

Theorem 1 (the dominant diagonal theorem). Strictly diagonally dominant matrices are nonsingular.

Proof. Applying the first step of Gauss reduction gives an SDD matrix by Lemma 1. The new $(2, 2)$ entry is nonzero (Exercise 1). Hence the pivot for the next Gauss step is the leading entry of the $(n-1)$ by $(n-1)$ matrix obtained by deleting the first row and column. Lemma 1 shows that the result is again SDD.

Proceeding in this way for $n-1$ steps gives a triangular matrix with nonzero diagonal entries (Exercise 1). Therefore it is nonsingular.   □

Now consider the matrix $A - \lambda I$. If λ is large, then $A - \lambda I$ will be SDD and therefore nonsingular. Therefore large numbers cannot be eigenvalues of A. This argument is quite crude, of course; 'large' is not a very precise description. Sharpening up the argument gives the following.

Theorem 2 (Gershgorin's theorem). Every eigenvalue λ of an n by n matrix A satisfies

$$|\lambda - A_{ii}| \leq \sum_{j \neq i} |A_{ij}| \quad \text{for some } i \text{ in the range } 1, \ldots, n \qquad (2)$$

Proof. The matrix $\lambda I - A$ is SDD if $|\lambda - A_{ii}| > \sum_{j \neq i} |A_{ij}|$ for every i. Hence if (2) is not satisfied, then $\lambda I - A$ is SDD and therefore nonsingular, so λ is not an eigenvalue. If λ is an eigenvalue, it follows that (2) must hold.   □

This theorem says that each eigenvalue must lie within a distance d_i of A_{ii} for some i, where d_i is the sum on the right of (2). Eigenvalues in general are complex numbers, which can be visualised as points in a plane, the complex plane or Argand diagram (see Appendix E).

Definition 2. For $i = 1, \ldots, n$, let $d_i = \sum_{j \neq i} |A_{ij}|$. The set $D_i = \{z \in \mathbb{C} : |z - A_{ii}| \leq d_i\}$ is called the ith **Gershgorin disc** of the matrix A. (A 'disc' here means the interior of a circle plus its boundary.)   □

For an n by n matrix there are n discs in the complex plane, each one centred on a diagonal entry of A. Every eigenvalue of A must lie in one of these discs. Given a matrix, it is easy to draw the discs and obtain useful information about the eigenvalues.

Notice that the theorem says that the eigenvalues lie in the union of all the discs. There need not be an eigenvalue in each disc, as the following example shows.

Example 1. The Gershgorin discs for

$$\begin{pmatrix} 0 & -2 \\ 6 & 7 \end{pmatrix}$$

are centred on 0 and 7 with radii 2 and 6 respectively. They are shown in Fig. 7.1 together with the eigenvalues, which are easily found from the characteristic equation. Both eigenvalues lie in the larger disc, and there are none in the smaller disc.   □

The discs in this example overlap. The following stronger version of Gershgorin's theorem implies that if discs do not overlap, then each contains an eigenvalue.

Theorem 3 (Gershgorin's theorem improved). A subset G of the Gershgorin discs is called a 'disjoint group of discs' if no disc in the group G intersects a disc which is not in G. If a disjoint group contains r nonconcentric discs, then it contains r eigenvalues.   □

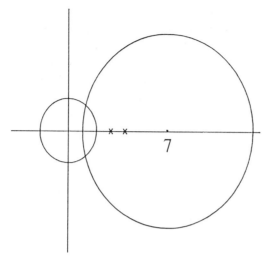

Fig. 7.1 – Gershgorin discs for $\left(\begin{smallmatrix} 0 & -2 \\ 6 & 2 \end{smallmatrix}\right)$.

For the proof of this theorem, see section 7E.

Example 2. For the matrix

$$\begin{pmatrix} 5 & 1 & 0 & -1 \\ 1 & 0 & -1 & 0.5 \\ -1.5 & 1 & -2 & 1 \\ -1 & 0.5 & 3 & -3.5 \end{pmatrix}$$

the disc with centre 5 and radius 2 is separate from the others. It forms a disjoint group containing just one disc, and therefore contains just one eigenvalue. Therefore there is at least one eigenvalue with positive real part. In section 8B we shall see that it is very useful to be able to decide quickly whether or not a matrix has an eigenvalue with positive real part; it tells you whether or not a dynamical system is stable. □

SDD matrices have other nice properties. The proof of Theorem 1 shows that zero pivots never arise in Gaussian elimination for SDD matrices, so row interchanges are not needed. In the language of section 4M, if A is SDD, then $A = LU$ where L and U are lower and upper triangular.

Now, section 4J showed that row interchanges may be needed for numerical stability even if no pivots vanish. However, for SDD matrices, it can be proved that row interchanges are not needed: Gaussian elimination without interchanges always gives accurate results, and roundoff errors do not grow disastrously in the manner of Example 4J-1. For further details see, for example, Golub and Van Loan (1983).

*7D   THE SPECTRAL RADIUS AND ITS APPLICATIONS

This optional section continues the discussion of section 4K, using eigenvalues to analyse iteration methods for solving linear systems.

We begin with Jacobi's method. As explained in section 4K, it solves the system $Ax = b$ by splitting A into its diagonal plus the rest. It applies when A is a nonsingular matrix with nonzero diagonal entries. The equations can then be multiplied by factors which make the diagonal entries of the matrix equal to 1; let $A'x = b'$ be the resulting system. Then $A' = I + C$ where C is a matrix with zero diagonal. We have

$$(I + C)x = b'$$

giving

$$x = b' - Cx \tag{1}$$

Jacobi's method starts with an initial vector $x^{(0)}$, and defines

$$x^{(r)} = b' - Cx^{(r-1)} \quad \text{for } r = 1, 2, \ldots$$

If X is the exact solution of (1), let

$$e^{(r)} = x^{(r)} - X \quad \text{for } r = 0, 1, 2, \ldots$$

Then $e^{(r)}$ is the error in the rth approximation; we hope that it is small when r is large. We have

$$e^{(r)} = b' - Cx^{(r-1)} - X = b' - C(e^{(r-1)} + X) - X$$

Since X satisfies (1), we have $b' - CX = X$, and

$$e^{(r)} = -Ce^{(r-1)} \quad \text{for } r = 1, 2, \ldots \tag{2}$$
$$= -C(-Ce^{(r-2)}) = (-C)^3 e^{(r-3)} = \ldots$$

so

$$e^{(r)} = (-1)^r C^r e^{(0)} \tag{3}$$

If $C^r \to 0$ as $r \to \infty$, then the error $e^{(r)}$ tends to zero and the approximations $x^{(r)}$ converge to the solution X. If C^r does not tend to zero, then the sequence will not converge, and the method fails.

The question of convergence hinges on the eigenvalues of C. Suppose that C has an eigenvalue λ with $|\lambda| \geq 1$. If $e^{(0)}$ happens to be an eigenvector corresponding to λ, then $e^{(1)} = -Ce^{(0)} = -\lambda e^{(0)}$, and $e^{(2)} = \lambda^2 e^{(0)}$, and so on, each $e^{(r)}$ being larger than the preceding one by a factor λ. Hence if there is an eigenvalue with magnitude ≥ 1, then it is possible for Jacobi iteration to diverge. The magnitude of the largest eigenvalue is clearly an important number.

Definition 1. The **spectral radius** of a square matrix A is the number $\max\{|\lambda|: \lambda$ is an eigenvalue of $A\}$. It is denoted by $\rho(A)$.   □

$\rho(A)$ is the radius of the smallest circle centred on the origin in the complex plane such that all the eigenvalues are inside or on the circle.

Theorem 1. For any square matrix A, $A^r \to 0$ as $r \to \infty$ if and only if $\rho(A) < 1$.

Incomplete proof. We prove the theorem under the assumption that A is an n by n

matrix with n linearly independent eigenvectors f_i with corresponding eigenvalues λ_i. Then $\{f_1,\ldots,f_n\}$ is a basis for $\mathbb{C}^n$. Let $c_1,\ldots,c_n$ be the components of any vector x with respect to this basis. Then

$$Ax = A\sum c_i f_i = \sum c_i A f_i = \sum c_i \lambda_i f_i$$

so

$$A^2 x = A\sum c_i \lambda_i f_i = \sum c_i \lambda_i A f_i = \sum c_i \lambda_i^2 f_i$$

Proceeding in this way gives

$$A^r x = \sum_i c_i \lambda_i^r f_i \quad \text{for } r = 1, 2, \ldots$$

Now if $|\lambda_i| < 1$ for all i, then each term in this sum tends to 0 as $r \to \infty$ and so $A^r x \to 0$. This is true for all x. If $x = e_j$, the jth standard basis vector with kth component δ_{jk}, then $A^r x$ is the jth column of A^r. Thus each column of A^r tends to 0, so $A^r \to 0$. This completes the proof in the case where A has n independent eigenvectors.

   The general case, where the eigenvectors do not form a basis, is more difficult. A proof will be given in Volume 2 (section 10D). □

We can now analyse the convergence of iterative methods.

Theorem 2 (convergence for Jacobi iteration). Jacobi iteration for $Ax = b$ converges if A is strictly diagonally dominant.

Proof. Equation (3) shows that the method converges if $C^r \to 0$, where $C = A' - I$ and A' is the matrix obtained from A by dividing each row by its diagonal entry (which is nonzero by Exercise 7C-1). Theorem 1 gives convergence if $\rho(C) < 1$. We now estimate $\rho(C)$ by Gershgorin's theorem.

   The diagonal entries of $C = A' - I$ are zero, so all the Gershgorin discs have centre 0. If

$$\sum_{j \neq i} |A'_{ij}| < 1 \quad \text{for } i = 1, \ldots, n \tag{4}$$

then all the discs have radius < 1, so all the eigenvalues are strictly inside the unit circle and $\rho(C) < 1$. But $A'_{ij} = A_{ij}/A_{ii}$, hence (4) is equivalent to

$$\sum_{j \neq i} |A_{ij}| < |A_{ii}| \quad \text{for } i = 1, \ldots, n$$

which holds if A is strictly diagonally dominant. □

   Now consider the Gauss–Seidel method for $Ax = b$. We split the matrix A into a part containing the entries on and below the main diagonal, and a part containing the entries above the main diagonal:

$$A = L + U$$

Here L is lower triangular and U is **strictly upper triangular**, that is, all its entries are zero except those above the main diagonal, so that $U_{ij} = 0$ if $i \geq j$.

The equation $Ax = b$ can now be written

$$Lx = -Ux + b$$

Take an initial guess $x^{(0)}$ and define a sequence of approximations $x^{(r)}$ by

$$Lx^{(r+1)} = -Ux^{(r)} + b \quad \text{for } r = 0, 1, 2, \ldots \tag{5}$$

If you write this equation out in terms of the entries of A and $x^{(r)}$, you will see that it is equivalent to the Gauss–Seidel equations of section 4K.

If the diagonal entries of A are nonzero, then L is triangular with nonzero diagonal, and is therefore invertible. So (5) gives

$$x^{(r+1)} = -L^{-1}Ux^{(r)} + L^{-1}b \tag{6}$$

This is similar in form to the Jacobi method, and its convergence can be discussed by similar methods

Theorem 3 (convergence for Gauss–Seidel iteration). Gauss–Seidel iteration for $Ax = b$ converges if A is strictly diagonally dominant.

Proof. The iteration (6) has the same form as the Jacobi method discussed at the beginning of this section, with $C = L^{-1}U$. It therefore converges if $\rho(L^{-1}U) < 1$.

The eigenvalues λ of $L^{-1}U$ are numbers λ such that $L^{-1}U - \lambda I$ is singular. Since $L^{-1}U - \lambda I = L^{-1}(U - \lambda L)$, it follows that $U - \lambda L$ is singular. Now,

$$U - \lambda L = \begin{pmatrix} -\lambda A_{11} & A_{12} & A_{13} & A_{14} & \cdots & A_{1n} \\ -\lambda A_{21} & -\lambda A_{22} & A_{23} & A_{24} & \cdots & A_{2n} \\ -\lambda A_{31} & -\lambda A_{32} & -\lambda A_{33} & A_{34} & \cdots & A_{3n} \\ \cdot & \cdot & \cdot & \cdot & \cdots & \cdot \\ & & & & \cdots & \\ \cdot & \cdot & \cdot & \cdot & \cdots & \cdot \\ -\lambda A_{n1} & -\lambda A_{n2} & -\lambda A_{n3} & -\lambda A_{n4} & \cdots & -\lambda A_{nn} \end{pmatrix}$$

We shall show that if A is strictly diagonally dominant, then $U - \lambda L$ is nonsingular when $|\lambda| \geq 1$.

The diagonal entry of the ith row of $U - \lambda L$ satisfies

$$|-\lambda A_{ii}| > |\lambda| \sum_{j \neq i} |A_{ij}|$$

because A is strictly diagonally dominant,

$$\geq |\lambda| \sum_{j=1}^{i-1} |A_{ij}| + \sum_{j=i+1}^{n} |A_{ij}|$$

if $|\lambda| \geq 1$. Hence if $|\lambda| \geq 1$, then $U - \lambda L$ is strictly diagonally dominant and therefore nonsingular (Theorem 7C-1). Hence $\rho(L^{-1}U) < 1$, and Gauss–Seidel iteration converges. $\qquad\square$

These theorems are only the beginning of the theory of iteration. The aim of this section is not to tell the full story, but to give some of its flavour, and to encourage you to find out more about this attractive subject, in which theoretical concepts and practical matters are so closely interwoven. See, for example, Todd (1978) and Burden and Faires (1985).

7E   PROOFS OF THEOREMS

Lemma 7C-1. If the matrix A is strictly diagonally dominant, so is the matrix obtained by one step of the Gauss reduction procedure, taking the $(1, 1)$ entry as pivot.

Proof. We first note two useful inequalities:

$$|x - y| \leq |x| + |y| \quad \text{for all } x, y \tag{1}$$

$$|x - y| \geq |x| - |y| \quad \text{for all } x, y \tag{2}$$

They follow from the fact that one side of a triangle cannot be longer than the other two sides put together. As noted in Appendix E, $|x - y|$ is the length of a side of a triangle in the complex plane whose other two sides have lengths $|x|$ and $|y|$, which gives (1). Putting $X = x - y$ and $Y = x$ in (1) gives (2).

Now, let B be the matrix obtained from A by a Gauss step with the $(1, 1)$ entry as pivot. Then

$$b_{ij} = a_{ij} - m_i a_{1j}$$

where

$$m_i = a_{i1}/a_{11} \tag{3}$$

The strict diagonal dominance condition on A is

$$\sum_{j=1}^{n}{}' |a_{ij}| < |a_{ii}| \quad \text{for } i = 1, \ldots, n \tag{4}$$

where $\sum'$ denotes a sum in which the term $j = i$ is omitted.

The first row of B is the same as that of A. We shall show that for each $i > 1$, the ith row of B satisfies the diagonal dominance condition.

As a result of the Gauss step, $b_{i1} = 0$ for $i > 1$. Hence

$$\sum_{j=1}^{n}{}' |b_{ij}| = \sum_{j=2}^{n}{}' |a_{ij} - m_i a_{1j}| \leq \sum_{j=2}^{n}{}' |a_{ij}| + |m_i| \sum_{j=2}^{n}{}' |a_{1j}| \quad \text{using (1)}$$

Adding the missing term in the second sum and subtracting it again gives

$$\sum_{j=1}^{n}{}' |b_{ij}| \leq \sum_{j=2}^{n}{}' |a_{ij}| + |m_i| \sum_{j=2}^{n} |a_{1j}| - |m_i||a_{1i}|$$

$$< \sum_{j=2}^{n}{}' |a_{ij}| + |m_i||a_{11}| - |m_i||a_{1i}| \tag{5}$$

by (4) with $i = 1$. The second term in (5) is $|a_{i1}|$ by (3). This fits into the first sum in (5), giving

$$\sum_{j=1}^{n}{}' |b_{ij}| < \sum_{j=1}^{n}{}' |a_{ij}| - |m_i||a_{1i}| < |a_{ii}| - |m_i||a_{1i}| \leq |a_{ii} - m_i a_{1i}| = |b_{ii}|$$

using (2). This is the strict diagonal dominance condition for the ith row of B. $\square$

Theorem 7C-3. A subset G of the Gershgorin discs is called a 'disjoint group of discs' if no disc in the group intersects a disc not in the group. If a disjoint group contains r nonconcentric discs, then it contains r eigenvalues.

Incomplete proof. This beautiful proof uses a fact which, though plausible, is not easy to prove.

For each real p between 0 and 1, let $A(p)$ denote the matrix obtained from A by multiplying all its off-diagonal entries by p. Then $A = A(1)$, and $A(0)$ is a diagonal matrix with the same diagonal as A. The Gershgorin discs of $A(p)$ have the same centres as those of A, and p times their radii.

Consider a disjoint group G of discs for $A = A(1)$. Let p decrease steadily from 1. The discs shrink as p is reduced, so for all $p < 1$ the discs in G remain a disjoint group.

Let $a_1, \ldots, a_r$ be the centres of the discs in G. Each a_i is one of the diagonal entries of $A(p)$ for each p. In particular, each a_i is a diagonal entry of the diagonal matrix $A(0)$, and is therefore an eigenvalue of $A(0)$. Hence $A(0)$ has eigenvalues $a_1, \ldots, a_r$ which are contained in zero-radius Gershgorin discs.

Now imagine p increasing gradually from 0 to 1. The discs grow steadily to their final size. The eigenvalues change as p increases, but at each stage they must lie in the union of all the discs by Gershgorin's theorem.

When $p = 0$, there are r eigenvalues in the group G of discs. As p increases, the entries of $A(p)$ change continuously, and it is plausible (and can be proved) that the eigenvalues also move continuously, that is, do not jump from one point to another without passing through intermediate points. It follows that the eigenvalues cannot leave the group G, so when $p = 1$ there are still r eigenvalues in G. ☐

PROBLEMS FOR CHAPTER 7

Hints and answers for problems marked [a] will be found on page 283.

Sections 7A, 7B

[a]1. For each of the following, find the eigenvalues and give a basis for each eigenspace:

(a) $\begin{pmatrix} 3 & 2 & 4 \\ 2 & 0 & 2 \\ 4 & 2 & 3 \end{pmatrix}$ (b) $\begin{pmatrix} 3 & 2 & 1 & 0 \\ 0 & 1 & 0 & 1 \\ 0 & 2 & 1 & 0 \\ 0 & 0 & 0 & 1 \end{pmatrix}$

[a]2. Prove that every square matrix has the same eigenvalues as its transpose, with the same algebraic multiplicity.

[a]3. Prove that the characteristic polynomial of an n by n matrix A is

$$p(x) = (\lambda_1 - x)(\lambda_2 - x) \ldots (\lambda_n - x)$$

where $\lambda_1, \ldots, \lambda_n$ are the eigenvalues of A listed by algebraic multiplicity, that is, with an eigenvalue of algebraic multiplicity m appearing m times in the list.

Deduce that the product of the eigenvalues of A equals $\det(A)$ (with the eigenvalues listed by algebraic multiplicity).

(For a small matrix, this gives a way of checking the correctness of eigenvalue calculations. See also Problem 6.)

[a]4. True or false?

(a) Complex eigenvalues of real matrices come in conjugate pairs. (In other words, if a matrix M with real entries has a complex eigenvalue $a + ib$, then $a - ib$ is also an eigenvalue of M.)

(b) A and A^T have the same eigenvectors.

(c) Two n by n matrices have the same determinant if they have the same characteristic polynomial.

(d) Converse of (c).

[a]5. A is an n by n orthogonal matrix. Using the results of Problems 2 and 3, prove the following:

(a) if λ is an eigenvalue of A, so is λ^{-1};

(b) if n is odd, then either 1 or -1 is an eigenvalue of A;

(c) if 1 is a simple eigenvalue, and -1 is not an eigenvalue, then $\det(A) = 1$;

(d) if -1 is a simple eigenvalue, then $\det(A) = -1$.

(For eigenvalue 1 the eigenvectors satisfy $Ax = x$; they stay fixed under the rotation. In odd-dimensional spaces, therefore, rotations are rotations about a fixed axis. This is not so in even-dimensional spaces; see Chapter 5, Problem 35. Eigenvalue -1 and determinant -1 correspond to reflections; see Chapter 6, Problem 17.)

6. The **trace** of an n by n matrix A is defined as the sum of its diagonal entries. It is denoted by $\text{tr}(A)$.

(a) Use the result of Chapter 6, Problem 6 to show that any n by n determinant equals the product of its diagonal entries plus terms containing products of no more than $n - 2$ diagonal entries.

(b) Hence show that the coefficient of x^{n-1} in the characteristic polynomial $\det(A - xI)$ is $(-1)^{n-1}\,\text{tr}(A)$.

(c) Hence show that the sum of the eigenvalues of A equals $\text{tr}(A)$, the eigenvalues being listed by algebraic multiplicity (explained in Problem 3). (This is a useful check on the correctness of eigenvalue calculations.)

(d) Use the result of Chapter 5, Problem 30 to prove that for any two n by n matrices A and B, $\text{tr}(AB) = \text{tr}(BA)$. Also prove it directly, without using eigenvalues.

(e) Extend the result of (d) to show that $\text{tr}(ABC..GH) = \text{tr}(HAB...G)$ for any $A, B, ... H$.

7. Given a matrix, you know how to find its characteristic polynomial. Sometimes things are the other way round: given a polynomial

$$p(x) = x^n + a_{n-1}x^{n-1} + a_{n-2}x^{n-2} + \ldots + a_1 x + a_0$$

one wants to find a matrix for which p is the characteristic polynomial.

The **companion matrix** for the polynomial p is the matrix with last row

$-a_0, -a_1, \ldots, -a_{n-1}$, with 1's in the $(j, j+1)$ position for $j = 1, \ldots, n-1$, and zeros elsewhere. Show that p is the characteristic polynomial for its companion matrix.

(For applications, see Problems 8 and 15; also Chapter 10, Problem 39)

8. Let E be the companion matrix (see the previous problem) of the polynomial $x^n - 1$. Show that E is a circulant (see Chapter 2, Problem 5). If A is the circulant with first row $a_1, a_2, \ldots, a_n$, show that $A = a_1 I + a_2 E + a_3 E^2 + \ldots + a_n E^{n-1}$.

If z is a number satisfying $z^n = 1$, show that z is an eigenvalue of E with eigenvector $(1, z, z^2, \ldots)$. Show that this is also an eigenvector of A, and find the corresponding eigenvalue.

If you are familiar with the complex nth roots of 1, prove that every n by n circulant has n linearly independent eigenvectors. (For another proof, see Chapter 12, Problem **00**).

9. Let $C_n(x)$ be the characteristic polynomial of A_n, the n by n matrix with $(A_n)_{ij} = 1$ if $|i - j| = 1$, and zero otherwise. Show that $C_2(x) = x^2 - 1$ and $C_3(x) = x(2 - x^2)$. By expanding the characteristic determinant by its first row, show that $C_n(x) = -xC_{n-1}(x) - C_{n-2}(x)$. Hence write down the characteristic polynomials of A_4 and A_5.

To solve for the eigenvalues is not easy. But they can be found by the following trick. Show that for a suitably chosen t, the n-vector $(\sin(t), \sin(2t), \ldots, \sin(nt))$ is an eigenvector of A_n with eigenvalue $2\cos(t)$. Show that there are n possible values of t, and hence determine all eigenvalues of A_n.

Use the above results to find the eigenvalues of the n by n matrix M_n defined by $(M_n)_{ij} = b$ if $|i - j| = 1$, $(M_n)_{ii} = a$ for all i, and all other entries zero.

10. A $2n$ by $2n$ matrix M is called **symplectic** if $M^T J M = J$, where J is a matrix with all entries zero except for n identical blocks arranged down the diagonal, each block being

$$\begin{pmatrix} 0 & 1 \\ -1 & 0 \end{pmatrix}$$

Show that a 2 by 2 matrix A is symplectic if and only if $\det(A) = 1$. Deduce that if k is an eigenvalue, then so is $1/k$. In the $2n$ by $2n$ case, show that every symplectic matrix is nonsingular, and that if k is an eigenvalue, then so is $1/k$.

(Hint for the last part: if v is the eigenvector, consider the vector Jv.)

11. $A = B + iC$ is a complex matrix; B and C are real matrices, with $B_{rs} = \text{Re}(A_{rs})$ and $C_{rs} = \text{Im}(A_{rs})$ for all r, s. Prove that if A is nonsingular, then there is a real number x such that $B + xC$ is nonsingular.

(Hint: $\det(B + xC)$ is a polynomial in x; prove that its degree is positive.)

[a]12. Use the result of Theorem 5I-1(i) to prove the following.

(a) Every eigenvalue of an orthogonal matrix has modulus 1.

(b) If a square matrix A is obtained from an orthogonal matrix by deleting certain rows and columns, then all eigenvalues λ of A satisfy $|\lambda| \leq 1$.

Sections 7C, 7D

[a]13. Show that a real n by n matrix has n real eigenvalues if it has n non-intersecting Gershgorin discs.

14. Use Gershgorin's theorem to show that every eigenvalue of a stochastic matrix (defined in Example 2C-4) has modulus less than or equal to 1. Use Chapter 5, Problem 32 to deduce that the spectral radius is 1.

15. An nth-degree polynomial can be reduced to the form $p(x) = x^n + a_{n-1}x^{n-1} + a_{n-2}x^{n-2} + \ldots + a_0$ by dividing by the coefficient of x^n. Prove that all zeros of p satisfy $|z| \le 1$ if $|a_0| + |a_1| + \ldots + |a_{n-1}| \le 1$.
(Hint: use Problem 7).

[a]16. Prove that if the spectral radius of A is less than 1, then $I - A$ is invertible, and $(I - A)^{-1} = I + A + A^2 + \ldots$. (Here the statement $B = I + A + A^2 + \ldots$ means $B - (I + A + \ldots + A^r) \to 0$ as $r \to \infty$.)

17. Problem: given a set of points (x_j, y_j), construct a smooth curve passing through them. Solution: use the dominant diagonal theorem, in the following way.
   Chapter 3, Problem 32 joins the points by straight line segments. A smooth curve can be obtained by using higher-degree polynomials.
   Given an interval $[a, b]$, and a set of points x_j with

$$a = x_0 < x_1 < \ldots < x_n = b$$

define a **cubic spline** on $[a, b]$ to be a function f satisfying the following conditions. In each interval $[x_{i-1}, x_i]$, f is given by a cubic polynomial; the coefficients of the cubic may be different in different intervals; f, f' and f'' are continuous across the boundaries $x_1, \ldots, x_{n-1}$; and $f''(a) = f''(b) = 0$. The points $x_1, \ldots, x_n$ are called the **knots** of the spline.
   The following steps prove that given $n + 1$ numbers $y_0, \ldots, y_n$, there is exactly one cubic spline passing through all the points (x_i, y_i).
   Take $f(x) = y_j + b_j(x - x_j) + c_j(x - x_j)^2 + d_j(x - x_j)^3$ for $x_j \le x \le x_{j+1}$ for $j = 0, 1, \ldots, n - 1$, where the b's, c's and d's are constants to be determined. Then $f(x_i) = y_i$ for $i = 0, \ldots, n - 1$. In other words, the spline f passes through the points (x_i, y_i).
   Write down the conditions that f, f', f'' are continuous at each knot, and $f''(x_0) = f''(x_n) = 0$. Assume for simplicity that all intervals $[x_{i-1}, x_i]$ have the same length. Eliminate the b's and d's, and obtain a system of equations for the c's. Show that its matrix is strictly diagonally dominant, and deduce that the c's, and therefore all the other constants, are determined uniquely by the given numbers y_j.
   Deduce that the set of all cubic splines with knots $x_0, \ldots, x_n$ is a vector space of dimension $n + 1$.
   If you are energetic, do the general case in which the intervals have different lengths.
   (Cubic splines give very useful approximations to functions; see, for example, Strang (1986).)

16

8

Diagonalisation of matrices and quadratic forms

This chapter deals with an important application of matrix eigenvalues. There are two ways of approaching the subject. The following sections take a straightforward route, and if you want to reach useful results quickly, you should read them. But there is another approach, involving theoretical ideas developed in Volume 2, which leads to a better understanding in the long run. To take this longer route, do not read this chapter now, but start on Volume 2.

Section 8A reveals a new aspect of the eigenvectors of a matrix A: they can sometimes be used to write A in the form PDP^{-1} where D is a diagonal matrix whose entries are the eigenvalues. The usefulness of this formula is illustrated in section 8B, which shows how eigenvalues control the stability or instability of systems governed by sets of differential equations. Section 8C returns to the general theory, developing and simplifying it in the important special case where A is a symmetric matrix. The last three sections apply the theory to maxima and minima of functions of several variables, and to the geometry of conic sections and their three-dimensional analogues.

8A DIAGONAL FACTORISATION

One of the main themes of matrix algebra is the idea of combining many equations into one. Matrices were introduced in Chapter 2 as a way of writing p equations in q unknowns as a single equation. The inverse of an n by n matrix A can be viewed as a way of combining the solutions of the n equations $Ax = e_i$ into a single matrix A^{-1}, which can then be used to solve $Ax = b$ with any b. This section is based on a similar idea; we shall combine the eigenvalue equations for several different eigenvectors into a single matrix equation.

Consider, then, an n by n matrix A, with linearly independent eigenvectors $v_1, \ldots, v_k$

belonging to eigenvalues $\lambda_1, \ldots, \lambda_k$. Thus

$$Av_i = \lambda_i v_i \quad \text{for } i = 1, \ldots, k \tag{1}$$

The vectors v_i can be assembled into an n by k matrix, which we write using partitioned-matrix notation as

$$P = (v_1 \; v_2 \; \ldots \; v_k)$$

where v_i is the ith column of P. Then (1) becomes

$$AP = (\lambda_1 v_1 \; \lambda_2 v_2 \; \ldots \; \lambda_k v_k) = (v_1 \; v_2 \; \ldots \; v_k)D$$

where D is the k by k diagonal matrix with entries $\lambda_1, \ldots, \lambda_k$. Thus

$$AP = PD \tag{2}$$

This is a compact form of the eigenvalue equations for the k eigenvectors $v_1, \ldots, v_k$.

Now, suppose that A has n linearly independent eigenvectors, so that $k = n$. The matrix P is then n by n, with linearly independent columns. Hence it is invertible, and multiplying (2) on the right by P^{-1} gives

$$A = PDP^{-1} \tag{3}$$

Thus if an n by n matrix A has n linearly independent eigenvectors, it can be factorised in the form (3), where D is diagonal; the matrix A is called 'diagonally factorisable', or 'diagonalisable' for short.

Some care is needed here. We saw in section 5H that the solution of the eigenvalue problem depends on the field; a real matrix may have eigenvectors over $\mathbb{C}$ but not over $\mathbb{R}$. The field should always be specified.

Definition 1. A matrix A is called **diagonalisable over $\mathbb{K}$** if there is a diagonal matrix D and an invertible matrix P, with entries in the field $\mathbb{K}$, such that $A = PDP^{-1}$. The matrix P is called a **diagonalising matrix** for A over $\mathbb{K}$.

Example 1. Consider the matrix

$$A = \begin{pmatrix} 3 & -4 & 2 \\ -4 & -1 & 6 \\ 2 & 6 & -2 \end{pmatrix}$$

Its characteristic polynomial is

$$\begin{vmatrix} 3-x & -4 & 2 \\ -4 & -1-x & 6 \\ 2 & 6 & -2-x \end{vmatrix} = \begin{vmatrix} 3-x & -4 & 2 \\ 0 & 11-x & 2-2x \\ 2 & 6 & -2-x \end{vmatrix}$$

$$= (3-x)\{(x-11)(x+2) - 12 + 12x\} + 2\{8x - 8 + 2x - 22\}$$

Elementary algebra now gives the eigenvalues as $3, 6, -9$. The eigenvectors are easily found by solving $(A - \lambda I)v = 0$. They are $(2, 1, 2)$, $(2, -2, -1)$, $(1, 2, -2)$ for eigenvalues $3, 6, -9$ respectively. These three vectors are linearly independent because they belong to different eigenvalues. Hence A is diagonalisable over $\mathbb{R}$, and the matrices

P and D in the theory above are

$$P = \begin{pmatrix} 2 & 2 & 1 \\ 1 & -2 & 2 \\ 2 & -1 & -2 \end{pmatrix} \quad \text{and} \quad D = \begin{pmatrix} 3 & 0 & 0 \\ 0 & 6 & 0 \\ 0 & 0 & -9 \end{pmatrix}$$

To compute the diagonal factorisation, we must invert P. The standard method of section 4I gives

$$P^{-1} = (1/9) \begin{pmatrix} 2 & 1 & 2 \\ 2 & -2 & -1 \\ 1 & 2 & -2 \end{pmatrix}$$

We have now computed all the factors in the formula (3) for this matrix, and it is easy to multiply out the three matrices, and verify the correctness of the calculations.
We shall look more deeply into this example in section 8C.   □

What, you may ask, is so wonderful about the factorisation (3)? One answer: read the next section; it puts (3) to work on differential equations. Another answer: do the following exercise.

Exercise 1. If $A = PDP^{-1}$, then $A^2 = PD^2P^{-1}$, and for any positive integer r,

$$A^r = PD^rP^{-1} \qquad\qquad □ \quad (4)$$

Since the entries of D^r are just the rth powers of the eigenvalues of A, (4) gives a simple expression for the powers of A. When r is large, (4) takes much less computation than multiplying A by itself r times.
Other applications of diagonalisation will appear soon. But first we shall consider the natural question: which matrices can be diagonalised? The argument at the beginning of the section shows that if an n by n matrix has n linearly independent eigenvectors, then it is diagonalisable. Is this condition necessary? The following result shows that it is, and sums up the whole of the discussion.

Theorem 1 (diagonalisability). (a) An n by n matrix A is diagonalisable over $\mathbb{K}$ if and only if it has n linearly independent eigenvectors over $\mathbb{K}$.
(b) If $A = PDP^{-1}$, with D diagonal, then the diagonal entries of D are eigenvalues of A, and the columns of P are the corresponding eigenvectors.

Proof. Suppose first that there are n linearly independent eigenvectors. The second paragraph of this section proves that in this case the matrix is diagonalisable.
Conversely, suppose that $A = PDP^{-1}$ where D is diagonal. Multiplying by P on the right gives $AP = PD$. The ith column of this equation gives

$$Ap_i = p_id_i \qquad\qquad (5)$$

where p_i is the ith column of P, and d_i is the (i, i) entry of D. Since P is invertible, none of its columns p_i can be zero, hence (5) shows that p_i is an eigenvector belonging to eigenvalue d_i. The vectors $p_1, \ldots, p_n$ are the columns of an invertible matrix, and are therefore linearly independent. This completes the proof.   □

In principle, part (a) of Theorem 1 gives a way of deciding whether or not a matrix is diagonalisable. In practice, the condition is almost useless, because one cannot tell whether the eigenvectors are linearly independent until one has gone through the laborious process of finding them. However, there is one special case in which it is easy to see whether the condition of Theorem 1 holds.

Corollary 1. If an n by n matrix has n different eigenvalues over the field $\mathbb{K}$, then it is diagonalisable over $\mathbb{K}$.

Proof. The n eigenvectors belonging to the n eigenvalues are linearly independent by Theorem 5H-1, so Theorem 1(a) shows that the matrix is diagonalisable. $\square$

The following optional section applies these ideas to more or less practical problems. It shows how the distinction between diagonalisation over the real and over the complex numbers leads to clear differences in the behaviour of dynamical systems.

*8B   APPLICATION OF DIAGONALISATION TO DIFFERENTIAL EQUATIONS

This section is optional; it depends on no other optional section. It applies the theory of the preceding section to systems of differential equations of a kind which occurs in many branches of science and mathematics.

Consider functions $x_1(t), \ldots, x_n(t)$ satisfying the equations

$$x_1'(t) = a_{11}x_1 + a_{12}x_2 + \ldots + a_{1n}x_n$$

$$x_2'(t) = a_{21}x_1 + a_{22}x_2 + \ldots + a_{2n}x_n$$

$$\ldots\ldots\ldots\ldots\ldots\ldots\ldots\ldots\ldots\ldots\ldots\ldots\ldots\ldots\ldots$$

$$x_n'(t) = a_{n1}x_1 + a_{n2}x_2 + \ldots + a_{nn}x_n$$

where the a_{ij} are constants, and primes denote derivatives with respect to t. The variable t often represents time; the ith equation then expresses the rate of growth of x_i in terms of the values of the x_j's. It is useful to think of t as representing time, even when the equations arise in a different context; thinking of the x_i as changing with time helps one's intuition and imagination. Our mathematical conclusions will, of course, be independent of such intuitive pictures.

The system can be written in the form

$$x' = Ax \tag{1}$$

where $x(t) \in \mathbb{R}^n$ is the n-vector with components $x_i(t)$, and A is the real matrix with entries a_{ij}. Thus A is a constant matrix, and x is an n-vector which changes with t.

If A is diagonalisable, we can easily solve equation (1). The solution depends on whether A is diagonalisable over the real or the complex field.

Case I: A diagonalisable over the reals

In this case there is a real matrix P and a diagonal real matrix $D = \mathrm{diag}(\lambda_1, \ldots, \lambda_n)$, whose entries are the eigenvalues of A, such that $A = PDP^{-1}$. Thus (1) becomes

$x' = PDP^{-1}x$ or

$$P^{-1}x' = DP^{-1}x$$

Define a variable n-vector $y(t)$ by

$$y = P^{-1}x \qquad (2)$$

Then

$$y' = Dy$$

Since $D = \mathrm{diag}(\lambda_1, \ldots, \lambda_n)$, we have

$$y_1' = \lambda_1 y_1$$

$$y_2' = \lambda_2 y_2$$

$$\ldots\ldots\ldots$$

$$y_n' = \lambda_n y_n$$

This system of equations is said to be in 'diagonal form' because its matrix D is diagonal. The variables y_i are decoupled from each other; each one satisfies a simple equation in which the others do not appear. The solution is

$$y_r(t) = C_r \exp(\lambda_r t) \quad \text{for } r = 1, \ldots, n \qquad (3)$$

where the n numbers C_r are independent of t. If the initial values $y_r(0)$ are known, then (3) gives $y_r(0) = C_r$, so

$$y_r(t) = y_r(0)\exp(\lambda_r t)$$

In matrix notation:

$$y(t) = E(t)y(0)$$

where

$$E(t) = \mathrm{diag}(\exp[\lambda_1 t], \ldots, \exp[\lambda_n t])$$

Now we transform back to the original variables x. We defined y by equation (2); multiplying by P gives

$$x = Py$$

so

$$x(t) = PE(t)y(0) \qquad (4)$$

Finally, (2) gives $y(0) = P^{-1}x(0)$, so

$$x(t) = PE(t)P^{-1}x(0) \qquad (5)$$

This is a complete solution of the original equations, in terms of the initial values of the variables, under the assumption that A is diagonalisable over the reals. If the eigenvalue problem for A can be solved, then the solution can be written down explicitly.

But one can learn something from (5) even without any detailed calculation. The only t-variation on the right-hand side is in the entries of $E(t)$, which are of the form

$\exp(\lambda_r t)$. The x-variables are combinations of exponential functions of t. Each exponential either tends to zero (if the eigenvalue is negative), or is constant (if the eigenvalue is zero), or tends to infinity (if the eigenvalue is positive). It follows that the system cannot oscillate. Each $x_r(t)$ either tends to infinity or tends to a constant, depending on whether any positive-eigenvalue terms are present; if all eigenvalues are negative, then x is a linear combination of negative exponentials and tends to zero. We thus reach the following conclusion.

Theorem 1. If A is diagonalisable over the reals, then all solutions of the system $\dot{x} = Ax$ either tend to a constant, or tend to infinity exponentially as $t \to \infty$. If all eigenvalues of A are negative, then all solutions tend to zero. $\square$

It is remarkable that such a strong and physically significant result – no oscillations – follows from the simple assumption that A is diagonalisable over the reals. It shows how useful it would be to be able to recognise diagonalisable matrices at sight.

We also see that the eigenvalues of A have a physical meaning: positive eigenvalues correspond to growing solutions, negative eigenvalues to decaying solutions. The size of the eigenvalues gives the rate of growth or decay.

Modes of growth and decay

The machinery developed above can be interpreted as follows. Suppose that the initial conditions are such that all but one of the y's vanish at $t = 0$, so that $y_r(0) = 0$ for $r \neq k$, say. Then $y(0) = ce_k$, where c is a constant scalar and e_k is the usual basis vector with ith entry δ_{ik}. Equation (4) gives

$$x(t) = cPE(t)e_k$$

$$= cP \exp(\lambda_k t)e_k$$

But Pe_k is the kth column of P, which, as we saw in the last section, is the kth eigenvector of A, call it v_k. So

$$x(t) = c \exp(\lambda_k t)v_k \qquad (6)$$

This solution of the equation (1) is particularly simple: all entries of x are proportional to $\exp(\lambda_k t)$; their growth or decay is governed by the size of the kth eigenvalue. This special type of solution, in which all the variables x_r are proportional to the same function of t, is called a **normal mode** of growth or decay, according to whether λ_k is positive or negative.

There are n normal modes, one for each eigenvector. Equation (5) says that every solution is a linear combination of normal modes. Each mode is a simple exponential function of t, but the general solution, being a combination of exponentials with different exponents, can behave in quite a complicated way. The complication fades away for large t, and the system then behaves in accordance with Theorem 1.

The structure of a normal mode is described by equation (6). Each entry x_r is proportional to $\exp(\lambda_k t)$ times the corresponding entry of the kth eigenvector v_k. In the kth mode, the components of x grow (or decay) together, the vector x always being proportional to the kth eigenvector. The eigenvector thus has a physical interpretation: it gives the structure of a normal mode. For this reason, the diagonalising matrix P, whose columns are the eigenvectors, is sometimes called the **modal matrix**.

Case II: A diagonalisable over the complex numbers

All the above equations still hold in this case, but the interpretation is different when the eigenvalues and eigenvectors are complex.

The general solution is a linear combination of normal modes, each of which is described by equation (6). If the eigenvalue λ_k is real, then the mode grows or decays exponentially, depending on the sign of λ_k. Suppose λ_k is complex, with real and imaginary parts λ' and λ'', say. Then the corresponding normal mode is given by

$$x(t) = c\,e^{(\lambda' + i\lambda'')t}v = c\,e^{\lambda't}[\cos(\lambda''t) + i\sin(\lambda''t)]v \qquad (7)$$

This represents an oscillation with frequency λ'', combined with exponential growth if $\lambda' > 0$ and decay if $\lambda' < 0$. The normal modes are now **normal modes of oscillation** if the eigenvalue has a nonzero imaginary part.

The general solution is a combination of modes of type (7). Each mode is a complex function of t in general. But if the original equations represent a real physical system, then the solution must be a real function of t; the complex modes (7) combine in such a way that the imaginary parts cancel out.

If all the eigenvalues are purely imaginary, then all the modes consist of steady oscillations. If all eigenvalues have negative imaginary parts, then all the modes tend to zero as $t \to \infty$, so all solutions of the system die away. If there is an eigenvalue with positive imaginary part, then there is a solution which grows exponentially.

We have now proved the following.

Theorem 2. If the matrix A is diagonalisable over $\mathbb{C}$, then
  (a) if all eigenvalues have negative real parts, then all solutions of the equation $x' = Ax$ tend to zero as $t \to \infty$;
  (b) if all eigenvalues have zero real part, then the solutions oscillate;
  (c) if there is an eigenvalue with a positive real part, then there is a solution which grows with t, and the growth is oscillatory if that eigenvalue has a nonzero imaginary part. $\qquad\square$

This theorem summarises the discussion above. Like Theorem 1, it shows how the theoretical ideas of this chapter have a direct bearing on the behaviour of physical systems.

8C   SYMMETRIC MATRICES AND ORTHOGONAL DIAGONALISATION

This section deals with a type of matrix for which the diagonalisation problem is particularly simple.

Example 1. Example 8A-1 showed that the matrix

$$A = \begin{pmatrix} 3 & -4 & 2 \\ -4 & -1 & 6 \\ 2 & 6 & -2 \end{pmatrix}$$

has diagonalising matrix

$$P = \begin{pmatrix} 2 & 2 & 1 \\ 1 & -2 & 2 \\ 2 & -1 & -2 \end{pmatrix}$$

We also wrote down P^{-1}. Perhaps you noticed that it looks rather like P. In fact,

$$9P^{-1} = P^{\mathsf{T}} \tag{1}$$

We have already met matrices for which the inverse and transpose are closely related. Section 5I defined orthogonal matrices as real matrices M for which $M^{-1} = M^{\mathsf{T}}$. This is not quite the same as equation (1); but that is easily remedied. Let $Q = P/3$. Then

$$Q^{-1} = 3P^{-1} = P^{\mathsf{T}}/3 \quad \text{(by (1))}$$
$$= Q^{\mathsf{T}}$$

Thus Q is an orthogonal matrix. The diagonalisation formula for A can now be written

$$A = PDP^{-1} = (P/3)D3P^{-1} = QDQ^{-1}$$

We have found an orthogonal diagonalising matrix Q for A.   □

Orthogonal matrices make life simpler because they are so easy to invert: the inverse is just the transpose. There is a special term for matrices which have an orthogonal diagonalising matrix.

Definition 1. A matrix A is said to be **orthogonally diagonalisable** if there is an orthogonal matrix Q such that $A = QDQ^{-1} = QDQ^{\mathsf{T}}$.   □

Thus the matrix in Example 1 is orthogonally diagonalisable.

We now generalise from this example, and ask when, and how, a given matrix can be orthogonally diagonalised. The answers will emerge from the properties of orthogonal matrices given in Section 5I.

We saw there that a matrix is orthogonal when its columns are orthonormal vectors. But the columns of a diagonalising matrix for A are eigenvectors of A. Hence A is orthogonally diagonalisable if it has orthogonal eigenvectors. It is easy to verify that the eigenvectors in Example 1 are orthogonal.

The diagonalising matrix is constructed from the eigenvectors. But some care is needed. The columns of an orthogonal matrix are normalised, that is, have magnitude 1. So if the diagonalising matrix is to be orthogonal, it must be constructed from normalised eigenvectors. An eigenvector can easily be normalised, by dividing it by its magnitude (the square root of the sum of the squares of its entries); the resulting vector has unit magnitude, and forms a column of the orthogonal diagonalising matrix. We therefore have the following result.

Theorem 1 (orthogonal diagonalisation). A real n by n matrix A is orthogonally diagonalisable if and only if it has n orthogonal eigenvectors. The columns of the orthogonal diagonalising matrix are the orthonormal eigenvectors of A.

Proof. (a) Suppose that A has n orthogonal eigenvectors $v_1, \ldots, v_n$. Set $w_i = v_i / \| v_i \|$ for each i, then $\{ w_1, \ldots, w_n \}$ is orthonormal. The matrix with columns $w_1, \ldots$ is orthogonal, and is a diagonalising matrix for A since its columns are eigenvectors.

(b) Conversely, suppose that A is orthogonally diagonalisable, so that $A = QDQ^{-1}$ where Q is orthogonal and D is diagonal. By Theorem 8A-1(b), the columns are eigenvectors of A. They are orthonormal because Q is an orthogonal matrix. Hence A has n orthogonal eigenvectors.   ☐

Remark 1. There is another way of looking at the relation between Q and the orthogonal eigenvectors. The approach above was entirely algebraic. Theorem 5I-2 shows that the diagonalising matrix Q can also be interpreted geometrically: it rotates the standard basis into the set of eigenvectors of A.   ☐

Theorem 1 gives an answer, of sorts, to the question of when a matrix is orthogonally diagonalisable. But faced with a given matrix, how can one tell whether it has orthogonal eigenvectors or not?

Look at Example 1 again. The matrix A has a special property: it is symmetric, $A_{ij} = A_{ji}$ for all i, j. If you work out the eigenvectors of any other real symmetric matrix, you will find that they too are orthogonal.

Theorem 2 (symmetric matrices). Every real symmetric matrix is orthogonally diagonalisable.   ☐

This theorem is simple and beautiful. One can see at a glance whether an n by n matrix is symmetric; if it is, Theorem 2 guarantees that it has n orthogonal eigenvectors.

The proof of Theorem 2 is too hard to be included in this chapter. Fortunately, a proof is not really needed at this stage. Whenever we deal with a symmetric matrix, we shall work out the eigenvectors and find that they are orthogonal by sheer calculation; Theorem 2 is used only as a check on the correctness of the computation. In Volume 2 we shall prove the theorem, and look at some of its consequences.

Example 2. Finally, let us reconsider Example 1 from the standpoint of Theorems 1 and 2. Given the matrix A, how can it best be orthogonally diagonalised?

The first step is to find its eigenvalues and eigenvectors by the method of Example 8A-1. There is no way of avoiding this calculation; but one can check it, and perhaps spot arithmetical mistakes, by verifying that there are three orthogonal eigenvectors as guaranteed by Theorem 2.

The next step is to normalise the eigenvectors, dividing each by its norm. Each eigenvector listed in Example 8A-1 has norm $\sqrt{(1 + 4 + 4)} = 3$, so the normalised eigenvectors are $(2/3, 1/3, 2/3), (2/3, -2/3, -1/3)$, and $(1/3, 2/3, -2/3)$. The matrix with these as its columns is the orthogonal diagonalising matrix Q:

$$Q = \begin{pmatrix} 2/3 & 2/3 & 1/3 \\ 1/3 & -2/3 & 2/3 \\ 2/3 & -1/3 & -2/3 \end{pmatrix} \qquad ☐$$

We have concentrated on symmetric matrices in this section. Most matrices are not symmetric. Why spend so much effort on such a special case? The answer, of course, is that this case is of special importance, occurring naturally in many applications of linear algebra. For an example, turn to the next section.

8D   QUADRATIC FORMS IN TWO VARIABLES

This section and the next apply diagonalisation to the theory of functions of several variables. For simplicity we begin with functions of two variables only.

Maxima and minima of functions of two variables

One often needs to know where functions take their maximum and minimum values. For functions of one variable, elementary calculus gives a way of doing this. From the graph of a function f, it is clear that if f has a maximum or a minimum at a point x, and the graph is a smooth curve through x, then the first derivative f' vanishes at x. If $f''(x) \neq 0$, its sign gives a simple test for whether $f(x)$ is a maximum or a minimum.

For a function of more than one variable, the picture is more complicated. A function f of two variables can be represented by a surface in three-dimensional space, with equation $z = f(x, y)$. If the xy plane is horizontal, then the surface vertically above the point (x, y) of the plane is at a height $f(x, y)$ above the plane.

The surface can be imagined as a landscape, with $f(x, y)$ giving the height above sea level. Maxima and minima are peaks and valley bottoms. These are places where there is a small horizontal piece of ground; they are called **critical points**. Thus a critical point can be a peak, or a valley bottom, or a point such as the origin in Fig. 8.1, where the ground is horizontal but one can go either uphill or downhill from the critical point, depending on which direction one takes. The critical point shown in Fig. 8.1 is called, for obvious reasons, a **saddle-point**. We shall label a critical point by its x- and y-coordinates; thus (x, y) will refer to the point on the surface with coordinates $(x, y, f(x, y))$.

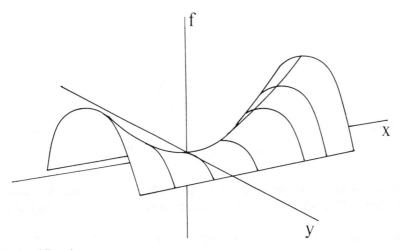

Fig. 8.1 – A saddle-point.

We now study the behaviour of f near a critical point. The notation is simpler if the critical point is at the origin. We can always transfer a critical point at (X, Y) to the origin by using $x - X$ and $y - Y$ as new variables. From now on we assume that this has been done, giving a critical point at $(0,0)$.

A smooth function f can be expanded in a series of powers of x and y:

$$f(x, y) = A + Bx + Cy + Dx^2 + Exy + Fy^2 + \ldots \tag{1}$$

where $A, B, \ldots$ are constants. Putting $x = y = 0$ shows that $A = f(0,0)$.

Now, close to the critical point, x and y are small, and so x^2 is negligible compared to x; all the quadratic terms in (1) are very small, and we have

$$f(x, y) \simeq f(0,0) + Bx + Cy$$

where $\simeq$ means 'is approximately equal to'. If $B \neq 0$, then the altitude $f(x, y)$ of the surface increases or decreases steadily as x increases through 0. But at a critical point, the surface is horizontal, f is locally constant, and we must therefore have $B = 0$. Similarly C must be zero.

We now go to the next approximation, and include the quadratic terms. If x and y are small, we have

$$f(x, y) \simeq f(0,0) + g(x, y) \tag{2}$$

where g is the quadratic part of (1):

$$g(x, y) = ax^2 + 2bxy + cy^2 \tag{3}$$

Here a, b, c are new names for the constants $D, E/2, F$ in (1). The reason for the factor of 2 will soon become clear.

A function of the form (3), involving only squares and products of the variables, is called a **quadratic form**. It can be rewritten in matrix notation as

$$g(x, y) = x(ax + by) + y(bx + cy) = (x, y)\begin{pmatrix} ax + by \\ bx + cy \end{pmatrix} \tag{4}$$

Write

$$u = \begin{pmatrix} x \\ y \end{pmatrix}$$

Then

$$\begin{pmatrix} a & b \\ b & c \end{pmatrix} u = \begin{pmatrix} ax + by \\ bx + cy \end{pmatrix}$$

and we have

$$g = u^T M u \quad \text{where } M = \begin{pmatrix} a & b \\ b & c \end{pmatrix} \tag{5}$$

Matrices and vectors in $\mathbb{R}^2$ give a natural and simple way of writing quadratic forms in two variables. You will notice that M is symmetric, so the theory of the last section applies.

The reason for the factor of 2 in the xy term of (3) is now clear. The xy term is split into two halves in (4); writing it as $2bxy$ avoids factors of $1/2$ in later work.

If the matrix M happens to be diagonal, with $M = \text{diag}(\lambda_1, \lambda_2)$, then (5) takes the simple form

$$g(x, y) = \lambda_1 x^2 + \lambda_2 y^2 \tag{6}$$

A quadratic form like this, with no terms of the xy type, is called a **diagonal quadratic form**, for the obvious reason that its matrix is diagonal. The properties of a diagonal form can be read off at a glance, as follows.

Suppose that λ_1 and λ_2 are both positive. Then $g(x, y) > 0$ for $(x, y) \neq (0, 0)$, and it follows from (2) that $f(x, y) > f(0, 0)$ for (x, y) near $(0, 0)$; in other words, f has a minimum at $(0, 0)$. A similar argument shows that if both λ's are negative, then f has a maximum, and if they have opposite signs, then the critical point is a saddle-point. Thus when M is diagonal, the signs of the diagonal entries tell us the nature of the critical point.

Of course, M will not usually be diagonal. But it is symmetric, so by Theorem 8C-2, it is orthogonally diagonalisable. Hence $M = QDQ^T$ for some orthogonal matrix Q. Equation (2) now gives

$$g = u^T QDQ^T u$$

which can be written as

$$g = v^T Dv \tag{7}$$

where

$$v = Q^T u \quad \text{and} \quad u = Qv \tag{8}$$

Note that the second equation in (8) follows from the first because $Q^T = Q^{-1}$ for orthogonal matrices.

Since the matrix D in (7) is diagonal, we have reduced the general case (5) to the simple diagonal form (6) by means of the transformation (8). This transformation is called **diagonalising** the quadratic form.

We noted in the last section that an orthogonal matrix represents a rotation. In terms of the rotated vector v, the quadratic form is given by an expression of the form of (6). The diagonal entries of D are the eigenvalues of M (Theorem 8A-1). So if both eigenvalues of M are positive, then f has a minimum; if they are both negative, then f has a maximum; and if they have different signs, then f has a saddle-point. Of course, rotating the vectors does not affect the nature of the critical point; a maximum remains a maximum after a rotation is applied.

You now see yet another aspect of eigenvalues: they are at the heart of maximum–minimum problems, as well as the theory of vibrations, and the geometrical structure of linear transformations. And atomic physics, and ill-conditioned systems of linear equations, as will be shown in Volume 2. And many other things besides.

8E   QUADRATIC FORMS IN n VARIABLES

We now extend the work of the previous section to n-space. It is quite straightforward. A **real quadratic form in n variables** is a function of the form

$$f(x) = \sum_{r,s=1}^{n} a_{rs} x_r x_s \tag{1}$$

where x denotes the real n-vector with entries $x_1, \ldots, x_n$, and the a_{rs} are real numbers. The terms in (1) with $r = s$ are the square terms, corresponding to ax^2 and by^2 in equation (3) of section 8D; the 'cross-terms' with $r \neq s$ in (1) correspond to the term $2bxy$ in 8D-(3). Notice that each cross-term occurs twice in (1); for example, $x_2 x_4$ occurs once with $r = 2$ and $s = 4$, and once with $r = 4$ and $s = 2$. Thus the coefficient of $x_2 x_4$ is $(a_{24} + a_{42})$. It is only the sum of a_{24} and a_{42} which appears in the quadratic form; their individual values do not matter. Therefore we can always make them equal without affecting the value of the form (1). So we shall assume that

$$a_{rs} = a_{sr} \quad \text{for all } r, s \tag{2}$$

Equation (1) can be written in matrix notation in the same way as for two variables. The **matrix of the quadratic form** (1) is the n by n matrix $A = (a_{rs})$, and we have

$$f(x) = x^T A x \tag{3}$$

The condition (2) means that A is a symmetric matrix.

Example 1. Consider the quadratic form in three variables:

$$x^2 + 2y^2 - z^2 + 3xy - yz$$

Its matrix is

$$\begin{pmatrix} 1 & 3/2 & 0 \\ 3/2 & 2 & -1/2 \\ 0 & -1/2 & -1 \end{pmatrix} \qquad \square$$

Quadratic forms in n variables are diagonalised in the same way as in the last section. By Theorem 8C-2, the matrix A is orthogonally diagonalisable, so

$$f(x) = x^T Q D Q^T x = y^T D y \tag{4}$$

where

$$y = Q^T x \qquad x = Q y \tag{5}$$

and $D = \text{diag}(\lambda_1, \ldots, \lambda_n)$, the λ's are the eigenvalues of A, and the columns of Q are the corresponding orthonormal eigenvectors.

Equation (4) can be written

$$f(x) = \sum \lambda_r (y_r)^2 \tag{6}$$

where y is related to x by (5). The entries of y are called **principal coordinates** for the quadratic form. Transforming to principal coordinates gives the simple expression (6), which is called a **diagonal** quadratic form, containing only squares and no cross-terms. Thus the transformation from x to y diagonalises the form (1).

The principal coordinates are Cartesian coordinates with respect to new axes, called **principal axes**, along the eigenvectors of A. This is easy to see. A point on the jth principal axis is, by definition, a point for which all entries of y are zero except the jth. From the second equation in (5), and the column lemma, it follows that x is proportional to the jth column of Q. This is the jth eigenvector of A. Hence the principal axes are along eigenvectors.

The diagonal form (6) reveals the behaviour of f quite clearly. If all eigenvalues of A are positive, then (6) shows that $f(x) > f(0)$ when $x \neq 0$, so f has a minimum at 0. Similarly, f has a maximum if all eigenvalues are negative, and the critical points are more complicated than simple maxima or minima when the eigenvalues are not all of one sign.

Example 2. Consider the quadratic form

$$f(x, y, z) = 3x^2 - y^2 - 2z^2 - 8xy + 12yz + 4zx \qquad (7)$$

Its matrix is

$$\begin{pmatrix} 3 & -4 & 2 \\ -4 & -1 & 6 \\ 2 & 6 & -2 \end{pmatrix}$$

Solving the characteristic equation gives eigenvalues 3, 6, -9 with eigenvectors $(2, 1, 2)$, $(2, -2, -1)$, $(1, 2, -2)$ respectively. These are the directions of the principal axes. We have

$$f = 3X^2 + 6Y^2 - 9Z^2 \qquad (8)$$

where X, Y, Z are coordinates with respect to axes along the eigenvectors given above. The relation between x, y, z and X, Y, Z is given by (5); in the present notation it gives $(X, Y, Z)^{\mathrm{T}} = Q^{\mathrm{T}}(x, y, z)^{\mathrm{T}}$, where Q is an orthogonal matrix whose columns are the normalised eigenvectors.

It follows from equation (8) that f has neither a maximum nor a minimum at $(0, 0, 0)$. If the origin is approached along the Z principal axis, f increases, but if the origin is approached in the XY plane, f decreases. In this way the principal axes lead to a clear description of the behaviour of the quadratic form.   □

It is now clear how diagonalisation helps in understanding quadratic forms. The next section carries the argument a stage further.

8F   CONIC SECTIONS AND QUADRIC SURFACES

This section looks at quadratic forms in two and three dimensions, from a geometrical viewpoint different from that of section 8D.

Conic sections

There are two ways of visualising a quadratic form in two variables. Section 8D considered a surface in 3-space, with equation $z = f(x, y)$. This technique does not easily extend to forms in three variables; it would need four-dimensional pictures, which are hard to draw. So we shall now take a different approach, in which a function of n variables is represented by an n-dimensional picture.

Consider first a form $f(x, y)$ in two variables. A two-dimensional picture can easily be derived from the surface of section 8D, by drawing a contour map. This means drawing the curves in the plane with equation $f(x, y) = K$, for various values of K.

Each curve is the set of points in the plane where the function takes the value K. The curves taken together give a map of the surface $z = f(x, y)$, and convey all the essential information about the function f.

The shape of the curves $f(x, y) = K$ may not be obvious at first sight. But we can always transform to principal axes in which the equation takes the form

$$aX^2 + bY^2 = K$$

where a and b are the eigenvalues, and X and Y the principal coordinates. This diagonal form is much easier to sketch. For one thing, it is symmetric about the X- and Y-axes (since changing the sign of X or Y in any solution gives another solution). The details are discussed in elementary analytic geometry. If a and b have the same sign, the curve is an ellipse whose axes are the X- and Y-axes; different values of K give ellipses of different sizes, nested inside each other without intersecting. If a and b have opposite signs, then the curve is a hyperbola. Both these curves are examples of **conic sections**, or **conics** for short, the curves in which a cone intersects a plane. If a or b is zero, we have a simple special case in which the curves are straight lines parallel to a principal axis, and the surface is plane.

In general, then, a quadratic form gives a family of conics, and the signs of the eigenvalues determine what type of conic. The principal axes are lines of symmetry of the curves; their directions are given by the eigenvectors of the matrix of the quadratic form.

Thus eigenvalues and eigenvectors give a useful way of drawing the curves corresponding to a general quadratic form. Of course, in the two-dimensional case one can manage without knowing about eigenvalues; the curves can be drawn by elementary methods. But the algebraic theory comes into its own in three and higher dimensions.

(By the way, if you are familiar with conics, have you wondered what happened to the parabola?)

Example 1. Consider the equation $7x^2 + 48xy - 7y^2 = 1$. What curve does it represent?

The matrix of this form is

$$\begin{pmatrix} 7 & 24 \\ 24 & -7 \end{pmatrix}$$

It is easily diagonalised: solving the characteristic equation gives eigenvalues ± 25, and the corresponding eigenvectors are $(4, 3)$ and $(3, -4)$. In terms of principal coordinates s and t, the equation becomes

$$s^2 - t^2 = 1/25$$

where the s- and t-axes are as shown in Fig. 8.2. This curve is easily drawn; it is the hyperbola shown in Fig. 8.2.   □

Quadric surfaces

A quadratic form f in three variables can be visualised as the family of surfaces $f(x, y, z) = K$ in 3-space. These surfaces are called **quadrics** (the word is a compound

Case (i): $K = 0$. Here we have $9Z^2 = 3X^2 + 6Y^2$. The intersection of this surface with a plane $Z = $ constant is an ellipse. Its intersection with the plane $X = 0$ is a pair of straight lines intersecting at the origin; similarly for the plane $Y = 0$. Thus the surface is a cone spreading out in the Z-direction, with an elliptical cross-section. See Fig. 8.4.

Case (ii): $K < 0$. The surface does not pass through the plane $Z = 0$, since $3X^2 + 6Y^2 = K$ can never be satisfied if $K < 0$. We have $9Z^2 = 3X^2 + 6Y^2 - K > 3X^2 + 6Y^2$, hence the surface lies inside the cone $9Z^2 = 3X^2 + 6Y^2$; it consists of two pieces, one lying inside each of the two half-cones $3Z = \pm\sqrt{(3X^2 + 6Y^2)}$. It approaches the cone asymptotically, because when X, Y, Z are very large, K is negligible, and the equation is close to the equation of the cone. See Fig. 8.4.

Case (iii): $K > 0$. Here the surface is in one piece, cutting the plane $Z = 0$ in an ellipse. We have $9Z^2 = 3X^2 + 6Y^2 - K < 3X^2 + 6Y^2$, hence the surface lies outside the cone $9Z^2 = 3X^2 + 6Y^2$. It approaches it asymptotically, by the same reasoning as for $K < 0$.

PROBLEMS FOR CHAPTER 8

Hints and answers for problems marked [a] will be found on page 284.

Section 8A

[a]1. Write down the matrices P and D which diagonalise the matrices of Examples 5H-1 and 5H-2 over $\mathbb{C}$. Verify in each case that $A = PDP^{-1}$. Also verify that the sum of the eigenvalues equals the sum of the diagonal entries (see Chapter 7, Problem 6).

[a]2. Find A^{17} where

$$A = \begin{pmatrix} -4 & 2 \\ -15 & 7 \end{pmatrix}$$

[a]3. (a) A is a real n by n matrix which is diagonalisable over $\mathbb{R}$. Prove that if $A^3 = 0$ then $A = 0$. Is this true for nondiagonalisable matrices?
   (b) Same as (a) with 0 replaced by I.

4. Is the 3 by 3 matrix with all entries 1 diagonalisable? If so, diagonalise it. Generalisations?

[a]5. A polynomial function f is defined by

$$f(t) = c_n t^n + c_{n-1} t^{n-1} + \ldots + c_1 t + c_0$$

where $c_0, \ldots, c_n$ are numbers. We apply f to square matrices A by setting $f(A) = c_n A^n + \ldots + c_1 A + c_0 I$. The factor I in the last term is needed to make it an n by n matrix, so that it can be added to the others.
   Prove that if A is diagonalisable, with $A = PDP^{-1}$, then for any polynomial f, $f(A) = Pf(D)P^{-1}$, where $f(D)$ is a diagonal matrix whose ith diagonal entry is $f(D_{ii})$. Hence evaluate $A^4 - 3A^2 + 2A - I$ and $A^3 - 63A + 162I$, where A is the matrix of Example 8A-1.

Prove that every diagonalisable matrix satisfies its own characteristic equation – that is, $p(A) = 0$ where p is the characteristic polynomial of A. (In fact, $p(A) = 0$ for any A, whether diagonalisable or not; this is called the Cayley–Hamilton theorem – see section 10D in Volume 2.)

6. Let $z = e^{2\pi i/3}$. Show that $z^3 = 1$ (so z is a complex cube root of 1), and $1 + z + z^2 = 0$.

Now consider a 3 by 3 circulant matrix M with first row (a, b, c) (see Chapter 7, Problem 8 for circulants). Show that $(1, 1, 1)$ and $(1, z, z^2)$ are eigenvectors, and write down a third linearly independent eigenvector. Hence write M in the form PDP^{-1}.

7. Tridiagonal matrices are defined in section 4K. This problem shows that they are diagonalisable if they are not too unsymmetric.

Consider a real tridiagonal matrix A for which the $(i, i+1)$ and $(i+1, i)$ entries have the same sign. Show how to find a real diagonal matrix D such that $D^{-1}AD$ is symmetric. Deduce that A is diagonalisable over $\mathbb{R}$.

Section 8B

8. Consider the nth-order differential equation

$$y^{(n)} + a_1 y^{(n-1)} + \ldots + a_{n-1} y^{(1)} + a_n y = 0$$

where y is a function of t, $y^{(r)}$ denotes the rth derivative of y, and the a's are constants. It can be reduced to the form studied in section 8B as follows.

Define an n-vector x by $x_1 = y$ and $x_r = y^{(r-1)}$ for $r > 1$. Show that $x' = Ax$, where A is the companion matrix (see Chapter 7, Problem 7) of the polynomial $P(z) = z^n + a_1 z^{n-1} + \ldots + a_n$. Hence the zeros of the polynomial P control the behaviour of the solutions.

Deduce that all solutions of the differential equation $y^{(3)} + 3y^{(2)} + 4y^{(1)} + 2y = 0$ tend to zero as $t \to \infty$.

(Hint: if you don't know how to find roots of a cubic $P(z)$, begin by plugging in integer values of z, to try and see where it changes sign.)

9. In the notation of section 8B, define the matrix $M(t, s) = PE(t - s)P^{-1}$ for all real s, t. Prove the following.

(a) $M(t, t) = I$, $M(s, t)M(t, u) = M(s, u)$, and $M(s, t)$ is invertible, with inverse $M(t, s)$, for all s, t, u.

(b) If x is a solution of $\dot{x} = Ax$, where $\dot{x} = dx/dt$, then $x(t) = M(s, t)x(s)$ for all s, t. In other words, $M(s, t)$ transforms the solution vector at time s into the solution at time t. (For this reason, M is called the 'transition matrix' for the system $\dot{x} = Ax$.)

(c) $\dot{M} = AM$, where $\dot{M}$ denotes $(d/dt)M(t, s)$.

(d) The equation $\dot{x} = Ax + b$, where b is a given vector function of t, can be solved as follows: the solution satisfying the condition $x(s) = X$ is

$$x(t) = M(t, s)X + \int_s^t M(t, u)b(u)\, du$$

(For more information on this very useful matrix, see, for example, Brockett (1970).)

Section 8C

^{a}10. Orthogonally diagonalise the following matrices:

$$\text{(a)} \quad \begin{pmatrix} 1 & 2 & 5 \\ 2 & -2 & -2 \\ 5 & -2 & 1 \end{pmatrix} \qquad \text{(b)} \quad \begin{pmatrix} 2 & -1 & -1 \\ -1 & 2 & -1 \\ -1 & -1 & 2 \end{pmatrix}$$

$$\text{(c)} \quad \begin{pmatrix} 2 & -1 & 0 & -1 \\ -1 & 2 & 0 & -1 \\ 0 & 0 & 5 & 0 \\ -1 & -1 & 0 & 2 \end{pmatrix}$$

^{a}11. True or false? (a) If A is orthogonally diagonalisable, then A is symmetric. (b) Converse of (a).

12. If A is a real symmetric matrix satisfying $A^3 = I$, prove that $A = I$. What if the restriction 'real symmetric' is removed?

13. Let X be the column vector $(a, b, c)^T$ where a, b, c are not all zero. Then $X^T X$ is a number and $X X^T$ is a square matrix M, say. Show that M is symmetric, $\text{rank}(M) = 1$, and the only nonzero eigenvalue is $X^T X$. What is the corresponding eigenvector?
   Let Y and Z be vectors of unit magnitude perpendicular to X and to each other. Show that they are eigenvectors of M. Hence diagonalise M.

14. Show that $a + b + c$ is an eigenvalue of the matrix

$$\begin{pmatrix} a & b & c \\ b & c & a \\ c & a & b \end{pmatrix}$$

What is the corresponding eigenvector? Find the other eigenvalues, and hence prove that $a^2 + b^2 + c^2 \geq bc + ca + ab$ for any real numbers a, b, c.
   (For a more systematic way of proving inequalities like this, see Problem 18.)

Section 8D

^{a}15. For each of the following functions, find whether it has a maximum, a minimum, or a saddle-point at the origin:
   (a) $2x^2 + 3xy - 2y^2 + x^3 - y^3$;
   (b) $xy + x^2 y^2$;
   (c) $x^2 - 2xy + 5y^2 + \sin^2(xy)$.

^{a}16. For each of the following quadratic forms, write down its matrix, find an orthogonal transformation which diagonalises it, and say whether it has a maximum, a minimum, or neither:
   (a) $2x_1 - 6x_1 x_2 + 10x_2$;
   (b) $x^2 + 8xy - 5y^2$.

Section 8E

[a]17. For each of the following quadratic forms, write down its matrix, find an orthogonal transformation which diagonalises it, and say whether it has a maximum, a minimum, or neither:

(a) $x^2 - 2y^2 + z^2 + 4xy - 4yz + 10zx$;

(b) $x^2 + y^2 + z^2 - xy - yz - zx$;

(c) $2xy + 2yz + 2zx$.

(For (a) and (b) see Problem 10.)

18. Use Problem 17(b) to give another proof of the inequality in Problem 14.

19. (a) Show that a diagonal quadratic form $\sum k_i(x_i)^2$ can be reduced to the form $\sum \mu_i(y_i)^2$ in which all the μ_i are ± 1 or 0 by a transformation of the form $x = Py$ where P is a nonsingular diagonal matrix.

(b) Hence show that every quadratic form can be expressed as a sum of squares (with coefficients ± 1) by a transformation $x = Ry$ where R is nonsingular (but not necessarily orthogonal).

20. Use the results of Problem 19 to show that if f and g are quadratic forms in n variables, and all eigenvalues of g are positive, then f and g can be simultaneously diagonalised. In detail: show that there is a transformation $x = Sz$ which transforms f into a diagonal form in the variables z, and at the same time transforms g into the form $z_1^2 + \ldots + z_n^2$.

21. Consider the values of a quadratic form $Q(x_1, \ldots, x_n) = x^T A x$ on the unit sphere $S = \{x : x^T x = 1\}$. Show that the maximum value of Q on S equals the largest eigenvalue of A, and occurs when x is the corresponding normalised eigenvector. Similarly, the minimum of Q on S is the smallest eigenvalue.

Hence prove that the largest eigenvalue of a symmetric matrix cannot be less than the largest diagonal entry. Similarly for the smallest eigenvalue.

(This idea is developed further in section 12C of Volume 2.)

22. (For readers familiar with multiple integrals and Jacobians.)

(a) If n-vectors X and Y are related by $X = PY$, where P is a nonsingular n by n matrix, show that the Jacobian of the change of variable from X to Y equals $\det(P)$.

(b) Evaluate $\iint \exp(-3x^2 + 2xy - 3y^2)\, dx\, dy$, integrated over the whole plane.

(Diagonalising the quadratic form reduces the problem to one-dimensional integrals of the form $\int_{-\infty}^{\infty} \exp(-at^2)\, dt = \sqrt{(\pi/a)}$.)

(c) If A is a real symmetric n by n matrix with all eigenvalues positive, prove that

$$\int_{-\infty}^{\infty} \ldots \int_{-\infty}^{\infty} dx_1 \ldots dx_n \exp(-x^T A x) = [\pi^n/\det(A)]^{1/2}$$

23. Use Chapter 6, Problem 12 to express quadratic forms as determinants.

Section 8F

[a]24. For each quadratic form f in Problem 16, use principal axes to sketch the curves $f(x_1, x_2) = k$, for various values of k.

[a]25. For each quadratic form f in Problem 17, sketch the graph of the equation $f(x, y, z) = k$, for various values of k.

26. Show that if $b^2 \neq 4ac$, the equation $ax^2 + bxy + cy^2 + dx + ey = k$ can be reduced to $aX^2 + bXY + cY^2 = K$ by a simple shift of axes $X = x + p$, $Y = y + q$ for suitable constants p, q. Why is the condition $b^2 \neq 4ac$ needed?

 Hence sketch the graph of the equation $x^2 + 8xy + y^2 - 4x + 2y = 10$.

27. For a quadratic equation in three variables, under what conditions can linear terms be removed in the same way as in Problem 26 above?

28. Prove that the principal axes of the quadric with equation $x^T A x = 1$ can be brought into line with the x_1, x_2, x_3 axes by rotating about a line parallel to the eigenvector of the diagonalising matrix of A.

Appendices

APPENDIX A   HOW TO READ MATHEMATICS

Mathematics is a peculiar subject. It is a precise logical structure, built of definitions, theorems, and the like. Yet there is a fabric of intuitive ideas behind the logical structure. The tension between the intuitive and the logical is perhaps one of the difficulties of the subject.

A page of mathematics may at first sight look unintelligible, full of meaningless definitions, theorems, and proofs. It is the underlying intuitive ideas that give meaning to the logical structure. The ideas and the logical structure are equally important; logic without intuition may be meaningless, but intuition without logic is unreliable. Mathematics develops by exploring new ideas in a tentative and imprecise way first, and then testing the ideas and proving or disproving them logically. When you read mathematics, distinguish the general explanations from the detailed logical exposition, and read each in the appropriate spirit.

Lemmas, definitions and all that

Mathematics is usually set out in a style which originated about two thousand years ago. It has survived because people have found it helpful; it packages the elements of the structure in easily identifiable units.

The basic facts are expressed as **propositions**. A proposition is a clearly stated fact, which the author proposes to prove.

Certain types of proposition have special names. A particularly important and memorable proposition is called a **theorem**. A proposition which is not very interesting in itself, but is needed in order to prove something else, is called a **lemma**. A proposition which follows immediately from the preceding proposition is called a **corollary** of that proposition.

The distinction between these types of proposition is not at all clear-cut. One author may call something a lemma while another calls it a theorem; it is a matter of personal taste.

All the propositions are deduced from certain basic ideas. These are set out in **definitions**, which introduce words for the basic ideas, and define their meaning. For example, Chapter 3 begins by defining the meaning of the phrase 'vector space', and proceeds to develop the properties of vector spaces in a series of propositions. Further definitions are introduced from time to time as the subject develops and new ideas arise.

The other main items of a mathematical text are **examples**. They are essential for a proper understanding of the theory. Textbooks may contain many examples. But experienced readers make up their own as they go along, in order to test their understanding. There is then less need of written-out examples. Similarly, an introductory textbook will contain a good deal of explanation, introducing and surrounding the formal structure of definitions and propositions. But more experienced readers need less verbose explanations, and advanced books are usually more concise.

The language of logic

Mathematics often uses familiar words in an unfamiliar and specialised way. This can be particularly confusing when the words are very simple. Consider the word 'some', for instance. The sentence 'some newspapers print the truth' carries a strong suggestion that while some do, others do not. But in mathematical writing, things are different. For example the statement 'some numbers in the set S are positive' implies nothing about whether S contains negative numbers. It is true when $S = \{1, 2, 3\}$ and also when $S = \{-1, 2, 3\}$.

Another word that often causes confusion is 'if'. The statement 'A if B' means that A follows logically from B. Thus we can say '$x > -1$ if x is positive'. This statement makes no assertion about the size of x; all it says is that if, somehow, we could establish that x is positive, then it would follow that $x > -1$.

It is amazingly easy to get confused by the word 'if'. The following statements all, by definition, mean the same thing.

$$\left.\begin{array}{l} \text{A is true if B is true} \\ \text{If B is true, then A is true} \\ \text{B is true only if A is true} \\ \text{B implies A} \end{array}\right\} \tag{1}$$

It is worth thinking this through slowly and carefully, with a concrete example in mind. For example, take A to be the statement '$x > 0$' and B the statement '$x > 1$'.

Statements can be turned upside down, in two different ways. One way of inverting a statement is to negate it. The **negation** of the statement 'A is true' is the statement 'A is not true'. For example, the negation of '$x > 0$' is '$x \leq 0$'. The negation of

$$x > 1 \text{ implies } y > 0 \tag{2}$$

is

$$x > 1 \text{ does not imply } y > 0 \tag{3}$$

The statement

$$x > 1 \text{ implies } y \leq 0 \tag{4}$$

is sometimes mistakenly thought to be the negation of (2). It is not; (4) gives definite information when $x > 1$, while (3), the correct negation of (2), says that no definite conclusion is possible in this case.

A logical statement of the type (1) above can be inverted in another way, by stating its converse. The **converse** of 'B implies A' is 'A implies B'. For any of the forms in (1), the converse is obtained by interchanging A and B. There is no general rule about whether the converse is true or not; the converse of a true statement may be true or it may be false. For example, '$x > 5$ if $x > 6$' is a true statement; its converse is '$x > 6$ if $x > 5$', which is false. On the other hand, '$e^x > 1$ if $x > 0$' is a true statement, and its converse '$x > 0$ if $e^x > 1$' is also true.

The converse of 'A is true if B is true' is the statement 'A is true only if B is true'; the latter is equivalent to 'B is true if A is true' according to (1). If statement (1) and its converse both hold, we combine them in the compound statement 'A is true **if and only if** B is true'. This means that each of A and B implies the other; they are logically equivalent to each other.

For a more detailed account of logic in mathematics, see Stewart and Tall (1977) and Binmore (1980).

Proofs and counterexamples

Mathematics is built up of general statements, usually of the form 'If... then...'. Some theorems say 'All X's have property Y', which is a variant of the same form: it is equivalent to 'If z is an X, then z has property Y'. A typical mathematical problem is to prove that such a statement is true, or that it is false; proving that a statement S is false is called **disproving** S.

When faced with a mathematical statement to be proved or disproved, the first thing to do is try to decide whether it is true. A good way to start is to try one or two simple examples. For instance, given the statement 'every quadratic equation has two roots', the obvious thing to do is write down some quadratic equations and test the statement. One soon finds the simple equation $x^2 = 0$ with only one solution. This immediately shows that the statement is wrong. An example like this, which disproves a general statement, is called a **counterexample**. To *prove* a general statement requires a general argument; but to *disprove* it, a single counterexample is enough. In this sense, disproving theorems is much easier than proving them.

Methods of finding proofs are discussed in Polya (1962); Solow (1982) and Stewart and Tall (1977) are useful too. There is no space here to discuss proof techniques in general, but the following paragraphs deal with one technique which often causes difficulty.

Proof by contradiction

A proposition of the form (1) above, 'if B then A', can be proved by assuming that B is true and then making a chain of deductions leading to A. But there is another method, called proof by contradiction. The plan is to start from the assumption that A is false, and show that it leads to a contradiction. It then follows that A cannot be false, so A is proved.

For example, consider the statement 'given any number of dots and any number of lines joining one dot to another, the number of dots with an odd number of lines emerging is even'. We prove this by contradiction, as follows.

Each line has two ends, each attached to a dot. Call a dot 'odd-type' if there is an odd number of lines attached to it. We shall prove by contradiction that the number of odd-type dots is even. Suppose that the number is odd. Then the total number of line-ends attached to odd-type dots is odd. There is an even number of line-ends attached to the even-type dots, hence the total number of line-ends is odd. This is impossible, since the number of line-ends is twice the number of lines, and therefore even. Thus the assumption that the number of odd-type dots is odd leads to a contradiction; so the assumption must be false, hence the number is even.

There is another proof technique which is not quite straightforward: the method of induction. It is discussed in Appendix B.

APPENDIX B   MATHEMATICAL INDUCTION

Mathematical induction is the name of a technique for proving certain kinds of proposition.

A 'proposition' is a statement which can be proved (or disproved) by mathematical means. Mathematical induction is a way of proving propositions involving the positive integers $1, 2, 3 \ldots$. For example, the statement 'the sum of the first n positive integers is $n(n + 1)/2$' is a proposition about the general positive integer n. We can name this proposition $P(n)$. Thus $P(3)$ is the statement '$1 + 2 + 3 = 3 \cdot 4/2$'.

Propositions of the type $P(n)$, for general integers n, can be proved by the following method.

The principle of mathematical induction. Let $P(n)$ be a proposition involving the positive integer n. If
(a) $P(1)$ is true, and
(b) for each positive integer k, $P(k)$ implies $P(k + 1)$,
then $P(n)$ is true for all n. □

This statement is justified as follows. Taking $k = 1$ in (b) we have $P(1)$ implies $P(2)$. But $P(1)$ is true by (a), hence $P(2)$ is true. Now take $k = 2$ in (b): $P(2)$ implies $P(3)$, but we have just proved $P(2)$, hence $P(3)$ follows. Repeating the argument shows that $P(4)$ is true, and similarly $P(5)$, $P(6)$, and all the other $P(n)$ are true.

Example. We illustrate the method by proving the following statement, which we call $P(n)$: the sum of the squares of the first n odd numbers is $n(4n^2 - 1)/3$.

First we show that condition (a) of the principle of induction is satisfied. A sum containing one term simply equals that term, so $P(1)$ says that $1^2 = 1 \cdot 3/3$. This is obviously true, so (a) holds.

Now we show that $P(k)$ implies $P(k + 1)$ for all k; that is, *if* $P(k)$ were true, *then* $P(k + 1)$ would follow. Here $P(k)$ is called the 'inductive hypothesis'. We assume it temporarily, to see whether it implies $P(k + 1)$. Since the kth odd number is $2k - 1$, the inductive hypothesis here is

$$1^2 + 3^2 + \ldots + (2k - 1)^2 = k(4k^2 - 1)/3 \tag{1}$$

for some integer k. The $(k + 1)$th odd number is $(2k + 1)^2$. Adding it to both sides of

(1) gives

$$1^2 + \ldots + (2k-1)^2 + (2k+1)^2 = k(4k^2-1)/3 + (2k+1)^2$$

and after some elementary algebra we have

$$1^2 + \ldots + (2k+1)^2 = (k+1)[4(k+1)^2 - 1]/3 \qquad (2)$$

But (2) is the statement $P(k+1)$, and we have deduced it from (1), which is $P(k)$. We have thus shown that if $P(k)$ were true then $P(k+1)$ would follow.

This completes the verification of conditions (a) and (b) of the principle of induction, and the truth of the statement for all n now follows.   □

Induction may look like a confidence trick at first sight. It begins by assuming $P(k)$, which resembles the logical crime of assuming what you are trying to prove. But if you think it through carefully, you will see that the method is sound. We assume the inductive hypothesis only temporarily, for the purpose of establishing (b). For further discussion see Polya (1962), Solow (1982) and Binmore (1980).

APPENDIX C   SETS

A 'set' means any collection of things. We take this as an intuitively obvious idea, and shall not try to define it formally. Analysing the idea of a set leads into deep waters; see, for example, Courant and Robbins (1941), Stewart and Tall (1977) and Binmore (1980). This appendix gives the elements of set theory, as far as is necessary for understanding linear algebra. The account is rather condensed; you might start by reading the first section only, and then return to the main text of the book, referring back to this appendix when necessary.

Basic notions and notations

One way of specifying a set is to list its members between curly brackets. Thus

$$A = \{1, 2, 3, 4\}$$

means that A is the set consisting of the four numbers $1, 2, 3, 4$. For any set X, we write $x \in X$ to mean that x belongs to the set X. For example, $2 \in A$ where A is the set defined above.

The order in which the members are listed makes no difference: $\{1, 2, 3, 4\}$ and $\{4, 3, 2, 1\}$ are just two ways of writing down the same set. So $\{1, 2, 3, 4\} = \{4, 3, 2, 1\}$. The list of members of a set may contain repetitions, but the repetitions do not mean anything. Thus $\{1, 1, 2, 3, 4\} = \{1, 2, 3, 4\}$.

Some sets have standard names. The set of all real numbers is called $\mathbb{R}$, and the set of all complex numbers is called $\mathbb{C}$.

Another way of specifying a set is by listing conditions for membership. For example, we write

$$\{x \in \mathbb{R} : 0 < x < 1\} \qquad (1)$$

for the set of all x in $\mathbb{R}$ which satisfy the condition $0 < x < 1$. The curly bracket $\{$ is pronounced 'the set of', and the colon is pronounced 'such that'; so (1) is to be read as 'the set of x in $\mathbb{R}$ such that $0 < x < 1$'.

A **finite set** is a set with a finite number of elements. An **infinite set** is one with infinitely many elements, such as the set of all real numbers. An infinite set cannot be specified by listing all its elements. But sometimes it can be indicated by means of dots; for example, $\{1, 2, 3, \ldots\}$ obviously means the set consisting of all positive integers.

The algebra of sets

The algebra of sets begins with the relation of equality. This is very simple: two sets are equal when they have the same elements. Thus, as we noted above, $\{1, 2, 3, 4\} = \{4, 3, 2, 1\} = \{1, 1, 2, 3, 4\}$.

We now consider combining sets, in the same way that elementary algebra combines numbers by multiplication and addition. There are two standard operations in set theory.

The **union** of sets S and T means the set obtained by putting them together. It is denoted by $S \cup T$. More formally, we have

$$S \cup T = \{x : x \in S \text{ or } x \in T\}$$

For example, $\{1, 2, 3\} \cup \{3, 4, 5\} = \{1, 2, 3, 3, 4, 5\} = \{1, 2, 3, 4, 5\}$.

The **intersection** of two sets S and T is their overlap, the set of all elements common to S and T. It is denoted by $S \cap T$. In other words,

$$S \cap T = \{x : x \in S \text{ and } x \in T\} = \{x : x \text{ belongs to both } S \text{ and } T\}$$

For example, $\{1, 2, 3\} \cap \{3, 4, 5\} = \{3\}$.

If two sets do not intersect, they are said to be **disjoint** from each other. Thus, for example, $\{1, 2\}$ and $\{3, 4\}$ are disjoint sets.

An awkward question now arises. What is $\{1, 2\} \cap \{3, 4\}$? We defined the intersection as the overlap, but there is no overlap here. In general the intersection of two disjoint sets contains no elements. A set with no elements is called an **empty set**. It may seem silly to call something a set if it has no elements. But it is no sillier than regarding 0 as a number; most people find that strange at first, but soon come to terms with it.

Warning. Do not confuse the empty set, which has no elements, with the set $\{0\}$, which has one element, namely, the number zero.   □

If A and B are finite sets, then $A \cup B$ and $A \cap B$ are obviously finite. Equally obviously, the union of an infinite set with any set is infinite, and the intersection of a finite set with any set is finite. The intersection of two infinite sets may be finite or infinite. Examples: if P is the set of all real numbers ≥ 0, and N the set of all real numbers ≤ 0, then $P \cap N = \{0\}$, a finite set with one element, and $P \cap P = P$ which is infinite.

If a set S is contained within another set T, we say S is a **subset** of T, and write $S \subset T$. Formally, $S \subset T$ if every element of S belongs to T. For example, if P is the set of all positive numbers, then $P \subset \mathbb{R}$. Notice that according to this definition, every set is a subset of itself: $T \subset T$ for every set T. A subset of T which is genuinely smaller than T is called a **proper subset** of T. By 'genuinely smaller' we mean $S \subset T$ and $S \neq T$. We do not try to compare the size of sets by counting the number of elements; many of the sets encountered in linear algebra are infinite, and comparing sizes of infinite sets leads to nonsense. (For example, both of the following statements are true: (a) half of the integers are even, so there are twice as many integers as there are even integers; (b) there are exactly as many integers as there are even integers. For a discussion of this absurd situation, see Stewart and Tall (1977), Binmore (1980) and the entertaining book by Rucker (1982).

Ordered sets

The elements of a set as defined above are not in any particular order. It may be necessary for practical reasons to list them in some order, but no significance is attached to the order in the list; $\{1, 2, 3\}$ is the same set as $\{2, 1, 3\}$.

However, we sometimes need to consider a list of things in some given order. An **ordered set of n elements** means a list of n items, written in a definite order. Thus $\{1, 3, 4, 2\}$ can be regarded either as a set, as discussed above, or as an ordered set in which 1 is the first element, 3 is the second, etc. Regarded as ordered sets, $\{1, 3, 4, 2\} \neq \{1, 2, 3, 4\}$.

Ordered sets may contain repetitions; thus the ordered set $\{1, 2, 2\}$ has first element 1, second element 2, and third element 2. Regarded as an ordered set, it has three elements. If we ignore the ordering and regard it as a plain set, it equals $\{1, 2\}$.

Sequences

A **sequence** is an ordered set with infinitely many elements. A sequence is defined by giving a rule which, for every positive integer n, specifies the nth element of the set. One often refers to the nth **term** of a sequence, rather than the nth element. For example, the rule 'for each n the nth term is $1/n$' describes the sequence $1, 1/2, 1/3, \ldots$. A sequence is sometimes distinguished from a set by using round brackets. For example, the above sequence is written $(1, 1/2, 1/3, \ldots)$.

Like an ordered set, a sequence may contain repetitions; for example, $(1, 1, 1, \ldots)$ denotes the sequence for which all elements are 1. On the other hand, the unordered set $\{1, 1, 1, \ldots\}$ contains just one element, and equals $\{1\}$.

The sequence $(a_1, a_1, \ldots)$ is sometimes denoted by (a_n). Thus $(1/n)$ stands for the sequence $(1, 1/2, 1/3, \ldots)$.

Warning. Some authors use the word 'sequence' to mean either a sequence or an ordered finite set. The phrase 'infinite sequence' is then used to mean what we call a sequence.

APPENDIX D   FUNCTIONS

This appendix skims briefly over material discussed at length in calculus books. If you have met the notions of continuity and differentiability before, the following will serve as a reminder. If you have not, then do not worry too much about details; only the broad ideas are needed to understand the examples in the text.

Real and complex functions

A **real function** is a rule which, given a real number, specifies another real number. For example, the sine is a real function; $\sin(x)$ has a definite real value for every real x. We write $f: \mathbb{R} \to \mathbb{R}$ to mean that a function f maps real numbers to real numbers. We write $y = f(x)$ to mean that f maps the number x into the number y; in this context x is called the **argument** of the function, and y is called the **value** of the function at the argument x.

We are also interested in functions which give a complex value for each real number. For example, e^{it} is a complex number for every real t. We say it is a **complex function**, or a function $\mathbb{R} \to \mathbb{C}$, mapping real numbers to complex numbers. In this book, the arguments of functions are always real numbers, though the values may be complex.

Every complex number c can be expressed as $c = a + ib$ where a and b are real. Hence a complex function $f(t)$ can always be expressed as

$$f(t) = g(t) + ih(t)$$

where g and h are real functions. Thus a complex function is equivalent to a pair of real functions. The functions g and h are called respectively the **real and imaginary parts** of f.

Sometimes a real function is defined only for certain values of its argument. For example, $\sqrt{[x(1-x)]}$ is real only if x is between 0 and 1. Write

$$[a, b] = \{x: a \le x \le b\}$$

This set is called the **closed interval** a, b. Then $\sqrt{[x(1-x)]}$ gives a real number for each x in $[0, 1]$, but not for each x in $\mathbb{R}$. We say that it is a function defined on $[0, 1]$.

Continuous functions

We define continuous functions as follows. A function f is called **continuous at a point** c if $f(x)$ approaches $f(c)$ as x approaches c. We say that f is **continuous on** $[a, b]$ if f is continuous at all points of $[a, b]$. Broadly speaking, a real function is continuous if its graph has no sudden jumps. A complex function f is continuous if its real and imaginary parts are continuous.

Most functions encountered in elementary calculus are continuous almost everywhere. It is not difficult to prove that the sum of two continuous functions is always continuous, and, more generally, every linear combination of finitely many continuous functions is continuous.

Differentiable and smooth functions

A real function f is called **differentiable on the interval** $[a, b]$ if its derivative $f'(x)$ exists for all x in $[a, b]$. This means that the graph of f has a well-defined tangent at all x in $[a, b]$. For example, every polynomial is differentiable on any interval. The function $f(x) = |x|$ is differentiable on $[a, b]$ if a and b are both positive or both negative. But f is not differentiable on intervals which include 0, because its graph has a sharp corner at 0 and there is no tangent to it at that point (see Fig. D1).

If the graph of a function has a sudden jump at $x = c$, as in Fig. D2, then there is no tangent at $x = c$, and the derivative does not exist at c. In other words, if a function

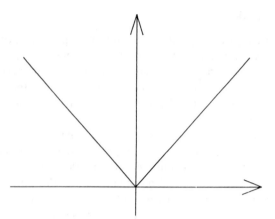

Fig. D1 – The graph of $f(x) = |x|$.

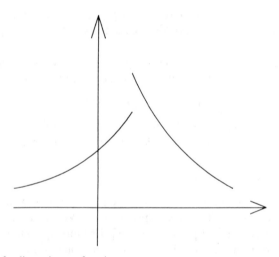

Fig. D2 – The graph of a discontinuous function.

is not continuous, then it is not differentiable. If on the other hand a function is continuous, it may or may not be differentiable. The function $f(x) = |x|$ is an example of a function continuous at $x = 0$ but not differentiable there.

A complex function is said to be differentiable on $[a, b]$ if its real and imaginary parts are. Again, every differentiable function is continuous, but not vice versa.

A function f is called **twice differentiable on $[a, b]$** if f is differentiable on $[a, b]$ and so is its derivative f'. Similarly, f is **n times differentiable on $[a, b]$** if it can be differentiated n times at all points in $[a, b]$.

A function f is called **smooth on $[a, b]$** if it can be differentiated any number of times. More precisely, f is smooth on $[a, b]$ if for every positive integer n, f is n times differentiable on $[a, b]$. For example, $1/x$ is smooth everywhere except at 0; it can be differentiated any number of times, giving higher and higher powers.

If f is n times differentiable, then f' is in general only $n - 1$ times differentiable. Let $C^{(n)}$ denote the set of all functions which are n times differentiable on some interval. Then differentiating a function in $C^{(n)}$ gives a function in $C^{(n-1)}$. On the other hand, the derivative of a smooth function is another smooth function. That is why the set of smooth functions is often convenient to work with: in the language of Chapter 5, it is mapped into itself by the operation of differentiation.

APPENDIX E   COMPLEX NUMBERS

This appendix gives a brief introduction to complex numbers along standard lines. For more details see Courant and Robbins (1941) and algebra textbooks. For an interesting alternative approach, see Chapter 22 of Feynman *et al.* (1964).

Informal approach

In ordinary algebra, the number -1 has no square root. The algebraists of the sixteenth century therefore invented a number of a new kind, written i (for imaginary), such that $i^2 = -1$. All the rules of elementary algebra apply to this number, except that its square is negative. Thus $i + i = 2i$, and so on. Any multiple of i is called an **imaginary number**.

Introducing the single new number i gives a whole family of imaginary numbers (multiples of i). These numbers give square roots for any negative number: for any a, the square roots of $-a^2$ are ia and $-ia$.

Real and imaginary numbers can be added, giving expressions such as $2 + 3i$. These are called **complex numbers** ('complex' because they are compounded of real and imaginary parts). Complex numbers are manipulated in the same way as ordinary numbers, with the extra rule $i^2 = -1$.

Every complex number can be reduced to the form $a + ib$, where a and b are real. For example, $1/i = i/i^2 = i/(-1) = -i$, which is in the form $a + ib$ with $a = 0$ and $b = -1$. Another example:

$$(2 + 3i)^2 = 4 + 12i + (3i)^2 = 4 + 12i + 9(-1) = -5 + 12i$$

Fractions such as $(2 + 3i)/(1 + 2i)$ can be reduced to the standard form by the following trick: if the denominator contains a factor $a + bi$, then multiply top and bottom by $a - bi$. For example:

$$\frac{2 + 3i}{1 + 2i} = \frac{(2 + 3i)(1 - 2i)}{(1 + 2i)(1 - 2i)} = \frac{2 + 3i - 4i - 6i^2}{1 - 4i^2} = \frac{8 - i}{1 + 4} = 8/5 + (-1/5)i$$

Objection. 'This theory is nonsense. There is no such thing as the square root of -1. Calling it i doesn't prove it exists. I could say 'let R denote a flying rhinoceros with purple stripes' but that wouldn't prove that such a thing exists, and I certainly couldn't deduce anything useful from it. Your imaginary numbers are well named. I will have nothing to do with them.'   □

This is a very reasonable objection. The following paragraphs answer it, by defining complex numbers in a way which leaves no doubt as to their existence.

Formal theory

To put the theory on a more logical foundation, we start again with a new definition of complex numbers. The definition will at first seem to have little connection with the square root of -1. But it will eventually lead to the $a + ib$ form of complex numbers given above.

We define a complex number as a pair of real numbers, in a definite order. We write $z = (a, b)$, meaning that z is the complex number consisting of the real numbers a, b. Thus $(1, 0)$ is a complex number, and $(0, 1)$ is a different complex number.

We define the sum of two complex numbers in the obvious way; for example, $(1, 2) + (3, 4) = (4, 6)$. The general rule is:

$$\text{if } z = (a, b) \text{ and } z' = (a', b') \text{ then } z + z' = (a + a', b + b') \tag{1}$$

Multiplication is more complicated. We define the product of two complex numbers by

$$(a, b)(a', b') = (aa' - bb', ab' + ba') \tag{2}$$

This may seem a strange definition of multiplication. But with a certain amount of work, one can show that for any complex numbers z, u, v we have $zu = uz, (zu)v = z(uv)$, and $z(u + v) = zu + zv$; all the rules of ordinary arithmetic apply when addition and multiplication are defined by (1) and (2).

Consider now pairs of the form $(x, 0)$. The rules (1) and (2) give

$$(x, 0) + (y, 0) = (x + y, 0) \quad \text{and} \quad (x, 0)(y, 0) = (xy, 0)$$

Thus combining two complex numbers of the form $(x, 0)$ gives another number of the same type, and the nonzero parts of the numbers add and multiply in the same way as in ordinary arithmetic. In other words, numbers of the form $(x, 0)$ behave just like real numbers x; they can be called real-type complex numbers. We shall write **a** for the real-type complex number corresponding to a real number a; thus

$$\mathbf{a} = (a, 0) \quad \text{for any real number } a$$

Now consider the complex number $(0, 1)$. Equation (2) gives

$$(0, 1)(0, 1) = (-1, 0)$$

Thus the square of $(0, 1)$ is the real-type complex number $-\mathbf{1}$. It is therefore sensible to write

$$(0, 1) = \mathbf{i}$$

It is the square root of $-\mathbf{1}$.

Now, for every ordered pair (a, b) we have

$$(a, b) = (a, 0) + (0, 1)(b, 0) = \mathbf{a} + \mathbf{ib}$$

Dropping the bold type, we see that every complex number can be written as $a + ib$ if the symbols are interpreted appropriately. The multiplication rule (2) agrees with the result of ordinary multiplication of $(a + ib)(a' + ib')$ if one sets $i^2 = -1$.

We have now shown that the ordered-pair definition gives the same structure as the informal approach. It has served its purpose of showing that i exists; from now on we shall use the more convenient $a + ib$ notation.

Complex algebra

Much of complex algebra is the same as the algebra of real numbers, but there are some new ideas.

If $z = x + iy$, the real numbers x and y are called respectively the **real part** and the **imaginary part** of z. We write $x = \mathrm{Re}(z)$ and $y = \mathrm{Im}(z)$. Beware: the 'imaginary part' of z means the real number y, not the imaginary number iy. The terminology is inconsistent, but everybody uses it.

The **modulus** of a complex number $z = x + iy$ is the real number $\sqrt{(x^2 + y^2)}$. It is denoted by $|z|$. Note that $x^2 + y^2 \geq 0$, so there is always a real square root. The symbol $\sqrt{}$ always denotes the *positive* square root (or zero if $x = y = 0$). The modulus $|z|$ is a non-negative number which can be thought of as the overall size of z, taking its real and imaginary parts into account.

The number $x + iy$ can be represented in a plane by the point with Cartesian coordinates (x, y). This representation is called the **Argand diagram** or the **complex plane**. Then $|x + iy|$ is the distance of the point from the origin.

Adding complex numbers and multiplying them by real numbers is very much like two-dimensional vector algebra. The modulus is the analogue of the magnitude of the vector. Thus $|z - a|$ is the distance between the points z and a in the Argand diagram; see Fig. E.1. For a given complex a and real positive R, $\{z: |z - a| < R\}$ is the interior of a circle with centre at a and radius R.

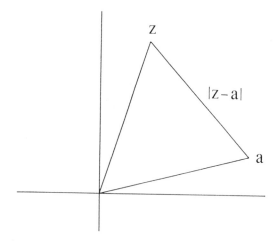

Fig. E1 – The distance between two complex numbers.

For any complex number $z = x + iy$, the number $x - iy$ is called the **complex conjugate** of z, and denoted by $\bar{z}$. In the Argand diagram, $\bar{z}$ is the reflection of z in the x-axis. It is easy to verify that

$$|z|^2 = z\bar{z}$$

and

$$\mathrm{Re}(z) = (z + \bar{z})/2 \qquad \mathrm{Im}(z) = (z - \bar{z})/2i$$

for every complex number z.

Re(z) and Im(z) are the Cartesian coordinates of z in the Argand diagram, and $|z|$ is the radial polar coordinate. The angular polar coordinate is called the **argument** of z. It is a real number θ such that

$$z = |z| \cos \theta + i|z| \sin \theta \tag{3}$$

Notice that the sine and cosine in (3) have the real number θ as their argument; they are the real functions of elementary trigonometry. But trigonometric and exponential functions of complex numbers can also be defined, by means of power series. In real calculus we have

$$\sin(u) = u - u^3/3! + u^5/5! - \dots$$

$$\cos(u) = 1 - u^2/2! + u^4/4! - \dots$$

$$e^u = 1 + u + u^2/2! + u^3/3! + \dots$$

for any real u. If u is complex, we define $\sin(u)$ to be the sum of the series above; similarly for cos and exp. The series converge for all complex u.

These series lead to a remarkable relation between the exponential and the trigonometric functions. The series for e^{iu} is

$$e^{iu} = 1 + iu - u^2/2! - iu^3/3! + u^4/4! + iu^5/5! - u^6/6! - iu^7/7! + \dots$$

because $i^2 = -1$, so $i^3 = -i$, $i^4 = (-1)^2 = 1$, and so on. Separating this series into its real and imaginary parts gives the beautiful formula

$$e^{iu} = \sin(u) + i\cos(u) \tag{4}$$

Equation (3) can now be expressed in the form

$$z = |z|e^{i\theta}$$

a neat expression for a complex number in terms of its modulus and its argument.

Finally, there is a useful expression for e^z in terms of the real and imaginary parts of z. If $z = x + iy$ we have $e^z = e^{x+iy} = e^x e^{iy}$, and by (4) we have

$$e^z = e^x[\cos(y) + i\sin(y)]$$

Thus the real and imaginary parts of e^z are expressed simply in terms of real exponential and trigonometric functions of the real numbers x and y. This formula is crucial to the application of eigenvalues to differential equations; see section 8B.

APPENDIX F   POLYNOMIALS

A **polynomial** is a function of the type

$$p(x) = a_n x^n + a_{n-1} x^{n-1} + \dots + a_1 x + a_0 \tag{1}$$

The numbers a_i are called the **coefficients** of the polynomial. Note that by definition a polynomial means a *finite* combination of powers; a series with infinitely many terms is not a polynomial.

The coefficients may be real or complex numbers. A **real** polynomial is one with real coefficients, a **complex** polynomial is one with complex coefficients. There is a more general theory of polynomials over any field; see, for example, Biggs (1985).

The **degree** of a polynomial is the highest power of x which it contains. If $a_n \neq 0$ in (1) then the polynomial p has degree n, but if $a_n = 0$ then the degree of p is less than n. A linear function such as $p(x) = ax - b$ has degree 1 if $a \neq 0$; if $a = 0$ we have a constant function, which is said to have degree zero.

Polynomial equations are generalisations of the quadratic equations of elementary algebra. Quadratic equations are solved by factorising. For any polynomial, finding a solution means finding a linear factor, as the following result shows.

The factorisation lemma. If p is a polynomial of degree n, and $p(a) = 0$, then there is a unique polynomial q, of degree $n - 1$, such that $p(x) = (x - a)q(x)$ for all x.

Proof. Given any number a, there is a unique polynomial q and a unique number r such that for all x,

$$p(x) = (x - a)q(x) + r \qquad (2)$$

This is easily seen by writing down the $n + 1$ equations that the n coefficients of q and the number r must satisfy; they can be solved one by one starting with the highest coefficient of q, giving unique values for the coefficients of q and for r.

If $p(a) = 0$, putting $x = a$ in (2) gives $r = 0$, proving the lemma.   □

Not every real quadratic equation has real roots; there is no real solution to the equation $x^2 + 1 = 0$, for example. But using complex numbers (see Appendix E), one can solve every quadratic equation by the elementary quadratic-equation formula.

Since solving quadratics requires the introduction of a new type of number, one might imagine that solving cubics, quartics, etc. would require other new kinds of number. However, a beautiful theorem of Gauss shows that complex numbers are sufficient to solve not just quadratics but every polynomial equation. This theorem is often given an imposing title.

The Fundamental Theorem of Algebra. For every nonconstant polynomial p there is at least one complex number z such that $p(z) = 0$.   □

In other words, every polynomial equation has a complex root, with the trivial exception of degree-zero polynomials. The proof of the theorem is difficult and subtle; see Courant and Robbins (1941) or Burkill (1970).

It follows from the Fundamental Theorem that every polynomial can be factorised.

The factorisation theorem. Every polynomial p can be expressed in the form

$$p(z) = C(z - a_1)(z - a_2)\ldots(z - a_{n-1})(z - a_n) \qquad (3)$$

where n is the degree of p, and $C, a_1, \ldots, a_n$ are complex numbers.

Note. The statement must be interpreted with common sense. If p has degree zero, then $p(z) = C$ for some C; this is regarded as a special case of (3) with $n = 0$, so that there are no linear factors.   □

Before proving the theorem, we note the following easy consequence.

Corollary. (a) A polynomial of degree $n > 0$ cannot vanish more than n times.

(b) If a polynomial $p(z)$ vanishes for infinitely many values of z, then $p(z) = 0$ for all z.

Proof of corollary. (a) The theorem shows that the polynomial has the form (3); if z is not one of the n numbers a_i, then $p(z)$ is the product of nonzero factors and is therefore nonzero. Hence $a_1, \ldots, a_n$ are the only points where p vanishes.

(b) p cannot have positive degree by (a), hence it is a constant, which must be zero since it vanishes for some z. $\qquad\square$

Proof of theorem. It follows from the Fundamental Theorem that there is a number a_1 such that $p(a_1) = 0$. By the factorisation lemma, $p(z) = (z - a_1)q(z)$ for some polynomial q of degree $n - 1$. Applying the Fundamental Theorem to q gives a number a_2 such that $q(a_2) = 0$, and the factorisation lemma shows that there is a polynomial r of degree $n - 2$ such that $q(z) = (z - a_2)r(z)$. Thus $p(z) = (z - a_1)(z - a_2)r(z)$. Applying the Fundamental Theorem to r, and proceeding in the same way, eventually gives $p(z) = (z - a_1)(z - a_2)\ldots(z - a_{n-1})w(z)$, where w is a polynomial of degree one, that is, a linear function. Hence w can be written in the form $w(z) = C(z - a_n)$, giving (3). $\qquad\square$

Finally we prove that the factorisation (3) is unique.

The unique factorisation theorem. For every nonconstant polynomial p there are unique numbers $C, a_1, a_2, \ldots$ such that (3) holds.

Proof. Firstly, C equals the coefficient of z^n in p, which determines C uniquely. Secondly, $a_1, \ldots, a_n$ are the roots of the equation $p(z) = 0$, which determines their values uniquely. If there are n different roots, then the numbers $a_1, \ldots, a_n$ are uniquely determined apart from their order. If there are less than n different roots, then some of the numbers $a_1, \ldots, a_n$ are the same, and we must prove that the number of times any given root appears in the list $a_1, \ldots, a_n$ is uniquely determined.

Suppose we have two different factorisations

$$p(z) = (z - a)^j q(z)$$

and

$$p(z) = (z - a)^k r(z)$$

where neither q nor r contains any factors of $(z - a)$. Subtracting, we have

$$(z - a)^j q(z) - (z - a)^k r(z) = 0 \quad \text{for all } z$$

If $j < k$, then

$$(z - a)^j [q(z) - (z - a)^{k-j} r(z)] = 0 \quad \text{for all } z$$

Hence $q(z) - (z - a)^{k-j} r(z) = 0$ for all $z \neq a$, and therefore for all z by the corollary above. Hence $q(z) = (z - a)^{k-j} r(z)$, a contradiction because q contains no factors of $(z - a)$. Hence we cannot have $j < k$. Similarly we cannot have $k < j$, so $k = j$, which completes the proof of uniqueness. $\qquad\square$

APPENDIX G   FIELDS

Number fields

The set of real numbers is the most familiar example of a field; the set of complex numbers is another. But the idea of a field is more general than this.

Consider, for example, the rational numbers. They are defined to be numbers of the form p/q, where p and q are integers with $q \neq 0$. Thus, for example, $2/3$, $-9/2$, $7 \,(= 7/1)$, and $0 \,(= 0/1)$ are all rational numbers, while $\sqrt{2}$ is not.

The sum of two rational numbers is another rational number. (Proof: $a/b + c/d = (ad + bc)/bd$, which is the ratio of two integers.) Likewise, multiplying or dividing two rational numbers gives another rational number. Thus the set of rational numbers has the property that applying the basic operations of arithmetic to members of the set gives members of the same set. This is the defining property of a field.

Definition 1. A **number field** is a set F of real or complex numbers with the property that the sum and the product of any two members of F belong to F, and the negative and the reciprocal of every nonzero member of F belong to F.

Examples 1. The sets of real numbers, complex numbers, and rational numbers are all fields. The set of integers is not a field because the reciprocal of an integer is not generally an integer. The positive numbers do not form a field because their negatives do not belong to the set. The imaginary numbers do not form a field because their products are real and do not belong to the set.   □

Example 2. The number $\sqrt{2}$ is irrational (i.e., not rational); for a proof, see Burkill (1962), for example. Consider the set S of all numbers of the form $a + b\sqrt{2}$ where a and b are rational numbers. A simple calculation shows that the sum and the product of two such numbers is another number of the same type. The reciprocal of $a + b\sqrt{2}$ can be expressed in the form $(a - b\sqrt{2})/(a^2 - 2b^2) = A - B\sqrt{2}$ where A and B are rational numbers; hence the reciprocal of every nonzero number in S also belongs to S. Note that the denominator $a^2 - 2b^2$ cannot vanish (unless $a = b = 0$) because $\sqrt{2}$ is irrational.

Abstract fields

Just as we generalised from 3-vectors to abstract vector spaces, we can generalise from fields of numbers to abstract fields. Broadly speaking, a field is any set in which addition and multiplication are defined so that sums and products have the usual properties. In detail, the definition is as follows.

Definition 2. A **field** is a set F such that

  (1) there is a rule which, given any a and b in F, determines an element of F called $a + b$, satisfying

    (a) $a + b = b + a$ for all a, b in F,

    (b) $a + (b + c) = (a + b) + c$ for all a, b, c in F,

    (c) there is an element of F, called zero and written 0, such that $0 + a = a$ for all a in F,

(d) for each a in F there is an element b such that $a + b = 0$;

(2) there is a rule which, given any a and b in F, determines an element of F called ab, satisfying

(e) $ab = ba$ for all a, b in F,

(f) $a(bc) = (ab)c$ for all a, b, c in F,

(g) there is an element of F, called unity and written 1, such that $a1 = a$ for all a in F,

(h) for each nonzero a in F there is an element c such that $ac = 1$,

(i) $a(b + c) = ab + ac$ for all a, b, c in F.   □

It is not hard to show that these axioms lead to all the familiar rules of arithmetic. For example, $0a = 0$ for all a in F; each element a has a unique negative $(-a)$ satisfying $a + (-a) = 0$; each element $a \neq 0$ has a unique inverse a^{-1} satisfying $aa^{-1} = 1$. Since $ab^{-1} = b^{-1}a$ (multiplication is commutative by axiom (e)), we can denote either of them by a/b.

Example 3. Consider the set $\{0, 1\}$. Since $1 + 1 = 2$ which does not belong to the set, the field axioms are not satisfied. But the addition rule of Definition 1 need not be the same as in ordinary arithmetic. We can define a new addition rule by $1 `+` 1 = 0, 1 `+` 0 = 1, 0 `+` 0 = 0$. It is easy to verify that with this addition rule, and ordinary multiplication, all the field axioms are satisfied. With the new addition rule, 0 and 1 still have the properties of 0 and 1 in ordinary arithmetic. The reciprocal of 1 is 1, of course, and since $1 `+` 1 = 0$, the negative of 1 is 1.

The addition rule can be described as follows: 'Add numbers in the ordinary way, and if the sum is greater than 1, subtract 2.' This procedure is called 'binary addition' or 'addition mod 2'. Generalising from 2 to any integer n gives the following.

Definition 3. Two integers a and b are **congruent modulo n** if their difference is a multiple of n, that is, $a - b = kn$ where k is an integer (positive, negative or zero). We write $a \equiv b \pmod{n}$; this formula is called a **congruence**.   □

It is easy to verify that for any fixed n, congruences behave like equations: if $a \equiv b$ and $b \equiv c$ then $a \equiv c$; if $a \equiv b$ then $ac \equiv bc$ and $a + c \equiv b + c$ for any integer c.

Example 4. For any fixed positive integer n, consider the set $S = \{0, 1, \ldots, n - 1\}$. Define an operation of 'addition mod n', written $+_n$, as follows: for each a, b in S, $a +_n b$ is the number in S congruent mod n to $a + b$. Multiplication mod n is defined in a similar way. Using these operations, all sums and products of members of S belong to S.

For $n = 2$ we have the field in Example 3. For $n = 3$ we also have a field. Since $1 + 2 = 3 \equiv 0 \bmod 3$, the negative of 1 is 2, and vice versa. Since $2.2 \equiv 4 \equiv 1 \bmod 3$, the reciprocal of 2 is 2; and the reciprocal of 1 is always 1.

But for $n = 4$ things are different. Axiom (h) fails for $a = 2$. We have $2.0 \equiv 0, 2.1 \equiv 2$, $2.2 \equiv 4 \equiv 0 \bmod 4, 2.3 \equiv 6 \equiv 2$. Thus there is no number in $\{0, 1, 2, 3\}$ whose product with 2 is 1 (mod 4). So 2 has no reciprocal, and the set is not a field for $n = 4$.

For $n = 5$ everything works properly, and we have a field. But for $n = 6, 8, 9, 10, 12, 14, 15, \ldots$ it fails in the same way as for $n = 4$. By now you will have guessed the following.

Theorem. The set $\{0, 1, \ldots, n-1\}$, with addition and multiplication mod n, is a field if and only if n is a prime number. □

For the proof, see Biggs (1985) or any textbook of abstract algebra. An application of these fields is given in Volume 2, section 9E.

Field extensions

Every vector space is based on a field; in this sense, linear algebra is an application of field theory. But their relationship is subtle; the following paragraphs outline a branch of field theory which is an application of linear algebra.

One field can be contained within another, in the same way that one vector space can be a subspace of another. The rationals, for example, are contained in the real numbers. The rational field is said to be a **subfield** of the real field, and the real field is said to be an **extension** of the rationals. Similarly the complex field $\mathbb{C}$ is an extension of $\mathbb{R}$, and $\mathbb{R}$ is a subfield of $\mathbb{C}$.

The set $\mathbb{C}$ of complex numbers has the property that the sum of two members of $\mathbb{C}$ belongs to $\mathbb{C}$, and the product of a member of $\mathbb{C}$ with a real number belongs to $\mathbb{C}$. Hence $\mathbb{C}$ is a vector space over the real field. This vector space is two-dimensional, as shown in Problem 3.24.

In the same way, if any field F has a subfield G, it can be regarded as a vector space over G. The dimension of this space is called the **degree** of F over G. Thus the complex field has degree 2 over the real field.

Example 5. Consider the field of Example 2; call it F. It is an extension of the rational field, which we shall call Q (for quotient: it is the set of quotients p/q of integers). Considered as a vector space over Q, F has a spanning set $\{1, \sqrt{2}\}$. This set is linearly independent. Proof: if $r + s\sqrt{2} = 0$ for rational r and s, and $s \neq 0$, then $\sqrt{2} = -r/s$ which is impossible because $\sqrt{2}$ is irrational; hence $s = 0$ which implies $r = 0$. This shows that the set $\{1, \sqrt{2}\}$ is a basis for F over Q. Thus F is 2-dimensional, and F has degree 2 as an extension of Q.

Example 6. We can extend the field F of Example 5 further by including multiples of $\sqrt{3}$: define G to be the set of all numbers of the form $x + y\sqrt{3}$, where $x, y \in F$. It is easy to verify that G is a field, with degree 2 as an extension of F (that is, dimension 2 when regarded as a vector space over F).

Now, G can also be regarded as an extension of Q. Every member of G is of the form $x + y\sqrt{3}$ where x and y belong to F; thus $x = a + b\sqrt{2}$ and $y = c + d\sqrt{2}$ where a, b, c, d are rational. Hence members of G have the form $a + b\sqrt{2} + c\sqrt{3} + d\sqrt{6}$. So G is the set of all rational linear combinations of $\{1, \sqrt{2}, \sqrt{3}, \sqrt{6}\}$, which is therefore a spanning set for G as a vector space over Q. The set can be proved to be linearly independent over Q, showing that G has degree 4 as an extension of Q. □

Example 6 illustrates the following elegant rule. If we have two successive field extensions, with H a subfield of G which is itself a subfield of F, then the degree of F over H is the product of the degrees of F over G and of G over H.

These ideas are fundamental to abstract algebra; but like most abstract mathematics, they were developed in order to solve a concrete problem, in this case the problem of solving polynomial equations. General formulas for solving quadratic, cubic and quartic equations have been known since the sixteenth century. The theory of field extensions was created by Galois in the early nineteenth century, as part of an attack on the general polynomial equation. Galois proved that there cannot be any general formula for solving equations of the fifth degree or higher.

The theory of field extensions also leads to the solution of one of the most famous problems in geometry: to trisect an angle, that is, to devise a procedure for using a straight edge and compass to divide any angle into three equal parts. The methods of Galois give a proof that there can be no such procedure. The proof is not difficult – see Courant and Robbins, for example. Amateurs who do not know this proof are still trying to solve the problem. We can now be sure that they will never succeed. Moral: abstract mathematics may have quite unexpected applications.

APPENDIX H   HISTORICAL NOTES

Some fields of mathematics are normally learnt in historical order. For example, most people first meet calculus presented in the spirit of Newton and Leibniz; it is developed and applied in an eighteenth century style; then one learns the nineteenth century theory of limits; and only very specialised and advanced courses deal with the twentieth century theory of calculus in abstract spaces.

Linear algebra is different. The abstract idea of a vector space has proved useful enough to be introduced at an early stage. But the subject developed quite differently; the inventors of linear algebra in the mid nineteenth century worked with matrices, not with vector spaces and linear operators. The following notes may help to give some perspective on the development of linear algebra. They are not intended as a balanced account of the history – such an account has yet to be written. Meanwhile, Kline (1972) is a useful general history of mathematics, and the periodicals *Archive for the History of the Exact Sciences* and *Historia Mathematica* may be consulted for recent historical research.

468–486 AD Zhang Quijian solves systems of three simultaneous equations by Gaussian elimination. He uses standard Chinese mathematical techniques, involving a board marked out in squares – not unlike a matrix.

1748 Colin Maclaurin studies sets of linear equations, which lead to determinants.

1762 Joseph Louis Lagrange's work in mechanics involves differential equations leading to the characteristic equation; the theoretical background of matrix eigenvalues does not yet exist.

1772 Alexandre Vandermonde gives a systematic treatment of determinants.

1809 Carl Friedrich Gauss publishes a book on celestial mechanics containing his elimination method for linear equations. The book also contains the method of least squares. An undignified controversy ensues, in which French mathematicians claim to have invented it first.

1815 Augustin-Louis Cauchy proves the rule for multiplying determinants, and introduces the modern notation.

1826 Cauchy develops the ideas of transformation to principal axes.

1843 Arthur Cayley studies the geometry of n-dimensional space, and William Hamilton works on quaternions, which will eventually lead to vector algebra.

1844 Herman Grassman publishes a difficult book (*Die Ausdehnungslehre*), which contains the idea of an abstract vector space but in such an obscure form that hardly anyone understands it; the ideal will become popular only at the end of the century.

1850 Joseph Sylvester coins the term 'matrix' for a rectangular array.

1855 Cayley studies matrices systematically. Charles Hermite introduces Hermitian matrices.

1875 William Kingdom Clifford tries to prove that every matrix which commutes with A equals a polynomial in A. Unfortunately this is not true, as will be shown by Sylvester.

1877 The astronomer Hill, studying the motion of the moon, uses infinite determinants for solving infinite sets of linear equations. A rigorous basis for this will not be given until 1885 (by Henri Poincaré).

1879 Georg Frobenius introduces the notion of rank, in terms of determinants.

1888 Giuseppe Peano publishes a book based on Grassman's idea of an abstract vector space. The idea still doesn't catch on.

1894 David Hilbert is led to the idea of a finite-dimensional vector space in his research in number theory. His work on integral equations (motivated by the physics of gases) leads to infinite-dimensional spaces.

1902 Julius Farkas works out the theory of linear inequalities; half a century later it will form the foundation of linear programming.

1906 Maurice Fréchet studies spaces of functions.

1908 Erhard Schmidt studies the geometry of abstract inner product spaces.

1918 Otto Toeplitz defines normal matrices, and proves that they are diagonalisable.

1920 Eliakim Moore introduces a generalised inverse matrix; the idea is not taken up by the mathematical community until it is rediscovered by Roger Penrose in 1955.

1925 Werner Heisenberg and Max Born formulate the laws of atomic physics (the quantum theory) in terms of matrices and eigenvalues. Their work stimulates the mathematician John von Neumann to work on the theory of eigenvalues of operators on infinite-dimensional inner product spaces.

1925 Erwin Schrödinger publishes a paper on the theory of colour vision in terms of vector spaces (at about the same time he is developing quantum mechanics in a way parallel to, but different from, Heisenberg's matrix mechanics).

1929 Stefan Banach sets out the general theory of the dual of a vector space, in the context of infinite-dimensional function spaces.

1931 Gershgoring publishes his theorem on eigenvalues. In the same year an international commission meets to choose a basis for the vector space of colour sensations.

1939 C. Eckart and G. Young obtain the singular value decomposition of an arbitrary matrix.

1939–1945 The second World War requires a great scientific and computational effort, which leads directly to the modern development of numerical linear algebra.

1947 Alan Turing analyses roundoff errors in the computational solution of linear systems. He introduces the idea of LU factorisation.
1961 J. G. F. Francis and V. N. Kublanovskaya invent the QR method for calculating eigenvalues.

Linear algebra continues to develop rapidly; see Horn and Johnson (1985) for an account of research in recent years.

Solutions to exercises

CHAPTER 1

1B-1. See Figure S1.

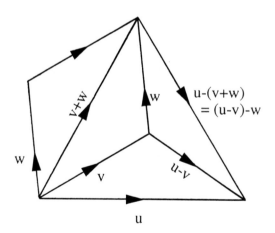

Fig. S.1

1D-1. Take three vectors in different directions in the same plane. Adding two vectors in a plane by the parallelogram rule gives another vector in the plane. Hence any combination of vectors in the plane is a vector in the plane, and no vector which is not in the plane can be a combination of vectors in the plane.

1E-1. If $\mathbf{b} = -\mathbf{c}$ then $\mathbf{a}*(\mathbf{b}+\mathbf{c}) = 0$, $\mathbf{a}*\mathbf{b} + \mathbf{a}*\mathbf{c} > 0$, if $\mathbf{a}, \mathbf{b}, \mathbf{c}$ are non zero.

1E-2. No. Their ratio is $\|\mathbf{a}\|/\|\mathbf{b}\|$.

1F-1. If x is any of the vectors, $x \cdot x = 1$, so $\|x\| = 1$. And the dot product of any pair is zero: every term in the sum has a zero factor.

CHAPTER 2

2B-1. $(A - B)_{rs} = A_{rs} + (-B)_{rs} = A_{rs} + (-1)B_{rs}$.

2B-2. (i), (ii) follow straight from Definition 2. (iii) For any i, j,

$$(C^T)_{ij} = ([A + B]^T)_{ij} = (A + B)_{ji} = A_{ji} + B_{ji} = (A^T)_{ij} + (B^T)_{ij}$$

2C-1.

$$\begin{pmatrix} -25 & -26 & -27 \\ 19 & 20 & 21 \\ 13 & 14 & 15 \end{pmatrix}$$

2D-1. False; see Example 2D-3.

2D-2. If A and X are diagonal, with diagonal entries a, b and x, y, easy calculation gives $AX = XA =$ the diagonal matrix with entries ax and by.

2D-3. $X^T Y = (X_1, X_2, \ldots, X_n)(Y_1, \ldots, Y_n)^T$ which is the 1 by 1 matrix $(X_1 Y_1 + \ldots + X_n Y_n) = X \cdot Y$ in vector notation. Similarly for $Y^T X$. But XY^T is an n by 1 matrix times a 1 by n matrix, which is the n by n matrix with (i, j) entry $X_i Y_j$ for $i, j = 1, \ldots, n$.

2D-4 Just work through the multiplication.

2D-5. Use A and B in Example 1, with $Y = X + B$ for any 2 by 2 matrix X.

2E-1. Do the multiplication.

2E-2. By Theorem 3, AB has inverse $B^{-1}A^{-1}$. Hence by Theorem 3 applied to (AB) and C, $(AB)C$ is invertible, with inverse $C^{-1}(AB)^{-1} = C^{-1}B^{-1}A^{-1}$.

CHAPTER 3

3D-1. If $k \neq 0$, not a subspace because doesn't include the zero element (namely, the function which maps every number to 0). If $k = 0$ it follows easily from Theorem 1.

3D-2. Follows from Theorem 1: the sum of two polynomials is a polynomial, and a constant times a polynomial is a polynomial.

3E-1. Must prove that each $s \in S$ is a linear combination of S. We have $s = 1 \cdot s$, which is a linear combination of S (with $n = 1$ in Definition 1).

3E-2. $(x, y, 0) = x(1, 0, 0) + y(0, 1, 0)$. This formula expresses each point of the plane as a linear combination of the set.

3E-3. For any subspace S, must show $S = \mathrm{Sp}(S)$. We have $\mathrm{Sp}(S) \subset S$ because every linear combination of members of a subspace belongs to that subspace. And $S \subset \mathrm{Sp}(S)$ by Exercise 1. Combining these gives $S = \mathrm{Sp}(S)$.

3F-1. (a) Each element of S belongs to T, so if a linear combination of S equals 0, it is also a linear combination of T. (b) If $0 \in S$ then $1 \cdot 0$ is a nontrivial linear combination of S which equals 0.

3F-2. If $au + bw = 0$ then $(a, b, 0) = (0, 0, 0)$ hence $a = b = 0$; so $\{u, w\}$ is linearly independent. If $av + bw = 0$ then $(a, a + b, 0) = (0, 0, 0)$, so $a = 0$ and $a + b = 0$ giving $b = 0$; so $\{v, w\}$ is linearly independent.

3F-3. Is logically equivalent to Exercise 3F-1(a).

3G-1. $a(1, 0, 0) + b(0, 1, 0) + c(0, 0, 1) = (a, b, c)$ for all real a, b, c. Thus every 3-vector is a combination of the given vectors, so they span $\mathbb{R}^3$; and the equation shows that if the linear combination equals $(0,0,0)$ then $a=b=c=0$, so the vectors are linearly independent.

3G-2. (a) Does not include the zero polynomial, so not a subspace. (b) Follows easily from Theorem 3D-1. (c) Obviously a spanning set. Linearly independent because if $a+bt+ct^2+dt^3$ is the zero member of the space, it is a polynomial which vanishes everywhere, which is possible only if $a=b=c=d=0$ (see Example 2). Hence it is a basis. It has 4 elements, so $\dim(P_3)=4$. (d) The above arguments apply unchanged in the complex case.

3G-3. $\{1,t,\dots,t^n\}$ obviously spans the space, and is linearly independent by the argument of the previous exercise. Hence it is a basis, with $n+1$ elements, so dimension $=n+1$.

3H-1. For each n, P_n is an $(n+1)$-dimensional subspace, so the condition of Definition 2 is satisfied.

3I-1. The formula $b_j=1\cdot b_j$ expresses b_j as a linear combination of the b's with all coefficients 0 except the jth, which is 1. Hence the i-component of b_j is 1 if $i=j$, and 0 otherwise.

3I-2. In the notation of version 2, $v+w=\sum v_r b_r+\sum w_r b_r=\sum(v_r+w_r)b_r$. Thus the r-component of $v+w$ is v_r+w_r. Similarly $kv=k(\sum v_r b_r)=\sum(kv_r)b_r$ showing that the r-component of kv is kv_r.

3J-1. Given $v\in V$ there is a unique $u\in U$, which gives a unique $w\in W$, and vice versa. And kv corresponds to ku which corresponds to kw; similarly for v_1+v_2. Hence an isomorphism from V to W.

3J-2. Let 0 and $0'$ be the zero vectors in V and V'. If $x\in V$ corresponds to $0'$ under the isomorphism, then $0x$ corresponds to $00'=0'$. Hence $x=0x$ by Definition 1(i), so $x=0$.

3K-1. $(1,0)$ and $(0,1)$ belong to the union but their sum $(1,1)$ does not.

3K-2. If $s\in S+X$ then s is the sum of two members of S, hence $s\in S$. Conversely, if $s\in S$ then $s=(s-x)+x$ for any $x\in X$, and $s-x\in S$ because $x\in S$, so $s\in S+X$.

3K-3. Let T be a complementary subspace of S, then $S\cap T=\{0\}$ by Theorem 2, hence T contains no member of S except 0, so $T\subset\{0\}\cup V\setminus S$. It is a proper subset because for any nonzero s in S and t in T, $s+t\in V\setminus S$ but $s+t\notin T$.

3K-4. $S\cap\mathrm{Sp}\{(1,c,0)\}=\{0\}$ since $k(1,c,0)$ has zero y-component only if $k=0$. We have $(x,y,0)=(x-y/c,0,0)+(y/c)(1,c,0)$, so every vector in the xy plane is a vector in S plus a member of $\mathrm{Sp}\{(1,c,0)\}$. Hence $\mathrm{Sp}\{(1,c,0)\}$ is a complement of S in the xy plane.

Again, $S\cap(yz\text{ plane})=\{0\}$. And $(x,y,z)=(x,0,0)+(0,y,z)$ says every 3-vector is a vector in S plus a vector in the yz plane, which is thus a complement of S in $\mathbb{R}^3$. Similarly, $S\cap\{(y,y,z)\}=\{0\}$, and $(x,y,z)=(x-y,0,0)+(y,y,z)$, hence $\{(y,y,z)\}$ is a complement of S.

CHAPTER 4

4A-1. At each stage the argument says 'let the leading entry be the kth' – this assumes that there is a leading entry, not true if the zero vector is allowed.

4D-1.

$$
\begin{pmatrix} 0 & 1 & 2 \\ 2 & 1 & 1 \\ 1 & 0 & 1 \end{pmatrix} \rightarrow \begin{pmatrix} 2 & 1 & 1 \\ 0 & 1 & 2 \\ 1 & 0 & 1 \end{pmatrix} \rightarrow \begin{pmatrix} 2 & 1 & 1 \\ 0 & 1 & 2 \\ 0 & -2 & -1 \end{pmatrix} \rightarrow \begin{pmatrix} 2 & 1 & 1 \\ 0 & 1 & 2 \\ 0 & 0 & 3 \end{pmatrix}
$$

4F-1. By Theorem 4B-1, the echelon form of A has no zero rows, so case 2 cannot apply. So must have case 1 or 3, in which solutions exist.

4F-2. Follows at once from the dimension theorem of Chapter 3.

4G-1. If there are no zero rows, then each row has a leading 1, so there are n leading 1's. So there is a leading 1 in each column, so columns are all zero except for the leading 1. Thus there are 1's down the main diagonal and zeros everywhere else.

4G-2. Follow the method of Example 1.

4I-1. $Ax = AMb = Ib = b.$

4I-2.

$$
\left(\begin{array}{ccc|ccc} 1 & 0 & 0 & 1 & 0 & 0 \\ a & 1 & 0 & 0 & 1 & 0 \\ b & c & 1 & 0 & 0 & 1 \end{array} \right) \rightarrow \left(\begin{array}{ccc|ccc} 1 & 0 & 0 & 1 & 0 & 0 \\ 0 & 1 & 0 & -a & 1 & 0 \\ 0 & c & 1 & -b & 0 & 1 \end{array} \right)
$$

$$
\rightarrow \left(\begin{array}{ccc|ccc} 1 & 0 & 0 & 1 & 0 & 0 \\ 0 & 1 & 0 & -a & 1 & 0 \\ 0 & 0 & 1 & ac - b & -c & 1 \end{array} \right)
$$

The inverse is thus a matrix of the same type with a, b, c replaced by $-a, ac - b, -c$.

4J-1. Gives a satisfactory solution: $x = y = 1.$

4L-1. Same as A with the last two rows interchanged.

4M-1. (a) See Exercise 4I-2; obviously works the same way for matrices of any size. (b) Just do the multiplication. (c) Applying the elementary operation to I puts a k in the (i, j) position and all other off-diagonal entries are zero. Hence the elementary matrix is lower triangular.

4M-2. The proof of Theorem 1 shows that the diagonal entries of U are the pivots: each pivot survives, unaffected by subsequent operations, into the final echelon form. The diagonal entries of D are the diagonal entries of U, as shown in the paragraphs above Theorem 2.

CHAPTER 5

5A-1. $f(2x, 2y) = 4f(x, y)$. If f were linear, $f(2x, 2y)$ would be $2f(x, y)$. We have $g[(x, y) + (u, v)] = g(x + u, y + v) = 2(x + u) - 3(y + v) = 2x - 3y + 2u - 3v = g(x, y) + g(u, v)$. A similar calculation gives $g[k(x, y)] = kg(x, y)$. Hence g is linear.

5B-1. $\operatorname{im}(a) = \{a(x): x \in \mathbb{R}^q\} = \{Ax: x \in \mathbb{R}^q\}$ = the set of all linear combinations of the columns = the column space. Hence $\dim(\operatorname{im}) = $ column rank. Example 5A-3 shows that a' is the a − map corresponding to A^T, hence the above argument shows that $\operatorname{im}(a')$ is the space of all linear combinations of the columns of A^T, which are the rows of A.

The matrix A has full column rank if and only if its columns span $\mathbb{R}^q$, which means that $\operatorname{im}(a) = \mathbb{R}^q$ which is the codomain of a.

5C-1. $f(x, y) = 0$ if and only if $3x - 2y = 0$ and $6x - 4y = 0$, which hold for any x if $y = 3x/2$. Hence $\ker(f) = \{(x, 3x/2): x \in \mathbb{R}\}$.

5C-2. Use the subspace criterion. If $u, v \in \ker(f)$, then $f(u + v) = f(u) + f(v) = 0 + 0 = 0$, hence $u + v \in \ker(f)$. Also $f(kv) = kf(v) = 0$ if $f(v) = 0$, so $kv \in \ker(f)$ if $v \in \ker(f)$. Hence $\ker(f)$ is a subspace.

5D-1. (a) $p = q = 2$, $\operatorname{rank}(A) = 1 < p$, no solution. (b) $p > q = \operatorname{rank}(A)$, unique solution. (c) $p = q$, $\operatorname{rank}(A) < p$, infinitely many solutions.

5E-1. If A is invertible, take any y and set $x = A^{-1}y$, then $a(x) = Ax = y$. There can be no other x' satisfying $a(x') = y$, since if $Ax' = y$ then $A^{-1}Ax' = A^{-1}y = x$, so $x' = x$. Hence there is a unique x satisfying $a(x) = y$; in other words, a is invertible.

5F-1. If $f(v) = 0v = 0$ for some nonzero v, then nullity $(f) > 0$, hence f is not invertible by Theorem 5E-1(a).

5I-1. Every square matrix with orthonormal columns has orthonormal rows, and vice versa. The reason is not quite obvious. See next exercise.

5I-2. If $A^T A = I$ then A^T is a left inverse of A, hence A is invertible by Corollary 4I-1. Hence $A^T A A^{-1} = I A^{-1}$, so $A^T = A^{-1}$. Therefore $I = A A^{-1} = A A^T = (A^T)^T A^T$; this shows that A^T is orthogonal. Hence the columns of A^T are orthonormal; they are the rows of A. This proves the result in Exercise 1.

5I-3. Every orthogonal matrix is invertible (see Exercise 2), and the set is closed under multiplication because if A, B are orthogonal, then $(AB)^T(AB) = B^T A^T A B = B^T B = I$, showing that AB is orthogonal. Hence it is a subgroup of the group of all invertible n by n matrices.

5K-1. $(fg)(x) = f(g(x)) = f(Bx) = ABx$. Hence the effect of the map fg is to multiply x by AB.

5K-2. In Example 2D-1, A and B have rank 1, but $\operatorname{rank}(AB) = 0$, not the product of the ranks of A and B.

CHAPTER 6

6B-1. (a) $\det(I) = 1 \cdot 1 - 0 \cdot 0 = 1$. (b) $ad - bc = 0$ if $a = c$ and $b = d$.

(c) $\begin{vmatrix} ax + bp & ay + bq \\ u & v \end{vmatrix} = (ax + bp)v - (ay + bq)u = a\begin{vmatrix} x & y \\ u & v \end{vmatrix} + b\begin{vmatrix} p & q \\ u & v \end{vmatrix}$

Similarly for row 2.

6B-2. Work out and plot the images of the vertices. Then draw lines joining them (since linear transformations map straight lines to straight lines, these lines are the images of the boundaries of the unit square).

6B-3. There might be inconsistencies which our technique does not reveal. To see how this might happen, suppose a fourth clause is added to Definition 1, saying that $\det(A) = 1$ if A has two identical columns. Then all the results in Theorem 1 and Example 1 would seem to hold, since they are deduced from parts (a), (b) (c) of Definition 1; yet the theory would be self-contradictory.

6D-1. Let P_N be the statement 'if the numbers i and j are separated by N numbers, then they can be interchanged by an odd number of adjacent interchanges'. P_1 is proved in the text. Suppose P_N holds. Then if i and j are separated by $N + 1$ numbers, interchanging i with its neighbour leaves i and j separated by N, hence they can be interchanged by an odd number of adjacent interchanges; now interchanging j with its neighbour restores the original order but with i and j interchanged. The number of adjacent interchanges is $2 +$ an odd number, so P_{N+1} follows. This completes the induction.

6F-1. Consider Example 4I-1:

$$\text{adj}(A) = \begin{pmatrix} 12 & -13 & -7 \\ -3 & 5 & 2 \\ -3 & 2 & 2 \end{pmatrix}$$

Det (A) is easily calculated by expanding by the first column, giving -3. Thus $\text{adj}(A)/\det(A)$ agrees with the result obtained in section 4I by the Gauss–Jordan method.

To invert a 5 by 5 matrix, about 300 operations (addition or multiplication) for the Gauss method, about 700 for the adjugate-matrix method.

CHAPTERS 7 AND 8

7B-1. The characteristic equation is $(1 - x)^2 = 0$, so 1 is an algebraically double eigenvalue. Solving for the eigenvectors (x, y) gives $y = 0$, and the eigenspace is $\text{Sp}\{(1, 0)\}$. This is 1-dimensional, so the geometric multiplicity is 1, not equal to the algebraic multiplicity.

7C-1. $|A_{ii}| >$ something which is either positive or zero. Hence it is positive.

8A-1. $A^2 = PDP^{-1}PDP^{-1} = PDIDP^{-1} = PD^2P^{-1}$. Similarly, $A^3 = AA^2 = PDP^{-1}PD^2P^{-1} = PD^3P^{-1}$, and so on.

Hints and answers to selected problems

Complete solutions to the problems in this volume can be obtained from the author (School of Mathematics, University Walk, Bristol BS8 1TW, England), for £2 (within Britain), or £3 (including postage overseas).

CHAPTER 1

1. Let X be the point with position vector $\mathbf{x} = t\mathbf{u} + (1-t)\mathbf{v}$. Then $\mathbf{x} - \mathbf{v} = t(\mathbf{u} - \mathbf{v})$, which is in the direction VU if $t > 0$ and UV if $t < 0$. Hence XV is parallel to UV, so X is on line UV. Taking $t = 0$ gives V; $t = 1$ gives U; $0 < t < 1$ puts X between U and V; $t < 0$ puts X on the far side of V from U.

3. (a) $\mathbf{x}$ and $\mathbf{y}$ in the same direction; (b) $\mathbf{x}$ and $\mathbf{y}$ in opposite directions with $\|\mathbf{y}\| < \|\mathbf{x}\|$; (c) reduce to (a) by setting $\mathbf{x}' = \mathbf{x} - \mathbf{y}$, $\mathbf{y}' = \mathbf{y} - \mathbf{z}$. Answer: true when $\mathbf{x}, \mathbf{y}, \mathbf{z}$ are position vectors of points X, Y, Z on the same straight line, with Y between X and Z.

4. $\mathbf{p} + r\mathbf{d}/\|\mathbf{d}\|$.

5. The plane through P which contains vectors $\mathbf{a}$ and $\mathbf{b}$.

6. If $\mathbf{a}, \mathbf{b}, \mathbf{c}$ all lie in one plane, then the set is a plane parallel to it through P. Otherwise, it is the whole of 3-space.

7. $\mathbf{p} + \mathbf{q}$ has N and E components 1.65, 1.126, hence magnitude 1.86, direction 34° E of N.

8. X in plane ABC means X is on the line joining C to some point on line AB. Use result of Problem 1 twice, and get $s + t + u = 1$. Point X is inside ABC if s, t, u are all between 0 and 1.

9. If $\mathbf{r} \cdot \mathbf{a} = k$, then the projection of $\mathbf{r}$ on $\mathbf{a}$ is $k/\|\mathbf{a}\|$; hence a plane at distance $|k|/\|\mathbf{a}\|$ from O perpendicular to $\mathbf{a}$.

10. $\mathbf{x} = -(\mathbf{x} \cdot \mathbf{b})\mathbf{a} = k\mathbf{a}$, say. Hence equation satisfied when $k\mathbf{a} = -k(\mathbf{a} \cdot \mathbf{b})\mathbf{a}$. Clearly satisfied by $\mathbf{x} = k\mathbf{a} = 0$. If $k\mathbf{a} \neq 0$ then $\mathbf{a} \cdot \mathbf{b} = -1$; this means there is a nonzero solution only when **a** and **b** satisfy $\mathbf{a} \cdot \mathbf{b} = -1$. In this case, $k\mathbf{a} = -k(\mathbf{a} \cdot \mathbf{b})\mathbf{a}$ holds for all k, so $\mathbf{x} = k\mathbf{a}$ satisfies the equation for all k when $\mathbf{a} \cdot \mathbf{b} = -1$.

12. (i) $\{(x, y, z, -x - y - z): x, y, z \in \mathbb{R}\}$; (ii) $\{(0, y, z, w)\}$; (iii) $\{(0, y, z, -y - z)\}$.

13. (a) Projection of S is the unit circle together with all points inside it. The projection of the t-slice is a circle of radius $\sqrt{(1 - t^2)}$; starts as a point, grows to radius 1 and shrinks down to nothing again. (b) (i) The unit sphere in $\mathbb{R}^3$ together with all points inside it. (ii) A sphere of radius $\sqrt{(1 - t^2)}$ if $|t| \le 1$; otherwise, the empty set.

CHAPTER 2

1.   (a) $(7, -7, 6)$ (d) $\begin{pmatrix} 2 & 2 & 2 \\ 1 & 4 & 2 \end{pmatrix}$ (e) $\begin{pmatrix} 0 \\ 3 \\ 3 \end{pmatrix}$

Rest undefined.

2. (a) True. (b) False (also need same number of rows). (c) True. (d) False, e.g.

$$\begin{pmatrix} 1 & 2 & 3 \\ 4 & 5 & 6 \end{pmatrix}$$

6.   (h) 1   (i) $\begin{pmatrix} 2 & 1 & 0 \\ 0 & 0 & 0 \\ 2 & 1 & 0 \end{pmatrix}$ (j) 2   (k) $(2 \;\; 2)$

(l) $\begin{pmatrix} 2 & 5 & 2 \\ 4 & 0 & -1 \end{pmatrix}$ (m) $\begin{pmatrix} 9 & 14 \\ -2 & 3 \end{pmatrix}$ (p) $\begin{pmatrix} 5 & 0 \\ 0 & 5 \end{pmatrix}$

Rest undefined.

7. (a) True. (b) Counterexample in the text. (c) True. (d) False (holds only when A and B commute).

8. If $u = (u_1, \ldots, u_p)$ and A is a matrix with rows $A_1, \ldots, A_p$, then $uA = u_1 A_1 + \cdots + u_p A_p$.

10. Experiment with simple matrices, such as

$$A = \begin{pmatrix} a & b \\ 0 & c \end{pmatrix}$$

Easy to see that $A^3 = 0$ implies $a = c = 0$ implies $A^2 = 0$. The result is true in general; you can prove it by tedious calculations, or wait until Chapter 10.

11. $\begin{pmatrix} 0 & 2t \\ 3t & 2t \end{pmatrix}$ for any t

12. Just work out the products and use trigonometric identities.

13. False (holds only when A and B commute).

14. True. In general, $(A + B)^n$ is the sum of 2^n terms.

18. Work out the consequences of commuting with all the matrix units.

22. (b) is false (unless A and B are square), the others are true.

23. $NM = I$ gives $ax + cy = 1$, $bx + dy = 0$, $az + cu = 0$, $bz + du = 1$; $MN = I$ gives

four similar equations. If $ad \neq bc$, solving the equations for $NM = I$ determines x, y, z, u. They also satisfy the equations for $MN = I$, hence M is invertible. If $ad = bc$, then manipulating the first two equations gives $b = d = 0$, and the second pair implies $a = c = 0$. But then $ax + cy = 1$ can never hold, and there is no solution. Hence no inverse in this case.

26. (a) False: each submatrix must be transposed. (b) False: consider

$$\begin{pmatrix} 0 & 1 & 2 \\ 0 & 3 & 4 \\ 0 & 0 & 0 \end{pmatrix}$$

It holds only if all blocks in A are of the same size.

28. $R = 0$, $Q = -A^{-1}BC^{-1}$.

CHAPTER 3

1. (a) Yes. (b) No (a real multiple of a positive number need not be positive). (c) Yes. (d), (e) No, for similar reason to (b).

2. (a) True for all except the trivial space of Example 3B-4(c). (b) True. (c) False (vectors cannot in general be multiplied together). (d) False: see Example 3B-4(b). (e) True. (f) False – real numbers belong to $\mathbb{C}$. (g) True. Proof: if $k \neq 0$ then $v = k^{-1}0 = 0$.

4. (a) and (b).

7. (a) and (c). The others fail the test that the sum of two members must belong to the set.

8. Both are right, in different ways. Chris's point is undeniably correct. Yet Pat has seen a deeper truth. Her point is that $\mathbb{R}^2$ is essentially a subspace of $\mathbb{R}^3$; more precisely, $\mathbb{R}^2$ is isomorphic to such a subspace (see section 3J for isomorphism).

10. We want all functions continuous on $[a, b]$ to belong to $C[0, 1]$; this holds if $[0, 1] \subset [a, b]$.

13. U is the xy plane; V is a plane containing the x-axis, at angle $45°$ to the y- and z-axes; $U \cup V$ consists of the two planes; $U \cap V$ is the x-axis. All except $U \cup V$ are subspaces.

14. (a) and (b) are false because $Sp(S)$ contains only finite sums of powers; (b) also seriously misquotes Taylor's theorem. (c) is wrong because there is no such thing as an 'infinite polynomial'. But the underlying idea is correct: $Sp(S)$ contains only polynomials (which have finitely many terms); V contains also power series with infinitely many terms.

15. $\{(1 - 3i, 2 + 3i, 0)\}$

17. (a) False. (b) True (u is a linear combination of v and w). (c) False (counterexample: $\{e_1, e_2, e_3, e_3\}$). (d) False, (e) See Exercise 3F-3.

19. (a), (b), (d). (Sketch proof for (b): suppose $a \sin t + b \cos t + ce^t + de^{-t} = 0$ for all t, then taking $t \to \infty$ gives $c = 0$, $t \to -\infty$ gives $d = 0$, then $t = 0$ gives $b = 0$, then $a = 0$.)

22. (a) False (e.g. $\{p, 2p, 3p, 4p\}$). (b) True because $Sp\{p, q, r, s\}$ is 4-dimensional. (c) True: $Sp\{p, q, r, s\}$ has dimension ≤ 4 here. (d) True. Proof: if it were linearly dependent, then a subset of it would be a smaller spanning set for $\mathbb{R}^4$, impossible because $\dim(\mathbb{R}^4) = 4$. (e) True (it is easy to construct such a set). (f) False (see Problem 14).

23. (a) No. (b) Yes. (c) No, because linearly dependent: $(1, 0, 0) + i(0, 1, i) = (1, i, -1)$.

24. Prove that $\{1, i\}$ is a basis, so that dimension $= 2$.

25. Set $S = \{v_1, \dots, v_k\}$. If S is linearly independent, it is a basis for $\text{Sp}(S)$, so $\dim(\text{Sp}(S)) = k$. If not, then removing vectors which depend on the others will give a basis of $< k$ vectors.

26. For any linearly independent sets X and Y, $X \cup Y$ is linearly independent if $\text{Sp}(X) \cap \text{Sp}(Y) = \{0\}$.

28. A linear combination of $\{e_1, e_2, \dots\}$ has only finitely many terms, and is therefore a sequence with only finitely many nonzero entries. Hence $(1, 1, 1, 1, \dots)$, for example, is in S but not in $\text{Sp}\{e_1, e_2, \dots\}$.

34. $i, -i/2, (2 - 3i)/6$.

35. (a) $2, -3, 3, -1$; (b) $0, 1, -1, 1$ (by method of Example 3I-3).

36. One answer is

$$\begin{pmatrix} a & b & c \\ d & e & f \end{pmatrix} \text{ corresponds to } (a \quad b \quad c \quad d \quad e \quad f)$$

There are many other possible answers.

37. (a) True. (b) False. This question is very subtle, and a full answer needs a deep understanding of infinite sets (see Courant and Robbins (1941), Rucker (1982) and Stewart and Tall (1977), for example). But the idea can be illustrated by the space F of all $\mathbb{R} \to \mathbb{R}$ functions and the space P of polynomials. We have shown that P is infinite-dimensional; so is F, since P is a subspace of F. The space F is much larger than P, since most functions are not polynomials. It can be proved that F contains so many more functions than P that there is no one-to-one correspondence between them. Thus there can be no isomorphism between these infinite-dimensional spaces.

39. (a) False, e.g., in $\mathbb{R}^2$ take $B = \{(1, 0)\}$, $C = \{(0, 1)\}$. (b) False, e.g., in $\mathbb{R}^3$ take $B = \{e_1, e_2\}$, $C = \{e_2 + e_3, e_2 - e_3\}$. (c) True. (d) True.

40. Chapter 2, Problem 4(d) shows that it is the sum of the subspaces, and part (c) shows that the sum is direct.

CHAPTER 4

1. (a), (b) are in echelon form; (b), (c), (d) are linearly dependent.

2. Regarded as an ordered set, $\{(1, 1, 1, 1), (1, 1, 1, 1)\} \neq \{(1, 1, 1, 1)\}$; see the note following Definition 4A-1.

3. (a) True. (b) False (there might be two zero vectors, which can be interchanged). (c) False. (d) False (e.g., unit matrix). (e) False (e.g., a triangular matrix might have first row zero). (f) True.

7. (a) $(5/7, -1/7, 5/7)$. (b) No solutions. (c) $\{(2t - u, t - u, 3t, u): u, t \in \mathbb{R}\}$.

10. Both true.

11. Soluble if and only if $b + c - 3a = 0$. Then the solutions are: $x_1 = a - (b + 3x_3 + x_4)/5$, $x_2 = (b - 2x_3 - 4x_4)/5$, x_3 and x_4 can take any values.

14. (a) False. (b) True; follows from the fact that rows $R, R', \dots$ of A are linearly dependent if and only if the corresponding rows $\bar{R}, \bar{R}', \dots$ of $\bar{A}$ are (because

$cR + c'R' + \ldots = 0$ if and only if $\bar{c}\bar{R} + \bar{c}'\bar{R}' + \ldots = 0$). (c) True (take y^T to be a nonzero row, then the ith row equals $x_i y^T$ for some number x_i). (d) A^T has row rank 1, so use (c). (e) False (e.g., (1 2)). (f) True.

15. (a) Both are 2.

16. If $Ax = b$ has solutions, then b is a combination of the columns of A. Hence a basis for the columns of A also spans the columns of $(A|b)$. If $Ax = b$ has no solutions, then b is not a combination of the columns of A. So if $\{a_1, \ldots, a_r\}$ is a basis for the column space of A, then $\{a_1, \ldots, a_r, b\}$ is linearly independent. Hence there are $r + 1$ independent columns of $(A|b)$, so col rank$(A|b) >$ col rank(A).

19. Yes. No.

20. Counterexample:

$$A = \begin{pmatrix} 1 & 0 \\ 0 & 0 \end{pmatrix} \qquad B = \begin{pmatrix} 0 & 0 \\ 1 & 0 \end{pmatrix}$$

23. (a) is invertible, the others are not.

24. (a) False, e.g. $I + (-I) = 0$. (h) If all entries of A are positive, then A^{-1} must have some negative entries. (i), (j) False, e.g.,

$$\begin{pmatrix} 1 \\ 0 \end{pmatrix}$$

has more rows than columns and is left invertible, giving counterexamples to both (i) and (j). The rest are true.

29. Use the fact that $\sum_1^n k^2 = n(n+1)(2n+1)/6$. Gauss elimination takes roughly $(1/3)n^3$ multiplications and the same number of additions. Back-substitution takes about $(1/2)n^2$ additions and multiplications, negligible by comparison.

30. Six Jacobi iterations give $(1.1175, 0.8494)$, six Gauss–Seidel iterations give $(1.1187, 0.8489)$. Exact solution: $(1.1186, 0.8475)$.

32.
$$\begin{pmatrix} 2 & 0 \\ 0 & 1 \end{pmatrix} \begin{pmatrix} 1 & 0 \\ 3 & 1 \end{pmatrix} \begin{pmatrix} 1 & 0 \\ 0 & 7/2 \end{pmatrix} \begin{pmatrix} 1 & -1/2 \\ 0 & 1 \end{pmatrix}$$

33.
$$\begin{pmatrix} 1 & 0 & 0 \\ -1/2 & 1 & 0 \\ 0 & -2/3 & 1 \end{pmatrix} \begin{pmatrix} 2 & 0 & 0 \\ 0 & 3/2 & 0 \\ 0 & 0 & 4/3 \end{pmatrix} \begin{pmatrix} 1 & -1/2 & 0 \\ 0 & 1 & -2/3 \\ 0 & 0 & 1 \end{pmatrix}$$

CHAPTER 5

1. (a), (c), (d), (e).

2. $(1, 0) = -i(i, 0)$ hence $A(1, 0) = -iA(i, 0) = -i(1, 0) = (-i, 0)$; $(0, 1) = (1/2)[(i, 2) - (i, 0)]$ so $A(0, 1) = (1/2)[A(i, 2) - A(i, 0)]$; and so on.

3. $A\{(x, y) + (X, Y)\} = A(x + X, y + Y) = (a[x + X] + b[y + Y], c[x + X] + d[y + Y]) = (ax + by, cx + dy) + (aX + bY, cX + dY) = A(x, y) + A(X, Y)$. A similar calculation gives $A\{k(x, y)\} = kA(x, y)$, so A is linear. (a) $1, 0, 0, -1$. (b) $-1, 0, 0, 1$. (c) $\cos(2\theta)$, $\sin(2\theta)$, $\sin(2\theta)$, $-\cos(2\theta)$. (d) $10, 0, 0, 10$. (e) $\cos\theta$, $-\sin\theta$, $\sin\theta$, $\cos\theta$.

5. Rubbish. The rule 'add 1' is pure guesswork, and must be wrong because it is not a linear map. $(0, 0)$ must go to $(0, 0)$ under a linear map.

6.

$$\begin{pmatrix} c & 0 & d \\ 0 & e & f \\ 0 & 0 & g \end{pmatrix}$$

7. (a) $\{(1, 1)\}$ is a basis for $\text{im}(A)$, $\{(1, -1)\}$ for $\text{ker}(A)$, rank $=$ nullity $= 1$. (b) $\{(3, -1, 1)\}$ is a basis for $\text{ker}(B)$, $\{(1, 0, 1, 1), (2, 1, 1, 0)\}$ for $\text{im}(B)$, rank 2, nullity 1. (c) Nullity zero, $\text{ker}(D)$ is zero-dimensional, no basis; $\{(1, 0), (0, 1)\}$ is a basis for $\text{im}(D)$, rank $= 2$.
8. (a) True: kernel $=$ domain, so all vectors map to 0. (b) False. (c) False. (d) False, e.g., $f: \mathbb{R}^2 \to \mathbb{R}$ defined by $f(x, y) = x$.
9. The kernel is the set of all knob settings which give zero dial readings. The image is the set of all possible dial readings.
15. Nullity$(A + B) \geq$ nullity$(A) +$ nullity$(B) - \dim(V)$.
16. n even, rank$(a) = n/2$, $a(a(x)) = 0$ for all x.
17. (a) False. (b), (c) True. (d) Doesn't make sense: d is in the wrong space. The rest are false; easy to construct simple counterexamples.
18. Nullity$(A) = 1$ if $k = n^2\pi^2$ ($n = 1, 2, \ldots$), $= 0$ otherwise.
21. (a) $A(x, y, z) = 0$ has nonzero solutions, not invertible. (b) If $Dp = q$ then $D(p + k) = q$ for any constant k. Hence for a given q, p is not unique; not invertible. (c) Notice that for any p, Jp vanishes at $x = 0$. (d) No, for same reason as in (c). (e) For a given polynomial $p(x) = a_0 + a_1 x + \ldots + a_n x^n$, find the unique q satisfying $DXq = p$. (f) No.
22. (a) Take a map $\mathbb{R}^2 \to \mathbb{R}^3$. (b) u exists because $W = \text{im}(f)$, and is unique if f is nonsingular by the linear operator theorem.
25. (a) No eigenvalues. (b) Eigenvalues $2 \pm i$.
26. Follows from the rank theorem of section 4H. M and M^T have the same rank, and hence the same nullity if M is square, by the rank plus nullity theorem. The rest is easy.
28. $A^2 v = A(Av) = A(\lambda v) = \lambda Av = \lambda^2 v$, etc.
29. Both are invariant.
31. (i) There exist values for f and r which stay the same from one year to the next. (ii) Stable equilibrium: any initial state can be expressed as a linear combination of the two eigenvectors, and the component for the p eigenvector decreases, so the state tends towards the 1 eigenvector. (iii) Population explosion. (iv) Population explosion unless the initial state is precisely the p eigenvector.
33. (a) Apply M to the standard basis. (b) Let P be the orthogonal matrix whose columns are the vectors in B, and Q the orthogonal matrix whose columns are the transformed basis vectors. Show that $MP = Q$ and deduce that M is orthogonal.
34. Along the vector $(12, -2, 3)$.
36. The calculation gives Fibonacci numbers (see Rouse Ball & Coxeter (1974)).
40. (a) See Problem 15. (b) True: apply (a) to rank$([f + g] + [-g])$. (c) Follows from (b) by symmetry in f and g. (d) False. Counterexample: $f = -g = i$, the identity.
41. $LR = I$, $RL(x_1, x_2, \ldots) = (0, x_2, \ldots)$.

CHAPTER 6

1. (a) 0. (b) -55.

4. Note that each 4-figure number is a linear combination of its digits, with coefficients 10^3, 10^2, etc.

5. (a) Not linear. (b) $\ker(f)$ is the span of the u's if they are linearly independent. If they are linearly dependent, $f = 0$.

8. $A^T = -A$, hence $\det(A) = -\det(A)$, so $\det(A) = 0$.

10. Add every other row to the first row, factorise, then subtract the first column from every other column. The second determinant can be turned into the first by multiplying the rows by suitable numbers.

11. $n^{n/2}$.

12. Expand by the last row, then by the last column.

19. True.

CHAPTER 7

1. (a) Eigenvalues $8, -1$; eigenspaces $\mathrm{Sp}\{(2, 1, 2)\}$, $\mathrm{Sp}\{(1, 0, -1), (0, 2, -1)\}$. There are many other equally good bases for these eigenspaces. (b) Eigenvalues $3, 1$; eigenvectors $(1, 0, 0, 0)$, $(1, 0, -2, 0)$; both eigenspaces are one-dimensional here.

2. Same characteristic polynomial, because $\det(A - xI) = \det(A - xI)^T$.

3. p has factors $\lambda_1 - x, \ldots, \lambda_n - x$, so $p = k(\lambda_1 - x)\ldots(\lambda_n - x)$. To prove that $k = 1$, consider the coefficient of x^n in $\det(A - xI)$: expanding by the first row, etc. shows that x^n appears via the product of the diagonal entries, and therefore has coefficient $(-1)^n$. For the last part, look at $p(0)$.

4. (a), (c) are true.

5. (a) λ is an eigenvalue of $A^T = A^{-1}$, with eigenspace S, say. Easy to show $As = \lambda^{-1}s$ for all $s \in S$. (b) Sum of all algebraic multiplicities (AMs) is n; if every eigenvalue satisfies $\lambda \neq \lambda^{-1}$, then sum of AMs of all eigenvalues is even. So n odd implies there is a λ such that $\lambda = \lambda^{-1}$. (c), (d) Use the last part of Problem 3.

12. (a) Use $\|\lambda x\| = \|Ax\| = \|x\|$ where x is an eigenvector. (b) By row and column interchanges (which do not destroy orthogonality), can arrange that A is the top left block of an orthogonal matrix, so that

$$U = \begin{pmatrix} A & B \\ C & D \end{pmatrix}$$

is orthogonal. If x is an eigenvector of A,

$$U\begin{pmatrix} x \\ 0 \end{pmatrix} = \begin{pmatrix} \lambda x \\ Cx \end{pmatrix}$$

Now use the fact that $\|Uz\| = \|z\|$ when U is orthogonal.

13. There is one eigenvalue in each disc. Their centres are on the real axis, so if $x + iy$ is a complex eigenvalue in a disc, then $x - iy$ is another eigenvalue in the same disc – contradiction.

16. $A - I$ singular would mean 1 is an eigenvalue – contradiction. Simple algebra gives $(I - A)\{(I - A)^{-1} - I - A - \ldots - A^k\} = A^k \to 0$ as $k \to \infty$, hence $\{\ldots\} \to 0$.

CHAPTER 8

1. (a)
$$P = \begin{pmatrix} 1 & 1 \\ -i & i \end{pmatrix} \qquad D = \begin{pmatrix} i & 0 \\ 0 & -i \end{pmatrix}$$

for Example 5H-1; similarly for the other. The answers are not unique. The diagonal entries of D are the eigenvalues, in any order; the columns of P are the corresponding eigenvectors; it does not matter which of the many eigenvectors you use.

2. Use equation (4) of section 8A.

3. (a) $A^3 = PD^3 P^{-1} = 0$ implies $D^3 = 0$, hence $\lambda^3 = 0$ for each eigenvalue of A, so $D = 0$ hence $A = 0$. Similar argument for (b). Counterexamples for the non-diagonalisable cases of (a) and (b):

$$\begin{pmatrix} 0 & 1 \\ 0 & 0 \end{pmatrix} \qquad \begin{pmatrix} 0 & 1 \\ -1 & -1 \end{pmatrix}$$

5. $f(A) = Pf(D)P^{-1}$ follows from equation (4) of section 8A. The first polynomial has 4-figure entries, the second is zero (this is no coincidence: see the Cayley–Hamilton theorem, section 10D).

10. (a) $\quad Q = (1/\sqrt{6}) \begin{pmatrix} \sqrt{3} & -\sqrt{2} & -1 \\ 0 & \sqrt{2} & -2 \\ \sqrt{3} & \sqrt{2} & 1 \end{pmatrix} \qquad D = \begin{pmatrix} 6 & 0 & 0 \\ 0 & -6 & 0 \\ 0 & 0 & 0 \end{pmatrix}$

(b) is the same kind of thing, with eigenvalues 3, 3, 0.

(c) $\quad Q = (1/\sqrt{6}) \begin{pmatrix} 0 & -2 & 0 & \sqrt{2} \\ \sqrt{3} & 1 & 0 & \sqrt{2} \\ 0 & 0 & \sqrt{6} & 0 \\ -\sqrt{3} & 1 & 0 & \sqrt{2} \end{pmatrix} \qquad D = \text{diag}(3, 3, 5, 0)$

11. (a) Easily follows from $A = QDQ^T$. (b) is Theorem 8C-2.

15. (a) Saddle-point (quadratic part is positive on the x-axis, negative on the y-axis). (b) Saddle-point (xy is positive in the first quadrant, negative in the second). (c) Minimum (quadratic part has two positive eigenvalues).

16. (a) Eigenvalues 1, 11: minimum. (b) Eigenvalues 3, -7: saddle-point.

17. (a) Matrix of Problem 10(a); neither maximum nor minimum. (b) Matrix of Problem 10(b); maximum (but not a strict maximum: $f(x, y, z) \geq f(0, 0, 0)$ for all nonzero x, y, z but not strictly greater; in the direction of the zero eigenvector, f is constant). (c) Eigenvalues 2, -1. Saddle-point-type behaviour.

24. (a) $k > 0$ gives a family of ellipses pointing along the line $y = -3x$. No curves for $k < 0$. (b) Hyperbolas, with half the plane filled with the $k > 0$ curves and the other half with the $k < 0$ curves.

25. (a) Hyperbolic cylinders, parallel to the vector $(1, 2, -1)$. (b) Circular cylinders with axis $(1, 1, 1)$ for $k > 0$, no solutions for $k < 0$. (c) Surfaces obtained by spinning a family of hyperbolas about an axis along $(1, 1, 1)$; general shape similar to Fig. 8.4.

References and bibliography

Biggs, N.L. 1985, *Discrete Mathematics*, Oxford U.P.

Binmore, K.G. 1980, *The Foundations of Analysis: a Straightforward Introduction. Book 1, Logic Sets and Numbers*, Cambridge U.P.

Braun, M. 1975, *Differential Equations and their Applications*, Springer, N.Y.

Brockett, R.W. 1970, *Finite Dimensional Linear Systems*, Wiley, N.Y.

Burden, R.J., Faires, J.D. 1985, *Numerical Analysis*, 3rd Edn, Prindle, Weber & Schmidt, Boston.

Burkill, J.C. 1962, *A First Course in Mathematical Analysis*, Cambridge U.P.

Burkill, J.C. & H. 1970, *A Second Course in Mathematical Analysis*, Cambridge U.P.

Courant, R., Robbins, H. *What is Mathematics*, 1941, Oxford U.P.

Coxeter, H.S.M. 1981, *Introduction to Geometry*, 2nd Edn, Wiley, N.Y.

Feynman, R.P., Leighton, R.B., Sands, M. 1964, *The Feynman Lectures on Physics*, Vol. 1, Addison-Wesley, N.Y.

Frisby, J.P. 1980, *Seeing*, Oxford U.P.

Gasson, P.C. 1983, *Geometry of Spatial Forms*, Ellis Horwood, Chichester.

Golub, G.H., Van Loan, C.C. 1983, *Matrix Computations*, John Hopkins University Press, Baltimore.

Graham, A. 1979, *Matrix Theory and Applications for Scientists and Engineers*, Ellis Horwood, Chichester.

Gregory, R.L. 1970, *The Intelligent Eye*, Weidenfeld & Nicolson, London.

Gregory, R.L. 1977, *Eye and Brain*, 3rd Edn, Weidenfeld & Nicolson, London.

Griffel, D.H. 1981, *Applied Functional Analysis*, Ellis Horwood, Chichester.

Horn, R.A., Johnson, C.A. 1985, *Matrix Analysis*, Cambridge U.P.

Hutson, V., Pym, J.H. 1980, *Applications of Functional Analysis and Operator Theory*, Academic Press, London.

Kline, M. 1972, *Mathematical Thought from Ancient to Modern Times*, Oxford U.P.

Kreider, D.L., Kuller, R.G., Ostburg, D.R., Perkins, F.W. *An Introduction to Linear Analysis*, Addison-Wesley, N.Y.

Land, E.H. 1977, The retinex theory of colour vision, *Scientific American* **237**, December.

Lipschutz, S. *Theory and Problems of Linear Algebra*, McGraw-Hill, N.Y.

Mandelbrot, B. 1982, *The Fractal Geometry of Nature*, Freeman, San Francisco.

Mason, J.C. 1984, *BASIC Matrix Methods*, Butterworth, London.

Minc, H. 1978, *Permanents*, Addison-Wesley, Reading, MA.

Mirsky, L. 1955, *An Introduction to Linear Algebra*, Oxford U.P.

Nering, E.D. 1963, *Linear Algebra and Matrix Theory*, Wiley, N.Y.

Polya, G. 1962, *Mathematical Discovery*, Wiley, N.Y.

Rouse Ball, W.W., Coxeter, H.S.M. 1974, *Mathematical Recreations & Essays*, 12th Edn, Toronto University Press.

Rucker, R. 1982, *Infinity and the Mind*, Birkhauser, Basel.

Rucker, R. 1986, *The Fourth Dimension*, Penguin, London.

Solow, D. 1982, *How to Read and Do Proofs*, Wiley, N.Y.

Stewart, I., Tall, D. 1977, *The Foundations of Mathematics*, Oxford U.P.

Strang, G. 1980, *Linear Algebra and its Applications*, 2nd Edn, Academic Press, N.Y.

Strang, G. 1986, *Introduction to Applied Mathematics*, Wellesley-Cambridge Press, Wellesley, MA.

Todd, J. 1978, *Basic Numerical Mathematics*, *Vol. 2*, *Numerical Algebra*, Academic Press, N.Y.

Wyszecki, G., Stiles, W.S. 1967, *Color Science*, Wiley, N.Y.

Index

Index

Mathematics and its Applications

Series Editor: G. M. BELL, Professor of Mathematics, King's College London (KQC), University of London